Women in Audio

T0264735

Women in Audio features almost 100 profiles and stories of audio engineers who are women and have achieved success throughout the history of the trade. Beginning with a historical view, the book covers the achievements of women in various audio professions and then focuses on organizations that support and train women and girls in the industry. What follows are eight chapters divided by discipline, highlighting accomplished women in various audio fields: radio; sound for film and television; music recording and electronic music; hardware and software design; acoustics; live sound and sound for theater; education; audio for games, virtual reality, augmented reality and mixed reality, as well as immersive sound.

Women in Audio is a valuable resource for professionals, educators, and students looking to gain insight into the careers of trailblazing women in audio-related fields and represents required reading for those looking to add diversity to their music technology programs.

Leslie Gaston-Bird (AMPS, MPSE) is an audio engineer, Governor-at-Large for the Audio Engineering Society, and member of the Recording Academy. She has worked for National Public Radio (Washington, D.C.), Colorado Public Radio, the Colorado Symphony Orchestra, Post Modern Company, and the University of Colorado Denver.

Audio Engineering Society Presents . . .
www.aes.org

Editorial Board
Chair: Francis Rumsey, Logophon Ltd.
Hyun Kook Lee, University of Huddersfield
Natanya Ford, University of West England
Kyle Snyder, University of Michigan

For more information about this series, please visit: www.routledge.com/Audio-Engineering-Society-Presents/book-series/AES

Women in Audio

Leslie Gaston-Bird

Routledge
Taylor & Francis Group

NEW YORK AND LONDON

First published 2020
by Routledge
52 Vanderbilt Avenue, New York, NY 10017

and by Routledge
2 Park Square, Milton Park, Abingdon, Oxon, OX14 4RN

Routledge is an imprint of the Taylor & Francis Group, an informa business

Library of Congress Cataloging-in-Publication Data
A catalog record for this book has been requested

ISBN: 978-1-138-31601-0 (hbk)
ISBN: 978-1-138-31599-0 (pbk)
ISBN: 978-0-429-45594-0 (ebk)

Typeset in Minion Pro
by Apex CoVantage, LLC

Visit the companion website: WomenInAudioBook.com

I dedicate this book to everyone who has shared their story with me. Thank you for your enthusiasm, spark, passion, and trust.

I would also like to dedicate this book to a number of people in my life:

To the loving memory of my mother, Frances Gaston.

To the memories of Joan Lowe, Marie Louise Killick, and Robin Coxe-Yeldham.

To Kellie Gaston Dixon, my sister, and Ryan Dixon, my nephew.

To my father, Berdell Gaston, who taught me how to thread a tape machine when I was a little girl.

To my mentors, Wayne Jackson, David Pickett, Pamela Laird, Judith Coe, the late Rich Sanders, and the late Roy Pritts.

To my daughter, Kira, who wants to be a veterinarian and a ballerina, and to my son, Zachary, who wants to be an automotive engineer, and who both have my unending support in whatever career path they choose.

And finally to my loving husband, Andy, who provided me with encouragement and watched the kids while I conducted interviews and finished the book.

Contents

Figures and Tables

Figures

Tables

Introduction

Women in Audio highlights the achievements of women in the field of audio engineering. The topics covered are women's audio organizations; radio and podcasts; sound for television and film; music recording and electronic music; hardware and software design; acoustics; live sound and sound for theater; audio for games; and education.

There is also a companion website, WomenInAudioBook.com, where you can find links to the organizations, research, videos, and sound recordings mentioned herein.

Throughout this book you will see names in **bold**. This is meant to help identify women who are being featured and who may appear more than once in the book. There are also "fun facts" that go into further detail about a concept or technology mentioned in the profile or subject.

At the end of the book is a glossary, which is a bit technical in nature; I refer the reader to resources in the "Recommended Reading" section to learn more about these various concepts.

The book starts with a brief history of audio with a focus on women's contributions to the field. Although I have endeavored to be thorough, I know for a fact there are some names missing. Ironically, although we lament the dearth and occasional invisibility of women in audio, there were too many to include in this book. If you have suggestions of women, events, and milestones to include, please contact me at WomenInAudioBook.com, as there could very well be another volume on the topic.

After the history chapter, subsequent chapters feature profiles of women who work in the field. Each has a unique story to tell: their childhood interests, how they began in the field, their own contributions to the industry, and their perspectives on being an audio engineer.

I have sought to be as inclusive as possible, reaching out to a diverse group of women from all over the globe. As such, this book makes a great addition to any curriculum struggling with the question of how to incorporate the subject of diversity in the field of recording arts, especially with regard to women and women in underrepresented groups.

Within each chapter, I have included a description of "tools of the trade" and provided some information on pursuing a career in each area.

This book is also meant to be inspirational. I have certainly been inspired and even moved to tears by some of these stories. I think you will be inspired, too; and if not, at the very least I hope you find the book to be informative.

Finally, the book is meant to be a resource for both men and women of all ages, races, gender identities, and nationalities.

Preface

I am not a woman in audio. I am an audio engineer.
Nor am I a "Black woman audio engineer."
I am an audio engineer who happens to be a Black woman.
—Leslie Gaston-Bird

However, in order to address the lack of representation in our field, it seems we must use the phrase "women in audio," even though all of the women I have spoken to and read about agree: given a choice, we'd rather just be thought of as "audio engineers," a luxury the men in our field are afforded without a second thought.

Gender-based designations such as these have been the case throughout history; perhaps more so in science, technology, engineering, and math, where women are underrepresented. In contemporary discussions in the media, we hear the phrases "women in medicine" or "women in physics."

Having been excluded from men's clubs and professions, over time women have needed to create their own networks, and there exists "the women's version" of almost every imaginable organization: "The American Association of University *Women*," "The American Medical *Women's* Association," and "The National Center for *Women* in Information Technology." The list goes on and on.

Writing this book has been an educational, revelatory, and emotionally moving exercise. It has also been exhausting, a true rabbit hole of discoveries as I have become familiar with the connections between women who work in each facet of the field, each of whom can introduce me to someone new or who mentions someone I may have overlooked, who in turn is connected to someone else. Although I'm terrible at remembering names and faces, I have found myself with almost total recall of every woman's name and face I have talked to and researched so far, almost as if we are in a special sort of school together – which, in a way, we are.

In fact, when I discovered that **Emily Lazar** mastered a project by **Fanny**, I exclaimed, "You gotta be kidding me!" It sums up perfectly how this book is all about *finding each other* and *connecting* in this new age of networking and social capital. Fanny was the first female rock group signed to a major label in the 1970s, with albums engineered by **Leslie Ann Jones**; Fanny's **June Millington** opened the **Institute for Musical Arts**, a forerunner to **Women's Audio Mission** (WAM), and almost 50 years later their work is being remastered by Lazar, the first woman mastering engineer to win a Grammy for Best Engineered Album (Non-Classical), who has coincidentally done tons of work with WAM.

Of the many women I have interviewed for this book, my experience finding **Joan Lowe** shortly before her death in 2019 was the most profound. (Lowe is also connected to Millington through Olivia Records.) It made me realize that so much of this history is rapidly slipping through our fingers, and capturing it to share with you, the reader, is an honor, a privilege – and a duty.

From: Joan Lowe
Subject: Re: Women in Audio
Date: November 8, 2018
To: Leslie Gaston

Hello, Leslie. Thanks for your interest in my past work in audio. I am about to reach my 90th birthday, so my experience ranges quite far back and into the era of heavy analog equipment. There were not many women working in the field, and when Olivia [Records] formed and wanted to learn about the recording industry to benefit women, I offered to help. If you would like to communicate further, I would be glad to talk with you. Telephone conversations are often somewhat difficult for my very senior and much-used ears, so I like to do as much as possible via email, at least in setting up basic questions.

I welcomed a group of women who wanted to learn how to produce, manufacture, and market records and volunteered whatever assistance I could to their endeavor. The recording business was so restricted at that time that the challenge was gladly accepted to help break through the old "rules." The experience with "Changer" was an incredibly inspiring one.

Don't hesitate to contact me further.
Joan Lowe

After sharing her story with me in a subsequent email, Joan Lowe passed away on February 8, 2019. I've shared the email with you in its entirety in Chapter 1.

The women I interviewed had various outlooks on the industry, among which I found a few themes: Some want to address the nature of our male-dominated industry, while others want to focus on their own career. Some women want to talk about the obstacles and about sexism, while others have encountered few, if any, obstacles and consider themselves "lucky" not to have experienced the worst sexist behavior. Most of the women interviewed have directly experienced or witnessed sexism. Many are excited to become mentors; others have retired and/or moved on to other careers and pursuits. All want to be recognized for their work as an engineer, not a "woman engineer."

Other common threads in the academic background (high school, college, and university) of my interview subjects include music, composition, math, electronics, physics, electrical engineering, and even environmental science in more than one instance. Overwhelmingly the two biggest traits I discovered among the women I interviewed and researched were *curiosity* and *determination*. Time and again, despite being told "you're not allowed because you're a woman" or "you won't succeed because you're a woman" and even with the lack of role models, some of the women featured in this book persisted, found their way, and succeeded. They were able to do this because their curiosity about the subject was the force driving them forward.

So I can't summarize a "universal experience" for all audio engineers who are women. I simply want to provide a compilation of history and role models for people: young and old, regardless of gender or gender identity. I want to show the "hidden figures" who have existed in our industry all along, who have sometimes been passed over in the retelling of our audio history. I want to show that we are involved in every aspect of audio you can think of, regardless of what the numbers show.

Why Wasn't My Favorite Woman Included? Where Is *My* Name?

As you browse the names in the book, you might wonder: What are the criteria for being included?

I have endeavored to focus on women in history who have earned notoriety for being the "first woman" to do something: start a record label, solve an equation, or win a Grammy or Oscar in a certain category. However, race and class can easily be overlooked because even those prestigious awards are not inclusive in nature. Therefore, I have also relied on my own experiences and networks to find women of color and ethnic minorities around the globe. This underscores the importance of women's audio networks (and their tireless leaders) in helping me to find a diverse group to write about and from which to compile our collective history.

I also wanted this book to be international in its scope. This added another layer of complexity, but the Audio Engineering Society has a rich community of professionals and educators who have helped put me in touch with institutions, men and women, and businesses who could help me find notable engineers who are women around the world. I also used social media – a lot!

If you or someone you know is an audio engineer who is a woman and you don't see your name in this book, please get in touch with me through the book's companion website, WomenInAudioBook.com. I'd love to see this book evolve as our industry evolves. Things have changed so rapidly in the past decade that the next edition of our book could paint a very different picture of our demographic.

What About Gender Identity?

This book features women, including transgender women and anyone who identifies as a woman. I have also been enlightened by transgender and gender-nonconforming audio engineers who have shared their experiences with me. "I'm not sure if the term 'women in audio' is supposed to include me," confided one young engineer. This was a heartbreaking moment for me but also a defining one, as I realized that being inclusive means being sensitive to the language and understanding the underlying issues. Another colleague related to me that a transgender woman working at a booth for a women's organization at an audio conference was harassed for being there by someone who said to her, "But you're not a woman!" This transgender woman is indeed a woman. She should be accepted at the Women's Audio Mission booth, and this harassment was unacceptable. If you have any concerns about the terms or phrasing used in this book, please let me know.

How Many Women Are Working in Audio? What Are the Numbers?

In the Wachowski sisters' film *The Matrix*, a young boy educates the hero, Neo, about a spoon-bending feat. The child tells him that it is impossible to bend a spoon and to realize the truth: that there is no spoon.

The statistics presented by the recent Annenberg Inclusion Initiative regarding inclusion in the music industry are presented here for you read and reflect upon. There are currently efforts being made to find these numbers for the audio industry as a whole. But although the number of women are depressingly low when compared to men working in the industry, there are nonetheless thousands of women working in audio. As I wrote this book, I found myself daunted by the prospect of interviewing an ever-growing and seemingly endless list of amazing women.

I then came to a similar realization as Neo and have adapted the child's advice to illustrate the profundity of the task at hand. I have adapted this as follows: "*Do not try to interview every woman in audio, for that would be impossible. Instead only realize the truth: There are no women in audio, only audio engineers who happen to be women.*"

If this comment rubs you the wrong way, don't worry: it should be obvious at this point, having dedicated the time and effort into writing on the subject, that I believe it is important to talk about "women in audio." However, as **Terri Winston**, founder and director of Women's Audio Mission, reminds all of us, "It's not about *you*; it's about the *next one*. It's about the *kid*. Does it matter that you're bristling at this when some young girl sees you up there and thinks, 'Oh my God, you're a badass! I want to be *you*'"?

To borrow a trending phrase, if "you can't be what you can't see," then surely it follows that "you *can be* what you *can see*." I hope you are inspired by whom you see on these pages.

To the Young Reader

If you are under the age of 18, perhaps you got this book from a teacher or your mother or father. Perhaps you found this book on your own. No matter how you found yourself holding this book, I am glad you are here!

I talked to many women about how they became interested in audio, and they all share one thing in common: they loved playing music and listening to music when they were children.

It wasn't easy for everyone to get an instrument; some of these women as young girls had to save money over several weeks to buy one. You might even be saving for a musical instrument, or for a computer so you can start making your own music using a software program.

Perhaps you just enjoy listening to music. You might find yourself curious about how the sounds are made. Maybe you use a computer to experiment with making your own sounds. There are certainly lots of different programs and apps to try. In some chapters, I list some "tools of the trade" that you can explore. You might even find some programs you haven't heard of.

Another thing that many of these women share in common is that as young girls, they liked math, physics, and science. Lots of girls used to take things apart to see how they worked. Some had parents who showed them how to build things or work in a studio; others read books or found a teacher to show them.

Not all of the girls enjoyed math or physics but had a knack for knowing when something sounded good and wanted to get musicians together to create a song and make a recording that could compete with anything on the radio (or Spotify, etc.).

As you look through the book, you may read that some girls were discouraged from learning math and science, either because they weren't allowed to or told that it's not "for girls." I hope this hasn't happened to you. If it has, seek out friends, teachers, and educational websites who can help you rekindle curiosity about these subjects. And of course, it's okay to say you just don't like math! The important thing is that no one *stops* you from wanting to learn.

"Follow your curiosity" is the advice of so many women in this book. That means, if there's something you want to know about, get your hands on as many books as you can and talk to people who are doing it. Do you want to know how to build a computer? Do you want to know how to make beats? Do you want to know how sound bounces around a room? Do you want to build your own studio? Do you want to make sound effects for a movie? This is what it means to be "curious," and you should keep looking for the answers to your questions . . . especially if they lead to more questions! You'll find at least some of the answers within this very book.

Some chapters also have sections called "Careers". I provided these so you can get a sense of what it's like to work in these fields, and what you need in terms of preparation.

I hope that somewhere in this book you can find something new to be curious about. Be sure to check out the "Recommended Reading" section at the end of the book, and browse the accompanying website, WomenInAudioBook.com, which has links to things you can listen to, watch, and experiment with.

I hope the stories you read inspire you. Most of all, I want you to "see what you can be."

Acknowledgments

Sue Barrett
Anna Bertmark
Dana Burwell
Carol Bousquet
Cossette Collier
Liz Dobson
Lorrie Evans
Karrie Keyes
Nicholas Klaus Larson
Theresa Leonard
Judy Clapp
Rob Jaczko, Carl Beatty, and Dakota Yeldham
Alissa Rams
Josh Reiss
April Tucker
Terri Winston
Cecilia Wu
Fei Yu
The Daphne Oram Trust
Images of Marie Louise Killick, the Sapphox stylus patent specification, and the Sapphox pamphlet are provided courtesy of the family of Marie Louise Killick and the website https://marielouisekillick.org.uk/
Calumet Regional Archives, Indiana University Northwest
Andrew Clayman, MadeInChicagoMuseum.com
Dr. Stacy L. Smith and the Annenberg Inclusion Initiative at the University of Southern California
The Gabriel/Mauro family

1   History

Women in Early Audio History

Ada Lovelace (England)

Ada Lovelace circa 1840

A DA LOVELACE IS widely regarded to be the first computer programmer, and she started long before the computer as we know it was invented.

Ada Lovelace's name at the time of her birth in 1815 was Ada Gordon. Her father was Lord Byron (George Gordon Byron), a famous poet and politician. Ada's mother, Annabella Milbanke, ensured that Ada was taught science, logic, and mathematics, which may explain Ada's love for machines as the industrial revolution unfolded around her. Her tutor introduced her to Charles Babbage, who was the Lucasian Professor at Cambridge (the same post that would eventually be held by the physicist and cosmologist Stephen Hawking). Her friendship with Charles Babbage endured for years. (findingada.com)

In 1834 she married William King, who became the Earl of Lovelace in 1838. As a result, Ada Gordon became "Lady Ada King, Countess of Lovelace," or Ada Lovelace for short. She kept working with Babbage, who had developed plans for a machine called the Analytical Engine. The machine caught the eye of Italian mathematician Luigi Frederico Menabrea, who in 1842 wrote about it in a paper called "Notions sur la machine analytique de M. Charles Babbage," which was published in the *Bibliothèque Universelle de Genève*, an academic journal. Babbage saw the paper and asked Lovelace to translate the article. She did so and wrote an addendum called "Notes from the Translator," which was comprised of a total of seven additional writings, lettered from A to G. In them, she takes it upon herself to clearly define "data" and "operations" (processing), further defining the term "operation" as "any process which alters the mutual relation of two or more things." The paper was released in English in 1843 as "Sketch of the Analytical Engine, invented by Charles Babbage Esq., By LF Menabrea of Turin officer of the military engineers, with Notes from the Translator" (Sterling, 2017).

In the book *Ada's Algorithm*, author James Essinger seeks to underscore that with her definition of the word "operation," she is giving birth to the science of computing and computer programming and is in fact now regarded to be the inventor of computer programming. Furthermore, she suggests for the first time that the device could have applications for music. She writes, "Supposing, for instance, that the fundamental relations of pitched sounds in the science of harmony and of musical composition were susceptible of such expression and adaptations, the engine might compose elaborate and scientific pieces of music of any degree of complexity or extent" (Essinger, 2014).

Because of her contributions to the field, in 2009 "Ada Lovelace Day" (October 8) was created in order to celebrate the accomplishments of women in STEM (science, technology, engineering, and math). The event was the brainchild of Suw Charman-Anderson, a social technologist in London. After reading a 2006 study called "Someone Like Me Can Be Successful: Do College Students Need Same-Gender Role Models?" by Penelope Lockwood in the *Psychology of Women Quarterly*, Suw realized that she had a hard time naming other women working in the same field. She then created a challenge, which she posted on Pledgebank. She promised, "I will publish a blog post on Tuesday 24th March about a woman in technology whom I admire, but only if 1,000 other people will do the same" (Charman-Anderson, 2009). She ended up getting almost 2,000 pledges (Phillips, 2011). The site, www.findingada.com, has been going strong for ten years as of this publication, and their social media pages highlight profiles of women in technology and a number of events held in celebration of International Women's Day.

Ada Lovelace portrait by Margaret Carpenter (c. 1836)

Fun Facts: What Is an Algorithm?

An **algorithm** is an ordered sequence of mathematical operations that are used to make complex equations easier to solve. Audio processing, such as equalization, reverb, and compression, rely on algorithms to function. As a very basic example, in order to add reverb to an audio signal, a system must 1) read the audio signal, 2) calculate the value of the signal's amplitude, 3) add a certain number of delayed signals or echoes at incrementally decreasing levels, and 4) output the final signal, perhaps mixed with the original signal. One such reverberator was described in more detail by Manfred Schroeder of Bell Laboratories in his AES paper, "Natural Sounding Artificial Reverberation," at the end of which he acknowledges his indebtedness to "**Mrs. Carol Bird**" (Schroeder, 1961). Carol Bird McLellan also worked at Bell Labs and did programming for Schroeder and a few others (Noll, 2019).

Marie Sophie-Germain (France)

If you have ever used a plate reverb (or plate reverb plug-in), you might be interested to know that the resonances you hear can be predicted mathematically. In fact they can even be visualized by placing sand on a flat metal plate and applying various frequencies that cause the plate to vibrate. The first mathematician to predict what these resonances would be was **Sophie Germain**.

Marie Sophie-German was born during a revolutionary period in France. By the time she was a teenager, societal conditions had deteriorated so much that it was dangerous to leave the house. Instead she spent many hours working on math equations. This troubled her parents, who knew that girls and women were not permitted to study math, but seeing their daughter persist led her mother to eventually agree to allow Sophie to pursue her passion. However, she would not be granted entrance to her chosen institution, École Polytechnique, because she was a girl. Instead she had to rely on copies of the lecture notes, which the school made freely available (Baguley, 2018).

In 1808, there was an open call by the French Academy of Science for mathematicians to build on the work of Ernest Chladni, who described how metal plates would resonate at certain frequencies when a vibration was applied (for example, with a violin bow).

This could be observed by sprinkling sand on the plate and noticing how the sand would arrange itself as it vibrated. (You can also see this phenomenon in action by searching for videos on the internet with the terms "sand," "plates," and "resonance.") The challenge was to be able to come up with a formula to predict what these resonances would be. The problem was that the system of analysis that would make this possible had not yet been invented (Mozans, 1913).

Germain seized on the challenge and made a total of three attempts to solve the problem. She was the only person to submit an attempt in 1808, which did not win, and a second attempt made in 1810 earned her an honorable mention. On her third attempt, she won the prize. However, the Academy would not let her attend the *grand prix* ceremony. Nor did they publish the work, so she had to publish it on her own in 1821. This work was called "Recherches sur la théorie des surfaces élastiques" (Germain, 1821). Eventually it was Joseph Fourier (who developed the Fourier transform) who invited Sophie to join him for meetings at the Academy. Fourier and other important figures in audio history, including André-Marie Ampére (for whom the electrical measurement of *ampere* is named) and Carl Friedrich Gauss (who studied electricity and magnetism), were great supporters of Germain's work.

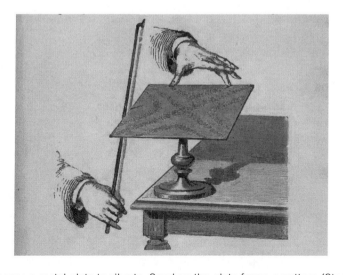

Using a bow to cause a metal plate to vibrate. Sand on the plate forms a pattern (Stone, 1879).

After her death from breast cancer, a posthumous degree was awarded to her, and now every year the French Academy of Science awards the Sophie Germain Prize in her honor (Baguley, 2018). Her work had tangible results: architectural engineers could use her formulas to predict the stress on metal objects, which contributed to buildings such as the Eiffel Tower. However, you will not find Germain's name on any of the four sides of the Tower, which was built between 1887 and 1889 and upon which are inscribed the names of 72 scientists who contributed their intellect to its construction. In 1913 author H. J. Mozans published a book, *Woman in Science: With an Introductory Chapter on Woman's Long Struggle for Things of the Mind*, within which he suggests Germain's name was omitted because she was a woman (Mozans, 1913).

Later, **Mary Desiree Waller** (1886–1959) would continue studying Chladni figures and recreate the figures Chladni discovered (*Female pioneers in audio engineering*, 2017). Her findings were published posthumously in 1961 in her book, *Chladni Figures, A Study in Symmetry* (Cymascope) (Mary Désireé Waller, 1962).

Fun Facts: Plate Reverb

You may have seen a "plate reverb" in your digital audio workstation's list of plug-ins. Before digital audio recording, audio engineers used metal plates to emulate the sound of reverberation. A transducer (such as a loudspeaker driver) is mounted upon a sheet of metal, and an audio signal is sent there (for example, from the mixing console's effects send). The metal would then resonate with the signal, creating an interesting effect that sounds like reverberation. Another transducer acts like a microphone pickup and returns the reverberant signal back to the console (for example, to the mixing console's effects return section). Engineers like **Lenise Bent** working in larger studios might have three or more physical plate reverb units – not the software plug-ins! – in a single mix.

For a student project, a metal plate is suspended with springs. Speakers are attached to the suspended metal plate.

Source: Courtesy of Chris Nation

Margaret Watts-Hughes (Wales)

In 1885, vocalist **Margaret Watts-Hughes** was experimenting with resonance and visualization. It is not clear if she were aware of Sophie Germain's work with metal plate resonances, but she was probably aware of the invention of the **phonoautograph** (1855) and the **phonograph** (1877) (see: Fun Facts).

In her article "Visible Sound," Margaret explains that she was looking for a way to indicate the intensity of a vocal sound. She experimented with a device she created and named the "eidophone" (pronounced "EYE-də-fōn").

She stretched a sheet of rubber over the end of a tube and used her voice to create amazing forms, which she gave names such as "daisy," "fern," and "pansy," and even more intricate patterns labeled "seaweed or landscape," "serpent," and "tree."

The article was published by *The Century Magazine* in 1891 and is followed by a response by **Sophie Bledsoe Herrick**, who authored scientific articles for *Century*. In her response, Sophie Herrick connects Watts-Hughes' work to Ernest Chladni's observations on metal plates (the same work that Sophie Germain would eventually describe mathematically). Herrick's response seems to address some critics who saw Margaret's work as amateurish. Herrick replies, "That her experiments are amateurish rather than scientific is no discredit, for she has opened up a new field into which the scientist may enter and reach results of great interest and value" (Herrick, 1891). In fact, **Grace Digney** uses the eidophone today for therapeutic purposes as well as art-making (see: Fun Facts).

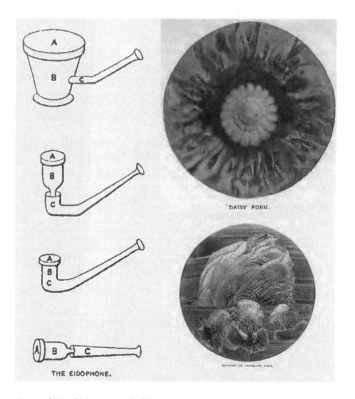

Possible configurations of the Eidophone (left); patterns created with the Eidophone (right)

Fun Facts: Grace Digney (Ireland) and a Modern-Day Eidophone

Around 2009, during research for her M.A. at the University of Ulster, visual artist **Grace Digney** discovered Margaret Watts-Hughes' work with the eidophone and has been using the technique to help develop singing abilities and aid with speech therapy for the hearing and hearing impaired. She captures the images with gummed (sticky) paper, which can also be felt and experienced by the visually impaired.

To make an eidophone, you can stretch a piece of latex rubber over a drum and secure it with a round frame and screws. You then feed a tube into the cavity of the drum and sing into it through the tube (see photos).

A multitude of patterns will form, which Digney captures and displays as artwork, as an "alphabet" or even chocolates so you can "eat your words." To find out more about Digney and her work, visit: gracedigney.com.

Grace Digney demonstrates a modern-day eidophone

Metal filings arrange themselves into concentric circles on the latex surface as it resonates with the voice.

An "alphabet" of patterns created by Digney
Source: Courtesy of Grace Digney

Eidophone patterns molded into a bronze tile by Digney
Source: Courtesy of Grace Digney

Frances Densmore (USA)

Dr. Frances Densmore was an ethnomusicologist born in 1867 who studied music at Oberlin College in Ohio, USA. She recorded the songs of Native Americans primarily using a graphophone and other tools in order to preserve their song, dance, and stories. This was complicated by the fact that her subjects were understandably hesitant to participate, since at time the United States government was working actively to eradicate Native culture; in particular, "religious and medicinal ceremonies were especially targeted under the Civilized Regulations of 1880 . . . only Native Americans could commit these offenses, which remained legal until 1936" (Kristin McFadden).

Densmore, however, strove to preserve their culture within this unfortunate social context. In the liner notes for "Songs of the Chippewa," Densmore writes, "These men realized that the old songs were disappearing and consented to record them when told 'their voices would be preserved in Washington, in a building that would not burn down'" (Densmore, 1950).

"My first recording equipment was an Edison phonograph which was then the best equipment available. The next summer, it was replaced by a Columbia gramophone with four heavy springs. At that time the Indians were not generally accustomed to phonographs and few, if any, had seen the making of records. One Chippewa woman, after hearing a record of her own voice, looked at the phonograph and exclaimed, 'How did it learn that song so quick? That is a hard song'" (Densmore, 1950).

Frances Densmore (Smithsonian Institution Archives)

Densmore with Blackfoot chief, Mountain Chief, during a 1916 phonograph recording session for the Bureau of American Ethnology.

Many of the recordings are housed in the Smithsonian Institution in Washington, D.C. They date from between 1907 and 1930. Dr. Densmore is credited with recording and editing the audio. Altogether, there are 2,385 cylinders in the collection, which were all recorded and edited by Dr. Densmore.

Fun Facts: The Phonautograph and the Phonograph

The first known device for recording sound was the phonautograph, which was patented in France by Édouard-Léon Scott de Martinville in 1857. However, it did not *play back* sound. It has been compared to a seismograph, which records vibrations from earthquakes (Voice, c. 2018). In April 1877, a method of playing back sound recorded by a phonautograph was conceived of in France by Charle Cros. However, he did not have a prototype, only the notes that he submitted to the French Academy of Sciences. He called his device the "paléophone." That same year (1877), Thomas Edison invented the phonograph, which was patented in 1878 (Service, 2017).

Fun Facts: The Graphophone

Not to be confused with the *gramophone*, the *graphophone* used wax cylinders to record audio. The gramophone recorded audio laterally (back and forth), while the graphophone recorded audio vertically (up and down). The one pictured here was used by **Frances Densmore** to record as early as 1897 (Williamson, 2018).

The Graphophone used by Frances Densmore

Aletha Mae Dickerson-Robinson (USA)

"The first time I handled a recording session, I brought in the Hokum Boys," remembers **Aletha Mae Dickerson-Robinson** (known then as Aletha Dickerson) of the session, which took place in 1928. "Among other artists I supervised were Ma Rainey, Ida Cox, Blind Lemon Jefferson, Will Ezell, Pace Jubilee Singers, Blind Blake, Meade Lux Lewis, Charley Spands [sic], Freddie Brown, Alex Hill's band, Laura Rucker and others whose names I can't recall. All these sessions were for Paramount" (Tuuk, c. 2017).

In the book, *The New Paramount Book of Blues: Elusive Artists on Paramount Race Records*, Alex van der Tuuk details Robinson's story, which he then summarized for an online article, "Paramount's Reluctant Recording Manager" (Ibid.).

Robinson had been J. Mayo "Ink" Williams' secretary. The two worked in Chicago while Williams acted as unofficial Paramount recording director. Williams was tasked with procuring talent and recordings for Paramount's "race records." Robinson typed out song lyrics and registered songs for copyright. Then in 1928, Williams left and joined the Vocalion company. With his departure, Dickerson became the recording manager. Tuuk writes, "As confirmation of her role, unofficial or otherwise, within Paramount's recording department, the 1930 Federal Census lists her as "Manager of Music Co." (Ibid.).

Tuuk's book includes written correspondences from Robinson in 1971, in which she continues her recollection of the first sessions she did for the Hokum Boys:

> This was the first Hokum Boys session for Paramount, and the first session I managed for Paramount. This session was in Chicago, many others were in the Gennett studios in Richmond, Indiana.
>
> That session, I discovered I was in charge. Nobody asked me. They just put me in charge. The trio consisted of Dorsey, Tampa Red and Alex. Dorsey was formerly pianist for Ma Rainey and, with Tampa Red, made Tight Like That for Vocalion, but weren't [sic] under any contract. Alex was the lead singer. Dorsey sang with him on choruses, composed the songs and had suggested the trio's name: Hokum Boys, cause this is pure hokum. The session he made for me was one record: "Selling That Stuff'."
>
> (Tuuk, c. 2017)

"Selling that Stuff" was released by Paramount Records as catalog number 12714-A in December 1928.

Women in Talking Pictures

Ursula Greville (England)

There was no sound for early films; today we call them "silent films." In an effort to marry sound with picture, there were a few different ways early pioneers tried to add sound to film. In one method, a record was played along with the picture. In the United States, "Vitaphone" employed this method.

In England, a similar method was invented called "Syncrophone." It combined a gramophone, radio, and motion picture recorder. Its associated distribution label, Synchrophone Record, was "the earliest label to have a female as recording expert, **Ursula Greville**, who was credited on recordings from 1934 and 1937" (Hoffmann, 2004) (Thomas).

Margaret Booth (USA)

Margaret Booth was a film editor before the days of "sound films." As part of the Academy of Motion Picture Arts and Sciences' Oral History project, Margaret Booth recalled how many of the first people to work on film with sound were women. Notable names include Blanche Sewell, who edited MGM's first sound film in 1928, Viola Lawrence who edited the first sound film for Goldwyn, and Jane Loring, who was reported by the

Los Angeles Times to have edited a "misspoken dialog" in the 1929 film, *Fast Company* (Hatch, 2013).

British Women in Film Sound

In her paper, "Learning to listen: histories of women's sound work in the British film industry," Melanie Bell laments that much of the work of women in film sound in the UK was "elided" ("omitted"). Women were "over-represented as Foley artists," she writes, noting that "roles such as Foley artist, for which women were employed, did not receive screen credits until the mid 1980s." This made the history of women in film sound prior to this time difficult to trace. Still, Bell brought to light the roles that women played in early British cinema. A database of union sound editors lists 4,050 sound technicians from 1930 to 1991. Of those, 248 women were listed, of whom only one was employed in the 1930s. Fifty-five of those women were brought on during the Second World War, and 68 were employed from after the war until 1980. The other 124 women joined between 1980 and 1991 (Bell, 2017).

Bell's paper highlights a few notable names: Beryl Mortimer, who began her career around 1955, was recognized as the "mother of Foley." Christine Collins began her career in 1956 and found success as a sound editor. Bell notes that it was documentaries, not feature films, where women made the most progress (Bell, 2017).

Meenakshi Narayanan (India)

In the article, "Meena Narayanan: She Broke Barriers to Become India's First Woman Sound Engineer!" author Sayantini Nath writes about Meenakshi (Meena) Narayanan, who was married to Ananthanarayanan Naryanan, a director of Tamil cinema (films spoken in the Tamil language) who established the first sound recording studio for film in the south of India: Srinivasa Cinetone. A. Narayanan brought Meena on as an assistant for a sound engineer named Poddar – even though she was still in high school. "Despite no prior knowledge," writes Nath, "Meena was keen to learn and quickly grasped the nuances of the art from Poddar. She assisted him in the sound recording of the first talkie produced in Southern Cinema – *Srinivas Kalayanam* (1934), thus earning the enviable status. She would practice with the mixing console tied to one end of her madisar saree." Film historian S. Theodore Baskaran, in his book *The Message Bearers* wrote, "It was unthinkable that at a time when cinema was considered a taboo, she made an entry into the film world and became a successful sound engineer." (Baskaran)

"There are a lot of problems when sound recording is done by those who have no idea of the language and trend in music," said Meena. "Since I wanted to rectify it, I paid attention to sound engineering and gained experience in two years." (Kolappan)

She went on to engineer several films: *Krishna Tulabaram* (1937), *Vikrama Shree Sahasam* (1937), *Tulsi Brinda* (1938), *Porveeran Maniavi* (1938), *Mada Sambrani* (1938), *Sree Ramanujan* (1938), and *Vipra Narayana* (1938) (Nath). She left the film industry after Srinivasa Cinetone was destroyed in a fire in 1939. The loss contributed to stress-related health issues for her husband who passed away shortly afterwards. Meena passed away in 1954 (ibid).

■ The Emerging Record Industry

As the technology for recording music and speech developed, the record industry was born. Recorded sound evolved from Edison's novelty, special-use wax cylinders (some of which were even used in pay-per-play "jukeboxes") to mass-duplicated (flat) records, which were perfected by Emile Berliner, a German immigrant to the USA. By 1910, record sales reached almost 30 million. Three companies mass-produced records: Edison, Columbia Gramophone Company, and Berliner Gramophone Company (eventually Victor records) were the first labels and referred to as "the Big Three" (Voice, c. 2018).

The invention and growth of radio started to change all of that in the 1920s, and the record industry felt threatened by music being piped into homes for free. Nonetheless, record sales still soared above 100 million units in 1927, followed by a swift decline in sales during the Great Depression.

Throughout this bourgeoning history, women and underrepresented groups were left out of telling their own stories. This resulted, for example, in the jazz music of African Americans being performed by white bands in the studio, which may have been nice for the genre (people who otherwise had no access to the music could hear jazz for the first time) and profitable for record companies, but left Black performers out of the equation. Even white performers doing blackface minstrel routines were given access to the airwaves. The broadcast control room was completely off-limits to African Americans until 1928 (Vaillant, 2002).

It wouldn't be until the 1930s–1950s when things slowly began to change for women, including **Helen Oakley Dance**, **Mary Shipman Howard**, and **Vivian Carter Bracken**.

Helen Oakley Dance (Canada)

"I was the first woman producer," claims Helen Oakley Dance in a 1998 video interview (Dance, 1998). (She may or may not have known the story of **Aletha Robinson**, who worked for Paramount in 1928.) She was inspired by early Black music, although she wasn't a trained musician, "but I did have ears," she emphasizes (Dance, 1998).

Helen Oakley Dance was born in 1913 and grew up in Toronto, Canada. She then moved to Detroit around the age of 18 and met Duke Ellington backstage at show. She pretended that she was referred by British jazz musician Spike Hughes but later admitted forging the letter. Despite meeting under false pretenses, Duke was nonetheless impressed by her, and they became friends.

She produced her first records featuring Paul Mares and Charles Lavere for the OKeh record label. In a 1998 interview, host Monk Rowe asked how she managed to arrange recordings with the artists she was interested in. "I don't really know," she replied. "I just did. I went into the studios and set up a time" (Dance, 1998).

Yale's archives show she also wrote for *Down Beat* magazine from 1934–1941; however, she is not listed as an author (Emily DiLeo [Ferrigno], 2019). In 1935 she started organizing jazz performances. "There's no dance floor," she recalls telling her team. "It's a concert." For one of these concerts, she brought Teddy Wilson, Benny Goodman, and Gene Krupa together, who were collectively known as the Benny Goodman Trio. The place was the Congress Hotel, and it was the first known time an interracial band had played together on stage (Jazz, 2006).

Helen Oakley Dance

For her recording work, she "hired the guys and told them what I wanted them to play, and stood in the control room, and decided whether it was happening or not" (Jrobinson, 2017). "I don't think of producing it, I think of it being produced *through you . . .* you can be helpful of course, but the main thing is that you are a *channel*" (Dance, 1998).

World War II

The Second World War had a major impact on the field of audio technology. During this time, women began to assume various technical roles as men left their homes to fight. Also during this period, the **Audio Engineering Society** was founded in response to the war's negative impact on the recording industry. The "discovery" of magnetic tape recording by John "Jack" Mullins in Germany during the war began a new era of recorded sound alongside vinyl records.

Women had been fighting for their right to join institutions of higher learning, and by this time in history, there was a small and slowly increasing number of women with university degrees. **Rear Admiral Grace Murray Hopper** earned a PhD in mathematics from Yale University in 1934 and worked as an associate professor at Vassar College until the war began. "She joined the US Naval Reserve as a lieutenant and worked for the Bureau of Ordnance Computation at Harvard where she learned how to programme the Mark I computer" (*Biography of Grace Murray Hopper*) (History).

However, not every woman with a degree would be able to put it to use – at least not at first. One such story involves **Joan Clarke**. Although Joan was a mathematician, she was hired to do clerical work. She worked at Bletchley Park in England, where over 10,000 people were employed, 75 percent of whom were women (*Bletchley Park names 'secret' World War II codebreakers*). Joan would eventually become a master code-breaker. Another woman, **Patricia Davies**, was tasked with listening to and deciphering radio communications. It was possible to intercept German radio communications due to the physical effects of the ionosphere, which bounced radio waves back to earth over long ranges (League).

Still another example of women being hired for posts beneath their skill set includes **Mary Shipman Howard**, who bought her own recording machine and started cutting her own records before 1940; that's around the time she applied for an engineer's job at NBC. Because "girls weren't being hired for that sort of thing," she was brought on as a secretary. Eventually as men left NBC to join the armed forces, she got the chance to prove herself (Record, 1948).

Of course, the war took many lives and imposed harsh conditions on citizens of Europe. **Marianna Sankiewicz-Budzyńska** was deported from her home country of Poland to Germany where she was sent to forced labor. She was later imprisoned by the Gestapo when it was discovered she was working with other POWs in secret. After surviving that horrific experience, she joined the Gdańsk University of Technology and would find success in the field of communications and electronics (Pożegnanie doc. dr inż. Marianny Sankiewicz-Budzyńskiej, 2018).

Helen Oakley Dance, who had already established herself as a record producer, joined the US Army in 1942 after the death of her brother in Dieppe, France, and was assigned to the OSS (the Office of Strategic Services, which preceded the formation of the CIA). She was deployed to Algiers, Morocco, to assist in "the disposition of US undercover operatives and radio technicians being sent to occupied countries" (Dance, 2001).

Mary Louise Killick, in a memoir published by her daughter, explains that she manufactured sound recording cutting equipment for the armed forces. "My cutters were used by the military and BBC war correspondents to record live reports on gramophone records on the field of battle," Killick says. "As the war advanced, demand for cutters and recording equipment increased and I found myself more and more interested in the quality of sound recording" (Killick, 2018),

Daphne Oram became a "balancing engineer" at the BBC in 1942 "when male technicians were away fighting" (Hutton, 2003).

During World War II, **Kay Rose** became a civilian film apprentice for the Army Signal Corps, the department of the Army that creates and manages communications. She helped create training films such as *Report from the Aleutians*, a John Huston documentary. Kay said of the experience, "I was eager, and I learned a lot" (Tucker, 2018).

The role of Black women in the USA during the war was largely unseen in popular media, but their numbers in the armed services were impressive: 600,000 in the wartime labor force, 4,000 in the Women's Army Corp, and 330 in the Army Nurse Corps, not to mention women fighting for civil rights and "performing artists who gained critical acclaim during the war years" (Honey, 1999). This is especially important when speaking about civil rights within the context of the music industry. Black people and women in particular were faced with painful struggles, such as when Lena Horne and her band had to sleep with a traveling circus because no hotel in Indiana would take them because of racist segregation laws (Honey, 1999). William Grant Still, a Black composer with access to Hollywood connections, was asked in another instance whether he could recommend a Calypso singer to a white composer who would study the music and adapt it for the screen – while the Calypso singer would be unpaid (Honey, 1999). If African American men and women weren't welcome in hotels, on the radio, or on the big screen, how could they ever get in the control room? Fortunately things would begin to change.

The Audio Engineering Society

The Audio Engineering Society (AES) was founded in 1948. The original rationale for its existence, as told by charter member Donald Plunkett, was that although the development of audio technology progressed rapidly after the end of World War II, there seemed to be more focus on radio, broadcasting, and electronics than with the quality of recorded sound (Plunkett, 1998). (In fact, throughout the Society's existence, discussions on "quality versus available bandwidth" continued to be quite common.)

Given the context of this book, it is very important to mention that Plunkett happened to be chief engineer for **Mary Howard Recordings** (See: **Mary Shipman Howard**) (Sutton, 2018).

Not only were AES' founders concerned about quality, supplies were short because of the war. This had a negative impact on business, so a small group of men formed the Sapphire Group in order to network and to find supplies. Until this time, many companies did not talk to each other, but the war meant that formerly competitive companies had to cooperate in order to survive. Hollywood, a major entertainment capital, became a hub for these discussions. The first AES meeting was in March 1948 and featured a presentation about a new loudspeaker system. AES' overseas outreach began when Warren Birkhead of Capital Records moved to Japan, and the Japan Audio Society became the first international AES section (Plunkett, 1998).

On its list of charter members and in a photo of the inaugural banquet of the West Coast Section from Donald Plunkett's "Reminiscences on the Founding and Development of the Society," there are no women present; however, over the years, that would change (albeit very slowly).

Mary Shipman Howard (USA)

In 1940, **Mary Shipman Howard** began working for NBC in Manhattan, New York, as a secretary. During World War II, she was promoted to engineer. She is recognized as the "first American woman to own and operate a successful modern recording studio" (Record, 1948). She opened her studio, Mary Howard Recordings, in 1946. Her clients included Ethel Waters and Alex Templeton, among others. She also started a record label, "MHR," featuring Waters, Lucille Turner, and Dale Belmont, to name a few (Record, 1948).

Donald Plunkett, one of the Audio Engineering Society's charter members, was her chief engineer. "She was a musician who understood musicians and understood a good deal about recording and how to marry the two – both the personalities of the musicians and

audio record

PUBLISHED BY AUDIO DEVICES, INC.

Vol. 4, No. 2   444 Madison Avenue, N. Y. C.   February, 1948

On-The-Spot Recordings Integral Part of Regular News Broadcasts at WOR

Listeners Given Quicker Eye-Witness Coverage of Special News Happenings

Equipped with a transcription library valued at half a million dollars and a crack staff of on-the-spot reporters, WOR-New York has perfected the use of transcriptions in news broadcasts to what probably is its most mature development. This development, increased since the war, results in more authentic broadcasts and gives listeners quicker eye-witness coverage of news events.

Larry Pickard, WOR writer, selects a disc from the station's huge file of on-the-spot recordings.

When a news story breaks, such as the search for the missing recluse, Langley Collyer, WOR reporters are sent to the scene wherever practicable to record descriptions of the event which are in turn inserted into regular news broadcasts. Reporter John Wingate, for example, was on hand when Collyer's body was discovered, described the event and raced his recordings back to the station so that WOR listeners might hear a complete story before the newspapers had hit the streets. During recent investigations of the House Committee on unAmerican Affairs WOR newscasts were supplemented with recordings of actual testimony given during the hearings.

The wedding of Princess Elizabeth furnishes another example of the way recording

(Continued on Page 2)

Don Plunkett, Chief Engineer of Mary Howard Recordings, adjusts one of the mikes in the spacious New York studio while an artist sits at the piano waiting patiently for Mary Howard's cue to begin. Inset: Recording's own, Mary Howard.   Photo by Murray Laden and Edward O'tern

The War Gave Mary Howard Her Big Chance to Make Good in Recording; She Did — And How!

Before the War, many jobs in American industry were considered "man-sized" positions and therefore . . . for men only. But the War and its tremendous drain on manpower soon gave the female a chance to "strut her stuff." And one such lady, who took full advantage of this opportunity to prove that it wasn't strictly a man's world after all, was Miss Mary Howard, daughter of a well-to-do New England family.

Mary Howard had a flair for good music and records particularly intrigued her. To satisfy her curiosity, she bought a recording machine and started on her own trial-and-error course in record cutting. Miss Howard's interest in recording steadily grew — and so did her recording equipment. And then . . .

Mary Howard came to New York in 1940 and immediately applied for an engineer's job at NBC. As girls weren't being hired for that sort of an assignment, Mary Howard had to be content with a secretary's position in the engineering department. Then, her big break came. NBC, losing man after man to the armed forces,

(Continued on Page 4)

Tempus Fugit!
Student Radio Writers

Yes, time is flying! Only a few more weeks for you high school and college radio writers to enter one of the two big radio script writing contests. Entries for SCHOLASTIC MAGAZINES' Script Writing Competition (co-sponsored by Audio Devices) for high school students positively must be received before midnight, March 5, 1948. The 1948 National Script Contest, also co-sponsored by Audio Devices and conducted by the Association for Education by Radio, closes March 30. So you haven't much time to win one of the many valuable cash prizes. Act now! For complete contest details write: (for high school students) William D. Boutwell, SCHOLASTIC MAGAZINES, 220 East 42nd Street, N. Y. C. (for college students) Dr. S. P. Lawton, AER Script Contest Chairman, U. of Oklahoma, Norman, Okla.

Mary Shipman Howard in *Audio Record*

the temperament of recording equipment," Plunkett recalled in a 1999 interview by Susan Schmidt Horning (Horning, cited in Tucker). Hers was also one of the first small studios to Ampex equipment for tape mastering (Sutton, 2018).

After retiring from the music industry, she became a dog breeder. Howard passed away in 1976.

Marianna Sankiewicz-Budzyńska

Her students called her "mom" or "aunty" as she protected them against martial law when it was enacted in Poland during the early 1980s. Her defiance was understandable: she had spent time in labor camps during World War II, then was sent to Stalag II B in Hammerstein, and imprisoned by the Gestapo. During the liberation in 1945, she was led by a Soviet patrol to be executed "because she knew three foreign languages and was again considered a spy. The execution did not happen only thanks to an unexpected allied bomb attack" (Chojnacki, 2018), (Karaś, 2018).

Having survived all of this, **Marianna Sankiewicz-Budzyńska** returned to Poland and began working for Polish Radio in Warsaw. She moved to Gdańsk and earned her master of science degree from the Faculty of Electronics at the Gdańsk University of Technology in 1950. She became assistant lecturer in the Department of Radio Broadcast Engineering. She went on to earn her PhD and later became the first woman in the postwar history of Gdańsk University of Technology to be elected vice rector for education (Karaś, 2018). She retired in 1992.

Together with her husband, Gustaw Budzyński, she cofounded a new department of Electrophonics, which become the Department of Sound Engineering. In 1995 she became AES Vice President for Central Europe and helped establish several new AES sections in Lithuania, Ukraine, Russia, Belarus, and a new student section in Poland. In 1998 she was presented with an AES Citation Award and an AES Fellowship Award in 2005. She passed away in 2018 at the age of 97 (Karaś, 2018).

Postwar Women in Audio

Kay Rose (USA)

Kay Rose moved to Hollywood in 1944 with letters of introduction from the professionals she worked with in the Signal Corps. She waited weeks for someone to call back, got down to her last five dollars, and was ready to ask her family for bus fare home. Rose took a streetcar to the San Fernando Valley – where she'd never been before – and got off at Universal. She asked the guard at the gate if she could call the editorial department to ask for a job as a picture assistant. He laughed and said, "That's not how you get a job – but because you're here, I'll let you do it." He used his phone to dial someone in the editorial department. The guard handed her the phone, and the man on the other end asked, "Are you any good?" She said "yes." And he hired her. What Rose didn't know was a half hour earlier the editor's assistant had just quit (Tucker, 2018).

Rose stayed at Universal as a picture assistant for two years until the GIs returned from the war. Rose described the crew:

> "The editorial crew consisted of one editor, one assistant, one effects editor and one music editor. Depending on the budget, the assistant also cut both music and effects – on a good job, you assisted and did one or the other, but not both – even if you didn't know how. If you were the assistant and sound editor on a film, you'd give the music editing to a friend and if he was the assistant and music editor on another film, he'd give the sound effects to you – that's how you worked it. These films were very small movies, like *I Was a Teenage Frankenstein*. . . . They had little money, so when the monster breaks out of the lab and is escaping on the street, the crew ran by with lights pretending to be cars, and I had to add motors, revs and skids."
>
> "I was taught by men. One of the mixers, Bob Glass, Sr. had me come back every day at lunchtime for a couple of weeks and he and the dialogue mixer, Mac Dalgleish taught me how to lay out dubbing sheets for mixers – in essence, how to cut dialogue and sound effects."
>
> (Rose, 2001)

In 1951, she married film editor Sherman Rose. Together they produced a sci-fi movie, *Target Earth*, for $60,000; produced the first educational children's television series in the 1950s; and had a child, Victoria ("Vickie"), in 1953. They divorced after five years. Vickie Rose remembers her mom (her primary caretaker) coming home, and her hands were black from working on a Moviola (Tucker, 2018).

In 1961, (Kay) Rose did a film called the *Pit and the Pendulum* for American International Pictures (AIP).

> "My sparse library was not adequate for the Pit and the Pendulum torture chamber, so I went to Universal to see about renting their Foley stage which, it turned out, was too expensive for AIP's modest budget. Waldo Watson, Universal's Sound Department head, said, 'You don't need Foley – just use Universal's library.' So I did – and made it all up as I went along. You see, working on those kinds of low budget, mostly independent pictures caused me to be more creative out of necessity."
>
> (Rose, 2001)

Rose's first big-budget film was in 1966 – a Western called *The Professional*. Later in 1973, Rose was working on *Cinderella Liberty* with Mark Rydell and needed an apprentice. She asked if her daughter, Vickie, could work. To her surprise, the head of editorial said it was okay if no one else was on the available list. There was only one name on the list, but it was someone with a history of conflict, so Vickie got the job (Tucker, 2018). (**Vickie Rose Sampson** would eventually teach **Leslie Ann Jones** as Jones transitioned from music recording to movies.)

In 1974, Kay worked on the film *California Split*, the first movie where dialog was recorded on 8-track, 1-inch tape.

> "That was such fun, making something work that had not been done before and that nobody knew how to do," she said. "California Split was recorded on 8-track

so that Bob [director Robert Altman] could have separations – separations by place (different rooms) or area (background or foreground) or various combinations. For example, one sequence had seven mikes set up in various rooms so that the actors could walk around and talk and never go off mike. Jim Webb was the production mixer – he had recorded rock concerts like Joe Cocker and others, so he was familiar with multi-track set-ups. . . . He tried to keep each actor on his own track but since almost everything was ad-libbed that didn't always work."

(Rose, 2001)

In 1984 Kay Rose became the first woman to win an Academy Award for Sound Editing for *The River*. Over the years, she would work on such famous films as *Comes a Horseman*, *The Rose*, *Ordinary People*, *On Golden Pond*, *California Split*, *Crimes of the Heart*, *The Milagro Beanfield War*, *Black Rain*, *Robocop 2*, *Switch*, *The Prince of Tides*, *For the Boys*, *Blake Edwards' Son of the Pink Panther*, and *Speed* (Oliver, 2002).

She was honored with a lifetime achievement award from the Motion Picture Sound Editors (MPSE) organization in 1993 and a career achievement award from the Cinema Audio Society in 2002. In October 2001, directors George Lucas and Steven Spielberg endowed the Kay Rose Chair in the Art of Sound and Dialogue Editing at the USC School of Cinema-Television (Oliver, 2002).

Rose passed away in 2002. "To me, the story rules. . . . Finding a balance between music and effects that supports the story in the best way possible has always been my focus. It was all a lot of fun. . . . I loved every minute of it" (Ibid.).

Vivian Carter Bracken (USA)

Vee-Jay Records was the first record label in the United States to release music by the Beatles. **Vivian Carter Bracken** founded the company with her husband Jimmy Bracken (the "J" in "Vee-Jay") (Hillery).

Bracken was born in 1921 in Tunica, Mississippi, and later moved to Gary, Indiana, which is just over the state line from Chicago, Illinois. In 1948, she won a contest for the "best girl disc jockey in Chicago." The prize was her own 15-minute radio show (Callahan, 2006).

She called her show "Livin' with Vivian" and played music by Black musicians, much of which was not available commercially (Ibid.). Kids could see her spinning records through the store window, and as a record played, she might come outside to "mingle with the kids to find out what they liked or disliked about each one." In 1953, she and her husband started Vee-Jay Records (Hillery).

Their first artist was a group called the Spaniels, who had attended the same high school as Bracken in Gary. She let them rehearse in her mother's garage and arranged for them to record at Chance Records in Chicago. After some time, Chance Records went out of business, and Vee-Jay recruited Chance's accountant, Ewart Abner, to handle management at Vee-Jay (Lane, 2011).

With the Vee-Jay label, the Spaniels achieved chart success with two singles, one of which was "Goodnight, Sweetheart." Over the next decade, Vee-Jay records released singles by John Lee Hooker ("Boom Boom"), Gene Chandler ("The Duke of Earl"), The Four Seasons ("Walk Like a Man" and "Big Girls Don't Cry"), The Dells ("Oh What a Night"), and

Vivian Carter at WWCA
Source: Courtesy of Calumet Regional Archives, Indiana University Northwest

Dee Clark ("Nobody But You"). The also signed the Pips (who later joined Gladys Knight on another label). Before long, Vivian Bracken was quite wealthy (Lane, 2011).

In 1962, Vee-Jay bought the rights to a song that was popular in Europe, "I Remember You" by Frank Ifield. "The British agent insisted they also take a quartet named The Beatles, unknown at that time in the United States" (Hillery). The agent referred to here was Paul Marshall, who worked for EMI Records affiliate, Transglobal. Interestingly, Vee-Jay's international rep at the time was Barbara Gardner, a Black woman who brokered this deal and went on to become the first known Black woman to own and operate her own ad agency (Clayman). The Beatles' album was called *Please Please Me*, but Vee-Jay omitted the single of the same name and called the album *Introducing . . . the Beatles* (various).

During this time, Capitol Records, who had initially turned down the Beatles, saw the band's popularity mushrooming and began planning to release new Beatles' material. As a result, Capitol released *Meet the Beatles!* just 10 days after *Introducing . . . the Beatles* was released in January 1964. But only just seven days after *Introducing . . .* was released, Capitol Records served Vee-Jay with a restraining order to stop them from distributing any more albums. However, Vee-Jay was allowed to keep selling into late 1964 and sold 1.3 million copies, going gold and platinum (Ibid.).

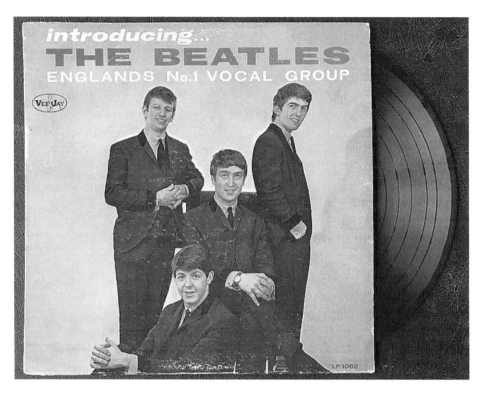

Vee-Jay Records' release of *Introducing . . . The Beatles* (1964)

Eventually, combined with some unfortunate business decisions and excessive spending habits by Abner, Vee-Jay Records went bankrupt, and in a few years, the IRS seized Bracken's record store in Gary for back taxes. Vivian Carter Bracken went back to radio, hosting a late-night talk show until 1982. She passed away in 1989 (Lane, 2011).

Cordell Jackson (USA)

When she was 12 years old, people – both men and women – used to tell Cordell Jackson that "girls don't play guitar." "Well, I do!" she shot back. She also played mandolin, piano, harmonica, and banjo (Conover, 1991) (Guitar Granny // Cordell Jackson Was Playing Rockabilly When Elvis Was a Babe, 1992).

In the article "Guitar Granny," she recalls how she came to be called the "first female recording engineer",

"After high school, I came up to Memphis and was working at an airplane factory: I played upright bass in the Fisher Aircraft band for about five years. Right after that, I wanted to have a record for the radio stations here. That's when I bought my very first recording equipment. In 1947 – it is documented – I cut my first record. At that point I had not started writing my own music. In '48, I started writing my first song. I started my own recording company in '56 and have been recording ever since – just developing little rhythm patterns and recording songs since then. I'm believed to be the first female engineer to produce music in America."

(Guitar Granny // Cordell Jackson Was Playing
Rockabilly When Elvis Was a Babe, 1992)

By now of course, we have established that there were other women and engineers before her, so perhaps we can say she is the *first woman engineer and producer for rock 'n roll records*. Jackson even wrote a book, *The Women Record Producers of Memphis*, but neither its manuscript nor any copies can be found (Branstetter).

Ethel Gabriel (USA)

Ethel Gabriel has produced over 2,500 albums and was the first woman record producer for a major label. She was the first woman to receive an RIAA Gold Record and would continue to win 15 Gold Records and 2 Platinum records during her career. She earned a Grammy for

Ethel Gabriel with Gold Records
Source: Courtesy of the Gabriel/Mauro family

Best Historical Album in 1982 for *The Tommy Dorsey/Frank Sinatra Sessions – Vols. 1, 2 & 3.* She produced *Cherry Pink and Apple Blossom White* by Perez Prado, which became a hit and introduced mambo to the USA. She went to Memphis and helped RCA sign Elvis Presley (Spevak, 2013).

Born in 1921 near Philadelphia, she was a dance-band leader in her teens and studied trombone, at one time playing with the Philadelphia Women's Symphony Orchestra. She graduated from Temple University and Columbia University (Ibid.). While at Temple, in 1940 she got a job as a secretary at RCA Records to earn money to help pay for school. She worked as a record tester doing quality control. "If it was a hit," she said, "I got to know every note because I had to play it over and over and over" (Kooomar, 2007). She worked her way up to A&R assistant "editing and breaking new records," and by the late 1950s, she had become the first woman A&R producer in pop, with artists such as Perry Como and Chet Atkins in her repertoire (O'Brien, 2004).

However, she recognized that women needed solidarity. "No one knew what networking meant," she said. "They were too busy making coffee for the bosses." So she began a networking series for women. Among the women to have benefited from these events and opportunities include Nancy Jeffries and Linda Moran, who went on to head record company departments; Jeffries became senior vice president of A&R at Elektra (O'Brien, 2004).

Gabriel worked for RCA for 40 years, finally becoming vice president of A&R in 1982. In 1984, a so-called "friend," Robert Anderson, convinced Gabriel to invest her life savings into a new music company. Shortly afterwards Anderson was sentenced to bank fraud in 1987, and the company never materialized. Gabriel lost all of her assets – even her gold records (Spevak, 2013).

In 1990 she wrote a fiery letter to the editor of *Billboard* magazine, responding to a male author who had written that there was no such thing as a woman producer because women didn't have "good ears" (see: sexism).

In 2013 Gabriel's nephew, Ed Mauro, successfully lobbied the RIAA to reissue her gold records, and Gabriel was presented with them at the Rochester Presbyterian Home, where she lived the rest of her life.

> "Even being here in one room, when I had an 11-room house," she said. "With an indoor swimming pool and a maid. The things that happen to you, you wonder, 'Why me?' People are made an example, so other people learn. I think the good Lord knows what he's doing, and why it happened."
>
> (Spevak, 2013)

In 2019 SoundGirls.org announced the Ethel Gabriel Scholarship. Also in 2019 SoundGirls began production on a documentary about Gabriel's career and impact on music. The crew for the project includes Caroline Losneck, Christoph Gelfand, **April Tucker**, and **Karrie Keyes**.

Marion Keisker MacInnes (USA)

Marion Keisker MacInnes, known in most research literature as Marion Keisker, was the first to record Elvis Presley. She was a secretary at Sun Records in 1953, whose office also doubled as a recording studio called the "Memphis Recording Service." Elvis came by to make a demo of two songs, "My Happiness" and "That's When Your Heartaches Begin." Sam Phillips, the owner, was not there that day, so Keisker made the recording.

Keisker recalls an exchange between herself and Presley that became legendary.

> I said, "What kind of singer are you?" He said, "I sing all kinds." I said, "Who do you sound like?" He said, "I don't sound like nobody." (Company)

Elsewhere in her career, Marion Keisker MacInnes was host of a radio show and an officer in the United States Air Force, and she became active in the women's rights movement (Company), becoming chapter president for the National Organization for Women in Memphis, Tennessee.

The 1960s

Wilma Cozart Fine

"**Wilma Cozart Fine** was one of the first high-ranking women executives in the recording industry," recalls her son as he accepted a Grammy Trustees Award[1] in her name in 2011. "Her career spanned and embraced recording and production innovation from the dawn of the LP era to the age of the CD. She loved classical music so much that she wanted to evangelize it; to produce works of enduring quality, and as the Mercury trademark said, 'living presence' so that people everywhere could share the joy she felt from listening to music" (Grammy.com, 2011).

Fine was the recording director for Mercury Records' Living Presence classical recordings during the mid-1950s and 1960s, and she was well known for her advice to "trust your ears" (Pearson, 2014).

Although she was careful not to call herself an engineer, as a producer it is clear that she had command of the technical concepts of recording, production, imagery, and quality. Interviewer Bruce Duffie asked her about this in a 1995 interview, and this was her response:

> "First of all, I'm not an engineer. I'm more a music person who works with the engineer. I'm what is called, today, the producer. . . . For these recordings we had a very wonderful team of people who made most of the catalogue . . . what we were trying to do was to let the listener hear the performance as it actually sounded in the concert hall."
>
> (Pearson, 2014)

In the same interview, she describes how they recorded using binaural recording and three-channel stereo recordings.

> "On the *Capriccio Italien* . . . we recorded a binaural master and a 3-track master because at that time we were testing whether we wanted to do 2-channel or 3-channel stereo recording. We did not have two three-channel tape machines because the first one of those was made for us.
>
> "If you notice, with the Mercury records, you always hear the center, but it isn't really there. It has been combined into the left and the right, and done in such a way

that the illusion is complete all the way across the stage. In other words, you don't have a hole there. You have the winds sitting back where they would when the orchestra was in its normal seating on the stage. I try to recreate exactly what that microphone picked up at the time of the session. . . . How close the mics are to each other is greatly influenced by the acoustics in the hall, so it's certainly not an arbitrary thing.

(Pearson, 2014)

She and her husband, C. Robert Fine, headed the classical music division for Mercury Records. Her husband was a recording engineer, and together they developed recording techniques in the 1960s, including recording onto 35mm film instead of traditional tape recorders, and they were among the first to mass-market stereo recordings. Their artists include Rafael Kubelik, Antal Dorati, and John Barbirolli; the composer and conductor, Howard Hanson; the Chicago Symphony Orchestra and the Detroit Symphony; the pianists Byron Janis, Gina Bachauer, and Sviatoslav Richter; and the cellist Mstislav Rostropovich (Kozinn, 2009).

The BBC Radiophonic Workshop

The BBC Radiophonic Workshop was established by **Daphne Oram** and Desmond Briscoe in 1957. It predated the invention of synthesizers and would have an impact on electronic music and sound effects for years to come.

Daphne Oram

On Friday, June 24, 2016, the London Contemporary Orchestra performed *Still Point*, **Daphne Oram's** piece for live orchestra and "live electronic manipulations," (almost 70 years after the piece was composed and 13 years after Oram passed away in 2003) at St John's Smith Square as part of the Southbank Centre's DEEP∞MINIMALISM festival (Dodero, 2016). On July 13, 2018, the piece was realized again at the BBC Proms in the Royal Albert Hall. That same evening, work by **Delia Derbyshire** was also featured in the concert. One of the people in the audience was **Suzanne Ciani**.

Composer and turntablist **Shiva Feshareki** felt it was her destiny to realize the piece, which she performed with James Bulley, two orchestras (one of which was "within the turntable manipulation," explains Feshareki), five microphones, "treated instrumental recordings," and echo and tone controls (BBC, 2018). In her thesis, "A quest to find 'real': September 2014 – February 2017," Feshareki explains:

> "The piece is scored for 'double orchestra', which in this case is a concept, as opposed to pieces like Michael Tippett's Concerto for Double String Orchestra (1938–9), which is quite literally that. After studying the materials I had gathered, I took the 'double orchestra' concept to mean an acoustic orchestra, and an orchestra made up of the orchestral material on the 78 RPM discs, the turntables, the microphones and the live- electronic manipulation. The word 'double' also refers to the fact that, in rehearsal, the orchestra is recorded in a 'dry' acoustic, and it is recorded again (or by a second orchestra) in a 'wet' acoustic."

(Feshareki, 2019)

Daphne Oram draws symbols using her "Oramics" technique
Source: Photo courtesy of The Daphne Oram Trust

Feshareki was interested in Oram's work and visited Bulley at the Goldsmiths Oram Archive. "*Still Point* was in the archive in a box full of loose sheets of manuscript," Feshareki recalls. "James showed me the box and suggested this could be the first work ever written for live electronics but still unperformed. When I realised this was a piece for turntables and orchestra I felt quite overwhelmed. I will never forget that moment. I promised James I would find this piece its world premiere," she says (Dodero, 2016). Feshareki kept her promise and mentioned the idea to Oliver Coates (DEEP∞MINIMALISM founder) who programmed the piece a few weeks later. "I am starting to feel that to complete the concept of the work, Oram somehow had it planned for a world premiere so long in the future! The piece was beyond groundbreaking for 1949; it was futuristic. It really has been a mind-bending experience" (Ibid.).

Daphne Oram was born in 1925 in Wiltshire, England. As a child, she wanted to build a machine that would produce any sound she desired. This machine was later realized in the form of her "Oramics" machine, built for that purpose (see photograph). The machine allowed her to draw a sound and play it back (Hutton, 2003).

During World War II, Daphne was hired at the BBC and developed a career as a balancing engineer. The development of magnetic tape-recording technology soon followed, and in the 1950s during her career at the BBC, she latched on to its potential not just for recording sound but creating and manipulating it as well. She wanted to borrow some equipment to experiment, but the senior engineer with whom she spoke with in the BBC department was dismissive.

"He reduced me to a very small height and finished by saying 'Miss Oram, the BBC employs 100 musicians to make all the sounds they require, *thank you*'. I went into broadcasting house late at night and collected from the listening rooms of all the studios around all the phonographs I could and started and I worked until I believe I looked rather grey."

(Chambers, 1979)

Fortunately she found a colleague at the BBC named Desmond Briscoe who was also interested in electronic music. They teamed up to write the score for a play called *Private Dreams, Public Nightmares* and which was subtitled *A Radiophonic Poem*. After a time, their electronic composition style caught on, and they obtained the resources to start the BBC Radiophonic Workshop (Hutton, 2003). However, she left just 10 months later. Dick Mills, who worked with Daphne in the workshop, recalls the departure.

"The BBC in its infinite wisdom even back in those days in the 50s seemed to have a thing about health and safety and they thought that anybody that was exposed to experimentation with weird sounds and things like that shouldn't work more than six months at the workshop. Now Daphne – this was her life's ambition. She'd gotten to this pinnacle of creativity not only in programme making but in development of a department and said, 'Well I'm sorry, but if you insist that I can't work here then I shall have to consider my position.'"

(Chambers, 1979)

Up until her departure, she had been composing "radiophonic art pieces" for the BBC, but she also had her own personal compositions from the 1940s. So after she left the BBC, she composed *Four Aspects*, which premiered in 1959. She also composed music for a film, *The Innocents*, in 1961 (Hutton, 2003).

"I'm interested in being able to manipulate sound to give every subtle nuance that I want. There seems to be no real notation system in electronic music. I wanted a system where I could graphically represent what I wanted and give that representation, that musical score and that to a machine and have from it the sound. I'm finding that what one has to do is to pick out each parameter separately. You want to be able to give a graph for how loud it is at that particular moment how the vibrato is giving a waver into the pitch. And so I have a number of film tracks going by and on this I draw the graphs for the pitch: I put what we call 'digital information': that is I can put a dot rather like a crotchet or quaver. I've been thinking about this for years actually I believe my father said when I was seven years old I was predicting that one day I would have a marvelous machine that would make any sound I wanted."

(Chambers, 1979)

For more information about Daphne Oram, visit DaphneOram.org.

Delia Derbyshire (England)

In 1937 when Delia Derbyshire was born in England, things were becoming slightly better for women in terms of gaining entrance to universities. During this time, Derbyshire supposed about 30 percent of admittees to universities were women. That number was the same in the fifties, when Derbyshire was making a decision about where to go to school. "Her degree was in math and music, and her approach to music was mathematical: she used graph paper and a slide rule," remembers **Dick Mills,** who worked with Daphne Oram and was there when Derbyshire began at the Radiophonic Workshop in 1962:

> "When she first came to the workshop, she rummaged along the bookshelves, and said, 'Oh! You've got this book'. And this book had a lot of tables in about frequencies. Anyway, she went through this book and she found *masses* of faults and mistakes in the tables, and she sat down and corrected all of those before she started."
>
> (Blake, 2009)

The Radiophonic Workshop existed before synthesizers, so Derbyshire was creating music with metal lampshades, wine bottles, sine waves, and white noise. Each sound was recorded onto analog tape, and multiple tape machines were used to play back the sounds. Derbyshire and her colleagues weren't called "composers." Rather they were called "assistants" – probably to pacify the orchestras who were creating music in the "traditional way" (Blake, 2009).

In an interview much later, she hints that comments about her music may have been gendered and comments about her role as a post-feminist in a pre-feminist era:

> "The music I was doing in the 60s was either too lascivious or too sophisticated. But, no; we're going back to the sexes thing. And so people would say to me, 'You must be an ardent feminist'! But looking back I think I was a post-feminist before feminism was invented."
>
> (Ibid.)

Her most well-known piece was the theme for *Doctor Who*. However, after her death, 267 reel-to-reel tapes were found in her attic. These archives can be accessed through the John Rylands Library at the University of Manchester in England.

In the spirit of continuing her legacy, Caro C (Caroline Churchill) was inspired to set up Delia Derbyshire Day with Ailis Ní Ríain and Naomi Kashiwagi in 2012. The organization was established as a registered charity whose mission is to "advance the education of the public in music technology and the history of British electronic music via the archive and works of Delia Derbyshire" and "advance the art of British electronic music via the archive and works of Delia Derbyshire." The group holds events and workshops in creating tape loops and making electronic music (DeliaDerbyshireDay.com).

Fun Facts: The Deliaphonica Game

The site, DeliaDerbyshireDay.com, has created an app so you can experiment with sounds in a similar way to Delia Derbyshire.

The game was created in honor of Derbyshire's eightieth birthday. When you launch the game, you can click on one of the squares to start or stop a "tape loop." The large squares at the top are loops, and the smaller squares at the bottom are "one-shot" samples that only play once per click.

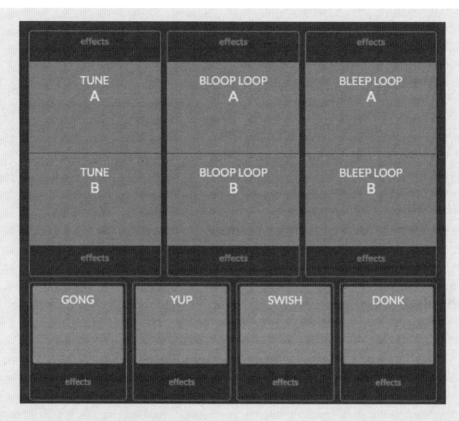

Delia Derbyshire Day "Deliaphonica Game"

Give it a try: go to https://deliaderbyshireday.com/deliaphonica-game/. The Deliaphonica game was developed in collaboration with David Boultbee, BREAD art Ltd.

Bell Labs

In his memoir, *Memories: A Personal History of Bell Telephone Laboratories*, A. Michel Noll recalls the innovations taking place at the facility in the 1960s. "During World War II, women were employed at Bell labs in what was called the girl's model shop, doing soldering and machine work to build models. However, at the end of the War, they were laid off as the men returned from service," he recalls (Noll, 2019).

Noll also remembers the business was very male-dominated, but there were programs to attract women to the company. "There was no conscious effort to avoid hiring female scientists – there simply were not many," he writes. "Bell Labs was beginning to make efforts to recruit women and to give them opportunities to work at Bell Labs during summers, with the hope to attract them to science and engineering." Some women, like **Carol Bird McLellan**, would eventually be promoted to Member of Technical Staff (MTS) (Ibid.).

In 1961, Bell Labs published a book called *Music from Mathematics*, which they sold with a phonograph record and booklet by Bruce E. Strasser. Max V. Mathews was a technical consultant on the project. Over time, Mathews would work with **Laurie Spiegel**, **Suzanne Ciani**, and **Elizabeth Cohen**. In fact, Spiegel composed pieces at Bell Labs, working with Mathews on computers like the DDP-224.

Bell Labs was also engaged with research in acoustics and speech processing. Manfred Schroeder acknowledges his indebtedness to "Mrs. Carol Bird" for her assistance in his seminal paper "Natural Sounding Artificial Reverberation," and indeed the two co-authored a paper with B. S. Atal on the "Early investigation of computer modeling of room acoustics," which became a forerunner of **ray tracing** (Brooks, 2003).

Ilene Busch-Vishniac was also at Bell Labs as a postdoctoral student and researcher, where she developed devices for microphones and earphones and addressed the "rain barrel" effect heard on conference calling systems (Busch-Vishniac, 1998).

1970s Second-Wave Feminism and Women's Music

The elements that define "women's music," according to author Bonnie Morris, include "a message that put female experiences and women's oppression at the center; woman-only shows wherever venues permitted the exclusion of men; publicity and staging that rejected a passive, commercialized female image and embraced powerful images of women; and an effort to keep the **business side of recording and distribution** in women's hands [emphasis added]" (Morris, 2015).

Virgo Rising

In 1973, producer **Mollie Gregory** got the idea to release an album that was performed, engineered, and produced by women. The album was called *Virgo Rising: The Once and Future Woman.*

"The album is a sort of private consciousness-raising session," the liner notes read, "songs to diaper babies by, songs to rivet or sculpt by, to drink or type or draw up the terms of your divorce by, to shelve books by, to make laws or run for office by, mow the lawn or fix the door by, songs to raise the heart when low, songs of sisterhood" (Various, 1973). The album was recorded at various locations in the western United States, not in "fancy studios" and with no overdubbing.

Singer Kit Miller was 17 at the time. In an interview with Sue Barrett, Miller reflects, "Virgo Rising was one of the first women's albums – all produced and performed by women, and all the songs were about women's rights. . . . It was part of a can-do spirit that had grabbed our communities – we could make our own record! . . . It was the early 1970s. Women's record companies like Olivia and Redwood didn't really exist until a few years later" (Barrett, 2008).

Joan Lowe was invited by Gregory to engineer the album. Lowe recalls transporting analog recording equipment around to make the recordings. "*Virgo Rising* was a real trip! Old-style analog recording equipment hauled all over to record the many artists on the album. Most sites were definitely not optimum for recording and some were live," she

Virgo Rising album cover, released by Thunderbird records
Source: Artwork by Josephine Cameron

recalled. "We recorded Charley's Aunts in their parents' living room in San Rafael with the raccoons they fed nightly out on the deck looking in and probably wondering what this noise was all about" (Ibid.).

Although the interview by Barrett took place several years later in 2008, Lowe remembers that it was one of a kind and perhaps the first of its kind. "I can't be sure about how many other woman-produced albums were out there at that time, but none like this," she says. "Some independent women were learning they could self-produce and that they did not have to wait for some record company to sign them on. There certainly were not many out there then. . . . I see the album as a definite marker in breaking through barriers to women and their ability to make music in the same way as had been traditional: being discovered by an A&R person for some label. Women learned they could manage the whole scene from studio to actual product" (Ibid.).

"The music business is hard," she says, "requiring dedication far beyond what one might expect. The person who thinks it's just a lot fun to get out there and tour from gig to gig, don't realize the extraordinary commitment it requires" (Ibid.).

June Millington and Fanny

"One of the most important female bands in American rock has been buried without trace: Fanny. They were one of the finest fucking rock bands of their time, in about 1973. They're as important as anybody else who's ever been, ever, it just wasn't their time. Revivify Fanny. And I will feel that my work is done."
            – David Bowie in an interview with Rolling Stone, 1999 (Berrett, 1999)

Any band would love to have this kind of celebrity endorsement, but to whom is this star-fueled command to "revivify Fanny" directed? To the record industry, to historians, to us, to women? Let's face it – although the late Bowie's work might be done (posthumously) when Fanny is revitalized, a woman's work is really *never* done.

Perhaps no one knows this more than **June Millington**, who founded the band Fanny with her sister Jean 50 years ago (as of this writing). June was born in the Philippines, and she and her sister found themselves struggling to make friends in the USA when they arrived in 1961. They faced horrible prejudice, but after playing at a talent show in junior high school, they received some positive attention. "Kids started coming up to us and telling us they liked it. So it dawned on us this was a way to make friends" (Albertoni, 2011).

In 1964, Millington and her sister started a band called The Svelts in high school. In the fall of 1968, they got a contract with Reprise Records. They were the first female rock group signed to a major label. After a few personnel changes, they renamed the band Fanny. They achieved Billboard chart success with singles such as "Butter Boy" (peaked at #29, 1975), "Charity Ball" (#40, 1971), "I've Had It" (#79, 1974), and "Ain't that Peculiar" (#85, 1972) (*Fanny: Chart History*).

Millington would produce and perform on records for Olivia Records artists Cris Williamson (*Changer and the Changed*) and Holly Near (*Fire in the Rain*), both of which were engineered by women (**Joan Lowe** and **Leslie Ann Jones**, respectively). Later when June decided to go solo, friend and engineer Tret Fure recorded her album, *Heartsong*, which was subsequently picked up by Olivia Records.

In the 1980s, Millington cofounded the **Institute for Musical Arts** and reunited with members of Fanny, reemerging as "Fanny Walked the Earth," whose story will be told in a documentary – and as a musical.

Olivia Records

Olivia Records was founded by Judy Dlugacz, Ginny Berson, Meg Christian, Kate Winter, and Jennifer Woodul in 1973. As the story goes, Berson and Christian had approached Cris Williamson, who was an established performer, to ask what Williamson thought about women's music and whether she would want to be interviewed on a new radio show, "Radio Free Women." As Berson and Christian revealed their affinity and passion for music through their conversation, Williamson asked, "Why don't you start a women's record company?" Within days, they did (Ibid.).

The name "Olivia" paid homage to a book published in 1949, called *Olivia: by Olivia*, written by Dorothy Bussy, which was an "evocative memoir of adolescent lesbian love" (Ibid.). The "Olivia collective" solicited new recordings and published their group's mission in a periodical for lesbian music journalism, called *Paid My Dues*. In it, they announced, "We are a group of lesbian/feminists who understand the need for women's music – music that speaks from all our experiences and towards all our thoughts and emotions. We intend to provide large numbers of women with access to the recording industry. We are interested in high quality music that is not oppressive to women" (Ibid.).

By 1974, the group's organizers located and recruited engineer **Joan Lowe**, who owned a studio in Oregon. Olivia Records recorded a single featuring Olivia cofounder Meg Christian and Cris Williamson, which made just $300. Wanting to reach a much larger audience, Christian sold the single during her tour and asked for donations for the next Olivia Records release, actions which subsequently brought in $15,000. The funds were used to record Christian's album, *I Know You Know*, which achieved $80,000 in sales (Ibid.).

On the album cover, Olivia Records refined its mission a bit, including the goal to "offer training in the technical, musical and other fields related to the recording industry," which was an early example of women's-only technical recording training.

In 1975, they were set to release their next album, *The Changer and the Changed*.

The Changer and the Changed

In 1975, Olivia Records produced their second, full-length album, *The Changer and the Changed*, featuring Cris Williamson. "With a voice famously described by Bonnie Raitt as sounding like 'honey dripping on a cello' . . . the album helped pave the way for latter-day Lilith Fairs and riot grrrls" (Gilbert, 2015). The album sold 250,000 over the next decade or so.

This time, they were able to connect with **June Millington** of Fanny. Other musicians included Margie Adam, Holly Near, vocalist and percussion player Vicki Randle (who later won a very public, featured position in the *Tonight Show* band), cellist Jackie Robbins, and banjo player Woody Simmons (Morris, 2015).

The engineer for this album was **Joan Lowe**.

Joan Lowe (USA)

Joan Lowe (Ross) was born in Pasadena, California, in 1928. She loved electrical engineering as well as other areas of science and technology. She worked for NASA at the MIT Lincoln Laboratory in Massachusetts.

Lowe was the engineer for *The Changer and the Changed*. She was also the engineer for *Virgo Rising: The Once and Future Woman* and later formed her own record label, Pacific Cascades Records. Lowe was kind enough to recount her experiences as an audio engineer to me shortly before she passed away in early 2019.

"There were no schools or classes in professional audio recording or audio work of any kind that I was aware of," she wrote. "When such training became available, it opened a new vocational opportunity which still suffered from preconceived concepts of hirable personnel.

"As far as the recording industry I had no single mentor or role model as such except briefly, and only now and then as I tried to work as an apprentice, often unpaid. It was clear that the production side of actual recording as well as much of the entire industry was tightly controlled. I knew I would have to do things very well and make my way carefully. I initially knew of no women in engineering or highly technical positions. I learned the hard way by observation coupled with my technical background and deep, loving interest in music in all its aspects. The only way to get started was as a 'gofer' doing errands of all sorts, dubbing, etc. This was my lot for a long time until steady perseverance and proof by small opportunities to contribute to actual recording finally came about. By the time I earned recognition, I was aware of but few women in engineering. One [of the women] that I knew of, but was not acquainted with, was in mastering at either Capitol or ABC, as I recall.

"My career as a recording engineer included working with labels such as Capitol and RCA but in a kind of individual assignment way, as needed. I freelanced now and then increasingly as I trended toward complete independence. I worked in a variety of types of recording: studio, location, film, but always primarily music. Recording as a large orchestra played while the movie ran was always fascinating to me. I was never the primary engineer on a session of that size, and only for a few smaller such assignments on location. The complexity was extraordinary which has changed a great deal since then with the development of digital over analog and direct input availability for many instruments which had to be miked previously.

"Olivia [Records] never had their own recording facilities, but always engaged a well-recognized professional studio for each separate project; at least, while I was still privy to their operations. I was basically an expense-paid volunteer by individual project, not staff, and left when I felt I had helped launch them in the ways of the recording industry, from studio, mastering, pressing management, album jacket design and printing, and all the details of getting an album ready for market. *Changer* was done in a beautiful studio in Alhambra, California, with well-equipped acoustic spaces and a large center studio and equipment. Just before we began the actual album, they installed a newly designed state-of-the-art 24-track board which was then the top of the line and a great help throughout. It had one bug that I had to cope with, but it was a seldom used component and did not present much of a problem by the time we got it straightened out.

"Most of the professional recording equipment of that era was made by Ampex and was analog, running at speeds of 15 or 30 ips [inches per second]. I used existing studios for work except for a certain amount of equipment that allowed me to edit, mix certain sources, copy, but also had a full complement of (very heavy) two half-track Ampex PR-10's in travel cases, a half-track Ampex 354–2 which was a 65-pound portable deck in a travel case, two matching Ampex mixers and lots of other equipment for location [recording]. These recorders were tube models, and the 354–2 had the best reputation for preference in the resulting true sound. I concur [with that assessment]. Almost all the studio equipment was professional Ampex. In freelancing or individual assignments, I also found that good microphones were a necessary part of my equipment. In addition to the basic standbys, I added a factory balanced, stereo pair of Neumann U-64's with their dual power supply and accessories to cover areas that benefited especially from the quality afforded by them.

"*Virgo Rising* would seem to fit the description of the first all-female production, but there may be another that qualifies about which we may know nothing, so I can't absolutely verify it. Certainly, it's the first that is somewhat known."

Joan went on to become the founder of Pacific Cascade Records.

"My involvement with Pacific Cascade Records is that I founded it and continued it for many years, finally closing it several years ago. I saw the influence and results of special people doing effective teaching and activity work which also covered some specialties in pre-school and other early childhood situations and wondered if that could be transferred via recording. I found excellent, proven, productive people whom I tested for viability from actual physical presence effect to recorded effect, some already known in entertainment and other fields, to add to my catalog. PCR won three special Parents' Choice Magazine awards.

Joan Lowe as a child c. 1936   Joan , age 20 (c. 1948)   Joan (late 1950's early 1960s)

Joan in the 1980s   Joan and her sound crew (around the 1980s)

Cris Williamson in a Jaguar XK140 that Joan restored   Joan Lowe in the 2000s

Engineer Joan Lowe through the years

PCR was available in the U.S. and Canada with some internationally. You could say it was a kind of 'designer' label for a special market.

"I became disenchanted with the commercial record industry from the aspects of treatment of studio staff/techs/engineers, abuses by producers, times of drug-saturated sessions and more. I then moved into my own business which was a mixture of location and studio work plus the establishment of Pacific Cascade Records which also put out a few regional country music 45's just for fun.

"Just a few comments which you might enjoy: Gone, I suppose, are the days of '**flangeing**' and locating physically by ear-and-hand the master for overdubbing and editing and some of the other hands-on skills an engineer had to have. Handling a relatively weighty reel of 2-inch tape and troubleshooting was a way of life. Utilizing some outboard equipment became an interesting test of keeping up with the latest designs and their idiosyncrasies. Perhaps, it's even more so now in the digital world. Basic engineering is much simpler in digital. It also offers a wider array of treatments. In my opinion, it produces a 'thinner' sound than analog, but there's no disputing its high quality. A good analog may offer a more true sound to the human ear, I've found, but like everything, it's all subject to individual opinion. Analog is subject to so much in erroneous intrusions. I did an album for a client at Different Fur in San Francisco, an excellent facility at that time. We finished the basic recording and took a few days' break. We returned to find some obvious errant ticks here and there on the 2-inch 24-track master that had not been evident when we finished the basic tracks. We had some very good ears on that project, and we all knew there was no problem. The tapes had been stored in studio in their usual very careful way, so no one had any answers for that. We worked around the problem carefully and got the mix done. No one could ever figure out specifically why that had occurred. Studio personnel were queried, but had no answers. The nature of analog recording, tape to head, etc., of course, adds opportunity for problems."

Joan passed away on February 15, 2019. She had a close circle of friends including Sue Barrett, who wrote the article on *Changer*, and Dana Burwell, who sent me these photos.

Woman Sound

"There was a time in my life when all my waking hours were spent working for political causes, but my love for music eventually led me into a career as a sound engineer. As I became established in that profession, I began to support political causes by donating my talents at rallies, marches, and fundraisers"(Sandstrom and Vetter, 2001).

In 1975, Boden Sandstrom and her partner, Casse Culver, founded Woman Sound, the first known, all-woman sound company (Boden Sandstrom papers, 1975–2015, 1975–2015).

Sandstrom was born in 1945. As a child she played trumpet and then the French horn. She played in the University Brass Ensemble and earned a bachelor's degree in English with a minor in music; later she earned a master of library science degree at the University of Michigan at Ann Arbor. During the Vietnam War, she was politically active and cofounded a group called Female Liberation in Boston, helping to create a journal called *The Second Wave* in 1971 (assuming the pen name Barbara Reyes). In 1970 she volunteered with the Greater Boston Peace Action Coalition (Ibid.).

In 1975 she moved to Washington, D.C., where Olivia Records was founded just two years prior; ironically it would be their farewell concert when Sandstrom had her epiphany.

> "While attending my first women's music concert, a farewell concert for the women of Olivia Records, the first women's record company, I saw a woman mixing the concert. In that instant, I knew what I wanted to do. . . . As a French horn player I had learned how to hear and fit in with all kinds of musical groups. Another strong academic interest was math and science. I instinctively knew that the role of an engineer would combine these two skills as well as serve as an outlet for musical expression that was missing in my life at that time. I felt empowered by the women's music – music by women written for women – I heard that night.
>
> (Sandstrom and Vetter, 2001)

The sound engineer was Judy Dlugacz, who introduced Sandstrom to Casse Culver, who was looking to train someone. Culver and Sandstrom started a business partnership, founding Woman Sound. Sandstrom went on to become a sole proprietor in 1978 and renamed the company "City Sound Productions." She did the sound for every event that National Organization for Women held in DC from 1975 to 1993 and later the March for Women's Lives in 2004. From 1976 to 1987, she did sound for the Michigan Womyn's Music Festival. She also worked with Sweet Honey and the Rock, Joan Jett, Cab Calloway, The Clash, and others (Boden Sandstrom papers, 1975–2015, 1975–2015).

Running her own sound business was far from lucrative; in fact, by 1988, she sold the business. However, she has no regrets. "I would not trade the experience of running Woman Sound for anything," she remarked. "It allowed me to achieve creative and personal fulfillment, and I was able to inspire many women to join a male-dominated profession" (Sandstrom and Vetter, 2001).

Not-So-Fun Facts: Imposter Syndrome

In their 1978 paper, "The imposter phenomenon in high achieving women: Dynamics and therapeutic intervention," Pauline Clance and Suzanne Imes describe "an internal experience of intellectual phoniness which appears to be particularly prevalent and intense among a select sample of high achieving women" (Clance and IMES, 1978).

It is bad enough that women who suffer from this phenomenon, known today as "imposter syndrome," feel that they are not intelligent and that they have "fooled" people; what's worse is the extreme manifestation of the syndrome in which they think their "high examination scores are due to luck, to misgrading, or to the faulty judgment of professors" (Ibid.).

It is important to note that not all women feel this way and that there is a conscious effort in today's environment of women networking in the profession to support and uplift women who doubt their abilities. Even women who do not experience the syndrome, including those interviewed and featured in this book (making no assumptions about who does or does not experience this), recognize the odds against women in the field, and many articulate the desire to "pull" other women up. This requires instilling a sense of self-confidence in those young and early-career women.

However, this positive atmosphere alone is not the cure for imposter syndrome: an enthusiastic "you can do it!" may fuel rather than diminish feelings of self-doubt. Instead, Clance and Imes recommended group therapy, in which "Mary" and "Jane," two fictional participants, have the opportunity to express disbelief that the other feels insecure. "Mary cannot believe that Jane thinks she is stupid. After all, Jane has a PhD from an outstanding university, is a respected professor, and is obviously bright. In a group setting, the ways in which an individual negates positive feedback and maintains her belief system **emerge in clear relief** and can be brought to the attention of the client [emphasis added]" (Ibid.).

Leslie Ann Jones (USA)

"At some point I decided two important things," **Leslie Ann Jones** recalls during her Heyser Lecture at the Audio Engineering Society Convention in 2017, "the first was not to be so quick to let people know who my father was. I was concerned there would be expectations about me and my performance, and I really wanted to stand on my own.

"The other was that I started to use my middle name, Ann. I realized that albums had credits and since my name could be male or female, I felt it was the only way I could distinguish myself as a woman. And that's when I started paying attention to gender" (Jones, 2017).

At any time during the last four decades, if anyone in the audio profession was asked to name a top engineer who happened to be a woman, there was one name almost *everyone* knew: Leslie Ann Jones. Her father, Spike Jones, to whom she refers in the first paragraph, was a famous musical satirist, and her mother was singer Helen Grayco. As such, Jones was raised in show business.

During the 1970s and 1980s as the women's music movement began to grow, Jones was establishing herself as a professional engineer in Los Angeles.

"I had two important experiences involving gender in positive ways. Many of the engineers at the Automatt [including myself] got a chance to work with the great Narada Michael Walden. Narada was just beginning his career as a producer and we all didn't work on the same projects. At some point I asked him 'why', and he said he cast the production team based on the best vibe for that artist and that session. Sometimes it was a male first engineer and a female second engineer; sometimes it was a female first engineer and a male second engineer. It was the first time I really thought about gender in that way and the fact that Narada appreciated the differences we each could bring not just because of our personalities or our skill level but also our gender. At that point I realized we were different and that was OK. I was being hired because I was a good engineer but also because I was a woman. It was up to Narada to decide how best I could contribute to his projects.

"The second experience was when I started working on women's music albums. Women's music was, as you can imagine, by women for women and about women with the express intention of having all the work done on records by women. Three of the labels Olivia, Redwood and Pleiades were all located in the Bay Area and it wasn't long before my old friend June Millington from the iconic women's rock band **Fanny** called to ask me to engineer a record for **Cris Williamson**. We had great fun making that record and I felt that it was great timing for us all. They were looking for a more professional recording experience and I was happy to provide that."

(Jones, 2017)

In 1978, Cris Williamson released her follow-up to *Changer and the Changed*, called *Strange Paradise*. **June Millington** was the producer, and Leslie Ann Jones was the engineer. The album was recorded at the Automatt in San Francisco. Then in 1981, Jones engineered Holly Near's *Fire in the Rain*, which was also produced by Millington and to which Cris Williamson contributed.

Jones was herself a guitarist. As a young woman, she played guitar in a band called Canticle during the "heyday of the Sunset Strip." They played clubs like the Troubadour and Hullabaloo, even got noticed by A&R, and even got help from Sonny & Cher's musical director. However, nothing ever came of it, and the record industry began to change: artists began to break away from company-owned recording studios like CBS and RCA in the 1960s and 1970s. Jones joined an all-female band and did gigs for her girlfriend's band as well. "I bought the sound system, and when that band broke up, I had to figure something out . . . my choices were: go back to school, start all over again as a guitar player, learn all the things I didn't know, or start fresh. And I chose to move forward, not backward. All those years in the studio recording as an artist, I never once thought about being a recording engineer. It never even occurred to me. Or even getting involved in sound, but once I put my fingers on faders and realized that I could affect the sound of a group or an artist without having the creative part of it – without being the artist – I was hooked (Jones, 2017).

"I formed a PA company with two male friends of mine, and we started doing live sound for other bands. We had this grand idea that we would eventually record bands live – I mean in those days, you know, bands could not afford to go into recording studios unless some A&R person signed them to do a demo or a spec deal or something. So, I bought a Tascam model 10 console and a 1/4" 4-track tape machine and set it up in my basement" (Ibid.).

Jones had also gotten a job at ABC Records. "I decided to ask for a job in the recording studio. I really wanted to be a producer and a manager. I had no thought about being a career engineer." However, Jones admired the producers at the time, like Peter Asher, who was a manger and producer. "So, I thought I should learn a little about engineering" (Ibid.).

She talked to the studio manager, Phil Kaye, and asked to be transferred to the studio. He was hesitant because he was "concerned about how all the guys in the studio would feel about having a woman in their midst," recalls Jones. But they agreed to try it out, and she was hired as a production engineer, making tape and cassette copies. When the opportunity came to step in as an assistant engineer when someone else couldn't come in, she grabbed it. It was around the time console automation was being introduced. "I would run out to the tech office and ask the tech guys questions about how to do this, how to do that . . . [they] must have thought I had a bladder infection since I had to keep excusing myself all the time," she jokes (Ibid.).

By 1977 she got her first album credit, *A Hard Core Package* by John Mayall. She had been an assistant engineer on one of his albums, and he invited her to be first engineer after noticing that she was such a hard worker: "you went for it and weren't afraid to get your hands dirty," he said to her (Ibid.) She also learned a valuable lesson from Roy Halee, an engineer/producer who was meticulous in his preparation. "I mean, it's like going on a road trip and you're supposed to check your oil and your tires and check your mirrors – how many of us really do that? But to sit there and make sure things are aligned, make sure all the mics are working make sure your reverb is at zero, you can just sit down, you don't have to worry; you can just be creative and go to work. and I've taken that with me my whole career" (Ibid.).

Over time, management at the studio changed, and Jones was let go. She ended up at the Automatt in San Francisco (which was actually Studio C of CBS Studios, where Roy Halee had worked). It was during this time she recorded *Strange Paradise*, mentioned above. In addition to her usual music recording duties, she mixed the music for the film score for *Apocalypse Now* (1979). She also recorded the first known digital multi-track session in San Francisco using the 3M digital machine for Carlos Santana's *The Swing of Delight* (1980). She also got involved with the Recording Academy, eventually becoming San Francisco chapter president.

After the Automatt closed, she worked with Frankie Beverly and Maze, among other artists, and then joined the staff at Capital Records in Los Angeles. In 1989, Studio A was reopened, and they began specializing in midsized television and film scores. Synchronization became a new skill Jones would add to her arsenal, and in order to learn more of the sound-for-film-and-television process, she took a class at the UCLA taught by **Victoria Rose Sampson**. Jones went on to win her first Grammys, first in 2003 for Kronos Quartet's *Berg: Lyric Suite* (which won the Grammy for Best Chamber Music Performance) and again in 2005 for *Good Night and Good Luck*, which won in the category Best Jazz Vocal Album and featured jazz vocalist Dianne Reeves.

Over time Jones became disenchanted with Los Angeles. "Even though I was born and raised there," she said, "it was the kind of town where you were defined by your next phone call; your self-worth was entirely dependent on making that audition and getting hired." She ended up pursuing a position running the Scoring Stage at Skywalker Sound, where Gloria Borders (who won an Oscar for Best Sound Editing on *Terminator 2*) was the general manager.

During the lecture, Jones admits, "Even now when I have been gone for a week or so I still pinch myself as I drive up to the tech building. How lucky am I?"

Leslie Ann Jones still works at Skywalker Sound. In 2000, she became the first woman ever to chair the Recording Academy (Grammy) Board of Trustees, and in 2018 she was inducted into the NAMM TEC Hall of Fame. Altogether she has four Grammys wins and six nominations: she, along with collaborator **Brandie Lane**, were the first women to win a Grammy for Best Engineered Album, Classical for their work on *Quincy Porter: Complete Viola Works* (which they shared with engineers Kory Kruckenberg and David Sabee).

And every summer, Jones still teaches young women at the Institute for Musical Arts' recording camp.

The 1980s

During the 1980s, the launch of MTV and the emergence of audio education programs began, making training accessible to women in an academic environment. **Lenise Bent** would become the first woman to win a platinum album for Blondie's *Autoamerican*. **Sylvia Robinson** started Sugarhill Records and produced the Sugarhill Gang's "Rapper's Delight." Women won in Grammy and Oscar categories for the first time, including **Joanna Nickrenz**, the first woman to win a Grammy for Classical Producer of the Year, and **Kay Rose**, the first woman to win an Oscar for Sound Editing for *The River*. **Janet Jackson** was the first woman nominated for a Grammy for Producer of the Year, Non-Classical with Jimmy

Jam and Terry Lewis for *Rhythm Nation 1814*. (Further milestones for this decade can be found in Appendix 1: Timeline.)

The Institute for the Musical Arts

> "Those of us involved in organizing the IMA (Institute for Musical Arts) stood enraptured as the IMA 'vision' unfolded before us," says Ann Hackler. "The buzz words of 'networking,' 'empowerment,' and 'role models' all took form on stage as prominent women musicians passed on their knowledge to, supported the work of, and shared the stage with student performers. All the while, the participants in the video workshop were capturing the event on tape."
>
> Post, L. 1989, *The Institute for the Musical Arts*,
> Hot Wire, Chicago (Jazz, 2006)

In the 1980s, the women's music movement began to dwindle. As **June Millington** puts it, women began to "process through a lot of issues. Women who had before only wanted to work with other women, started to shift their focus in the 80s; and some women wanted to make it in the 'real world'" (Millington, 2019).

Sometime earlier at an Olivia Records meeting in 1976, one topic of discussion was the effort to recruit more women of color. "[She] looked around the room, saw all the talent and thought to herself, 'who's going to take care of the business?'" (Tsui, 1991). By the 1980s Millington found herself wanting to escape from the sometimes heartbreaking, fickle experiences she had in Hollywood. Finally, around 1985, she decided to take action. "When I got tired of hearing myself complain about that, I decided to try to do that which I was complaining about" (Millington, 2019).

She had enough passion for the concept but needed help implementing her plan. She reached out to women, including activist Angela Davis, musician Linda Tillery, and guitarist/producer Roma Baran (who had worked with Laurie Anderson). "That's what IMA (the Institute for Musical Arts) is about – *doing* it, in a righteous way. But it's also hard" (Millington, 2019).

The Institute for Musical Arts was initially conceived of as a "non-profit teaching and performing arts organization" whose primary goal was to "support women, especially women of color, pursuing careers in music and in the business of music" (Tsui, 1991). They also offered recording studio classes.

"We didn't have a design engineer, we couldn't afford one; this was at a creamery that closed down in World War II. The place was resurrected in the 1970s in the main part of the creamery. There were two rooms and we turned it into a recording studio. We borrowed a board from my brother that was doing live sound, and I bought a 16-track. We started IMA in 1986 but we moved into the creamery in 1989–1990, and started the studio in 1992" (Millington, 2019).

By 1994, the IMA had produced "five or six albums in our recording studio with women who wouldn't have done it otherwise – or we would have had to wait for many years. . . . We make available all the training and technology a musician needs, plus offer workshops and networking information on booking and management." (In 1987, June Millington cofounded the Institute of Musical Arts [IMA] in Bodega, 1994.)

A multiracial group governed the Institute, with Ann Hackler, Davis, Baran, and Millington serving on the board of directors and a board of advisors that included Tillery, Leslie

Ann Jones, Bonnie Raitt, Audre Lorde, Gloria I. Joseph, Irene Young, Mary Watkins, Teresa Trull, and Cris Williamson. By 1989 the IMA was a fully functioning nonprofit institution providing services to more than 300 individuals in the Bay Area (San Francisco) and had "attracted upward of 3,000 people to fundraising events; received two resource development grants; had permanent office space donated to it; developed a pool of forty-five artists, technicians, and music business professionals willing to teach classes and workshops; compiled a mailing list of more than 2,000; received commitments in several major North American cities to sponsor IMA mini-institutes; and sponsored a workshop series and a mini-institute in the Bay Area" (Post, 1989).

In 2001, the Institute relocated to Goshen, Massachusetts. In addition to providing training for youth, the camp also hosts workshops and residencies for adult women musicians, "following the women's music model of developing a safe space in which to nurture female creativity" (Powers, 2015). The property is 25 acres and has two recording studios. "But we didn't have the opportunity to build our own studio. Literally, we added the studios onto the barn here," she laughs. "And that that took 7 years. But we did it" (Millington, 2019).

The studios were designed by Walters-Storyk Design Group (WSDG). Companies including Avid, Millennia Media, and Grace Design offered their support to IMA. **Leanne Unger**, a professor at Berklee College of Music, helped organize a console donation, and **Judy Elliot-Brown** of WSDG led the installation of the console (Female Future – Nonprofit Programs Train Women And Girls for Audio Careers, 2017).

The Recording Camp is taught by Leslie Ann Jones, who is now on the board of directors, along with Lee Ann Unger and Roma Baron.

Today, the IMA's Facebook Page asks visitors to ponder a few questions:

> What if we could all name three Native, Latin and Asian American musicians without thinking twice? What if it was ordinary to see a woman running the sound at a concert, or if female producers & engineers weren't anomalies? What if there were more women at the top of record companies?

The mission feels as relevant today as it did in 1986.

> "But you gotta get your hands on," Millington reminds us. "And if you have the aptitude, and you can get yourself your hands on the gear as runner, gopher, or 2nd [engineer], then that's where you start. It has to be a passion. And I think that in the '70s, going into the '80s, women didn't go after that passion because they didn't know they could.
>
> (Millington, 2019)

The IMA is still going strong and offering programming and accepting applications from girls between the ages of 17 and 24. You can find out more at www.ima.org.

Audio Engineering Society 1980

In 1980, Pamela Peterson wrote a paper for AES called "History of Women in Audio," which discussed the "dearth of women in audio" in the context of women in engineering fields and society in general. In 1976, only 1.8 percent of professional women listed their occupation as

TABLE 1.1 US Working Women Data 1.2. U. S. Working Women: A Databook 1950–1976, Bureau of Labor Statistics, Washington, D. C., 1977

Employment of women in selected occupations, 1950, 1970, and 1976

(Numbers in thousands)

Occupation	Both sexes			Women					
				Number			Percent of all workers in occupation		
	1950	1970	1976	1950	1970	1976	1950	1970	1976
Professional-technical	4,858	11,452	13,329	1,947	4,576	5,603	40.1	40.0	42.0
Accountants	377	711	866	56	180	233	14.9	25.3	26.9
Engineers	518	1,233	1,190	6	20	21	1.2	1.6	1.8
Lawyers-judges	171	277	413	7	13	38	4.1	4.7	9.2
Physicians-osteopaths	184	280	368	12	25	47	6.5	8.9	12.8
Registered nurses	403	836	999	394	814	965	97.8	97.4	96.6
Teachers, except college and university	1,123	2,750	3,099	837	1,937	2,198	74.5	70.4	70.9
Teachers, college and university[1]	123	492	537	28	139	168	22.8	28.3	31.3
Technicians, excluding medical-dental	102	339	897	21	49	122	20.6	14.5	13.6
Writers-artists-entertainers	124	761	1,099	50	229	381	40.3	30.1	34.7
Managerial-administrative, except farm	4,894	6,387	9,315	673	1,061	1,942	13.8	16.6	20.8
Bank officials-financial managers	111	313	546	13	55	135	11.7	17.6	24.7
Buyers-purchasing agents	64	361	376	6	75	89	9.4	20.8	23.7
Food service workers	343	323	505	93	109	177	27.1	33.7	35.0
Sales managers-department heads; retail trade	142	212	322	35	51	114	24.6	24.1	35.4
Clerical	6,865	13,783	15,558	4,273	10,150	12,245	62.2	73.6	78.7
Bank tellers	62	251	371	28	216	338	45.2	86.1	91.1
Bookkeepers	716	1,552	1,688	556	1,274	1,519	77.7	82.1	90.0
Cashiers	230	824	1,256	187	692	1,101	81.3	84.0	87.7
Office machine operators	143	563	726	116	414	535	81.1	73.5	73.7
Secretaries-typists	1,580	3,814	4,368	1,494	3,686	4,303	94.6	96.6	98.5
Shipping-receiving clerks	287	413	440	19	59	76	6.6	14.3	17.3

[1] Includes college and university presidents.

"engineer," up from 1.2 percent . . . in *1950*. Over that 26-year period, the number of women in engineering increased by just six tenths of a percent.

Because of the wide range of subdisciplines in engineering, Peterson explains that audio engineers might include recording studio owners, salespeople, and film sound mixers: in other words, differently than "engineers" as the Bureau of Labor and Statistics might define. Still, the numbers were jaw-droppingly low.

Peterson's paper also highlights notable women in audio at the time: Carlene Hutchins, Brownwyn Jones, Marion Downs, Karina Friend, Nadia Bell, Sarah Beebe, Laurel Cash, and Nancy Byers. A list of officers, governors, and editors from 1949–1998 shows the first woman to sit on the AES Board of Governors was Nancy Timmerman, an acoustician specializing in noise control who was elected to the board in 1984 (36 years after the Society's founding).

MTV: Music Television

MTV debuted on August 1, 1981, and was the arguably the biggest music industry disrupter since cassette tapes before it and Napster afterwards. It was a new way to promote music, sell soap, and entice teenagers with dreams of starring in their own music video.

Although I have not found a direct correlation between the rise of music television and the emergence of several schools of recording engineering, I do document the founding of several recording schools in the early- to mid-1980s, and it was, in this author's personal experience, one of the influences that led many young men and women (including myself) to seek a career "behind the glass."

Two artists that made use of the new medium, Prince and Michael Jackson, have an engineer in common: **Susan Rogers**.

Susan Rogers (USA)

Susan Rogers was working in Los Angeles as an electronic technician trainee with Audio Industries Corporation in Los Angeles, visiting studios and repairing consoles and tape machines. She had begun studying electronics after overhearing a colleague saying to someone "become a maintenance technician and you'll never be out of a job" (Barrett, 2001).

She did maintenance for Rudy Records, which was owned by Graham Nash and Stephen Stills. In 1983 she heard that Prince asked his management to locate a technician for him. She jumped at the chance and was hired by Westlake Audio for the gig (Campbell, 2018). She travelled to Minnesota to do the work. "The first thing [Prince] asked me to do when I came to Minnesota was to install an API console in his home studio and repair the multi-track" (Ibid.).

"After it was all up and running, he just kind of expected me to sit in the chair and do the recording. It dawned on me pretty quickly, 'This guy doesn't know that I'm not a [recording] engineer.' I thought, 'Let's just keep that secret then, shall we?'" (Campbell, 2018).

Although Prince eventually caught on that she wasn't experienced, he retained her. One benefit to having Rogers as an engineer/technician meant there was no downtime – if something broke, she fixed it. "I worked for him for five years . . . I went on tour, worked on three films, many videos, countless rehearsals and recording sessions and generally had the time of my life. It was sort of a cross between being in college and being in the circus" (Barrett, 2001).

Rogers worked for Prince from 1983 until 1987 and then worked for Michael Jackson in 1988. Eventually she pursued a PhD in psychology from McGill University and is now an associate professor in the Music Production and Engineering department at Berklee College of Music. She also spends her summers teaching at the **Institute for Musical Arts**.

■ The 1990s

During the 1990s, the number of schools with audio technology and engineering programs continued to climb. In 1991, Mariah Carey became the second woman nominated for a Grammy for Producer of the Year, Non-Classical, for *Emotions* with Walter Afanasieff. Cecelia Hall was the second woman to win an Oscar for Best Sound Editing for *The Hunt for Red October*, and Gloria Boarders won an Oscar for Sound Editing on *Terminator 2: Judgement Day*. Elizabeth Cohen became the first woman to serve as president for the Audio Engineering Society. (Further milestones for this decade can be found in Appendix 1: Timeline.)

AES' Women in Audio: Project 2000

In 1995, **Carol Bousquet** and **Robin Coxe-Yeldham** launched an initiative called "Project 2000 Working Group on Women in Audio," as it was formally known within the Audio Engineering Society. **Cosette Collier** of Middle Tennessee State University was also part of the group and served as a committee chair from 1999 to 2001. "The idea was to find out why there weren't more women in audio," Collier recalls. **Bousquet** was the group's chair, which set a five-year plan to research the issue (Collier).

Their first public meeting was held in October 1995 at the 99th AES Convention in New York. "It was packed," recalls Bousquet (Bousquet). An archived webpage also shows it was one of the best-attended events at the convention with over 150 attendees and that their luncheon sold out (Society, 1998). "We had a Q&A and the line was endless," added Bousquet (Ibid.). The panel featured leading women in the industry, some for whom Bousquet sought sponsors: Isabel Carter-Stewart (national executive director of Girls Incorporated); performance artist **Laurie Anderson**; **Robin Coxe-Yeldham** (Associate Professor at Berklee College of Music); record producer **Angela Piva**; and **Elizabeth Cohen** (then president-elect of the Audio Engineering Society).

Dr. Catherine Steiner-Adair was the keynote speaker at the event and gave some sobering and revelatory insight into the minds of young girls as they progress from eight years to fifteen years of age. She and her research colleagues followed a single group of girls during their development during this age range. She helped the audience connect to the language used by the girls when they were younger, which was confident, self-assured, and enthusiastic; to the language used by the same girls during their middle school and into the teen years, which was doubtful, confused, and hesitant. During this transition, when girls are around 10 to 12 years old, they were culturally indoctrinated into a behavior known by authors Carol Gilligan and Lynn Brand as "The Tyranny of Kind and Nice." Steiner-Adair explains,

> "And it's at this moment in their lives that girls get the message much more intensely than before, that they have to be *kind* and they have to be *nice*. And that it's much

more important, in fact, to be kind and nice than it is to tell the truth. It's more important for your safety in the world, to which there's a lot of truth. It's important if you want to stay connected to people. And it's *very* important if you want adult approval. This 'tyranny of kind and nice' comes historically from thousands of years of idealizing woman for being all-nurturing and caretaking, eternal mothers. The perfect woman or the perfect girl is the girl who doesn't have a mean bone in her body. She can just go into any situation and say whatever she thinks and feels, 'Cause it's always *so nice.*'"

(Steiner-Adair, 1995)

As a result, girls don't give straight answers to questions, instead testing the waters using phrases such as, "Do you want to know what I think?" "There is a perception that there's a right answer out there," Steiner-Adair explains. "That's *your* answer. And [the girl] has to figure out if you want *your* answer: 'Do you want me to tell you back *your* answer or do you want me to tell you what I really think?' . . . And then at ages 11, 12, and 13 that turns into 'I don't know. I dunno'" (Ibid.).

"And what happens at this age is secrets erupt into the lives of girls and women. You know, where you can't tell your friend that you're angry at her, but you tell everybody else . . . but you're angry at *her*. And this is horrible leadership training for girls and women. Horrible because it teaches them to deal with their own power covertly, not directly. And it pits girls against girls and women against women and sets the stage for the kind of Victorian statements that my grandmother grew up hearing all the time, which was 'don't trust other women'. 'Women are catty bitches'. 'They're backbiting. Very nasty'. 'They'll get you'. I do not think women are innately bitchy, but when you give women the message that they cannot deal directly with their anger, they will become catty. Anybody would, this is not about being female, this is about gender . . . this is about dis-empowering women from experiencing their own leadership and authority and taking it on directly. So girls struggle with, 'how can I be in a real relationship if I have to hide the truth of who I am?'"

(Ibid.)

Bousquet recalls that Laurie Anderson was the first panelist to speak after the opening remarks by Steiner-Adair and Isabel Carter-Stewart. "I turned to Laurie . . . she looked at me and her mouth hit the table. She said, 'you want me to go first after *that*?' She was overwhelmed with the information. Obviously she had given the topic thought throughout her lifetime but to hear it from those perspectives and how far reaching, and hard hitting, and how horribly we have been undermining girls so they don't have a fighting chance by the time they reach that point in their lives . . . and they are supposed to make these important decisions . . .?" (Bousquet).

Other women reacted strongly to the keynote address. "We had women in tears approaching us saying, 'if I had only heard this', 'do you know how important it was to hear this'. After the panel we moved to the next room for lunch so if they wanted to speak with the panelists they could gravitate to that table," Bousquet recalls.

But others were not so excited. "There was positive feedback . . . *except* from the AES. Whew! I was taken down for doing that because they said 'AES doesn't do that. AES is for technical information, PERIOD. Going forward there will be no more of this.'" But the Women in Audio project continued, and two years later they had a follow-up panel in New York. "And it was funny," says Bousquet, "when we opened up for questions at the end, one of the AES New York chapter members rushed to the front of the line, not even waiting his turn, shouting, 'You can't be talking about this stuff here!' And I had the wonderful satisfaction of telling him, 'Get in back of the line like everybody else please'. And so he wasn't going to, and I insisted. I said, '*you need to get in line like everyone else*'. Time ran out before I could give him the floor. I would have enjoyed giving him the floor. But 99% of the feedback was ultra-positive."

At the next AES conference in California, the Women in Audio Committee also gave out "Granny" awards to address the absence of women in audio technology at the Grammy awards, which was established in 1958. At this point in time, only three women had *won* Grammys for their technical work: **Joanna Nickrenz**, who along with Marc Aubort won for "Producer of the Year, Classical" in 1983; Judith Sherman in 1993 who won in that same category; and producer Ethel Gabriel, who won "Best Historical Album" with Alan Dell and Don Wardell for *The Tommy Dorsey / Frank Sinatry Sessions*. The first "Granny" award winners in 2007 were **Suzanne Ciani** (who also had five *Grammy* nominations) and **June Millington**, whose 1970s band Fanny was the first all-female rock band to be signed by a major label (Wolk, 2013). Millington also cofounded the **Institute for the Musical Arts**, which offered workshops in recording techniques for girls and women. (In 1987, June Millington cofounded the Institute of Musical Arts [IMA] in Bodega, 1994.) **Robin Coxe-Yeldham** also won a Granny award recognizing her as the "first lady of audio education in America."

In her leadership role, **Theresa Leonard** remembers the period of time near the dissolution of the AES Women in Audio Committee. "Someone stepped down from the Women in Audio position, and I did not step up to run it. I was doing my thing and I was busy, but I always insisted on panels and said to [my colleagues] we have to make sure women see themselves reflected on panels. I think I did my part, but at the time I didn't feel it wasn't run correctly; maybe because the larger AES was not open to it. Maybe there was something I could have done differently. At the time it felt like 'Men in Music', . . . so, 'Women in Audio'. At the convention it felt to me like we were separating women the same way we were separating students" (Leonard).

"We tried to figure out why women don't choose the profession, and it was a lack of role models," remembers Collier. "So we set up mentoring meetings with **Robin Coxe-Yeldham** to get women to sign up and go . . . but then the idea became, 'why should it be just for women?'" (Collier).

In 2000, Collier announced the committee's dissolution, saying, "Our research showed that the average number of women in recording or audio engineering programs was about 10 percent. The problem did not seem to be within the industry, but actually something more related to society and early education; the AES didn't feel that this was something a technical standards organization could effectively address" (Burgess, 2013).

Over time, it became difficult to find a chair for the committee, and eventually the committee disbanded.

AUDIO ENGINEERING SOCIETY

WOMEN IN AUDIO: PROJECT 2000

November 10, 1996, 10:00 – 12:00

> "NEVER DOUBT THAT A SMALL GROUP OF THOUGHTFUL, COMMITTED CITIZENS CAN CHANGE THE WORLD. INDEED, IT'S THE ONLY THING THAT EVER HAS."
>
> Margaret Mead

Women in Audio, a subcommittee of the Education Committee of the Audio Engineering Society, is committed to increasing opportunities for women in the audio industry. Today's special event will examine the challenges women face entering or progressing in the field of audio engineering; will recommend initiatives that will contribute to future growth of the number of women in audio; will focus on successful women who are presently in the field, and will discuss the roles that the AES, its' sections and individuals can play over the next five years that can improve women's active involvement in the industry. Ideas for future programs will also be discussed.

With special appreciation for the event sponsors:

Audio Technica USA, Inc.

Dolby Laboratories

David Moulton Professional Services

Eastern Acoustic Works

Gateway Mastering Studios

Parson's Audio

THAT Corporation.

A special thank you to

The Institute for the Musical Arts for videotaping and production services. IMA is a non-profit teaching, performing and recording facility dedicated to the advancement of women in music and its related businesses. It is located in Bodega, California. In addition to offering a bimonthly series of workshops and performances by female artists, IMA has a recording studio which serves as a teaching laboratory for artists and aspiring engineers and specializes in working with women on first and second album projects. The institute has begun to maintain an archive of women's music-related experiences for historical purposes.

For more information contact
Ann Hackler, Executive Director,
tel: 707/876-3004, fax: 707/876-3028

Note: This videotape will be made available through the
AES for educational purposes, thanks to the IMA.

*Recommended: The audio tape of the 1995 *Women in Audio: Project 2000* panel,
featuring Dr. Catherine Steiner-Adair of The Harvard Project on the Psychology
of Women and Girls Development and Laurie Anderson, performance artist, is
available from Conference Copy, Inc. at 717/775-0580.

Women in Audio Project 2000 Program handout

The 21st Century

Why Are There So Few Women in Audio?

In recent years, this question has been asked many times by various magazines, journals, and books. For example, I was interviewed for *The Atlantic* to give my perspective for an article, "Why Aren't There More Women Working in Audio?" (2017). I responded by saying the traditional image of a recording studio put in front of us has been "sexy lady singer/male producer" and quoted the musician Björk, who says she has to remind people she produces her own work. I also summarized the results of a 2016 paper that found that a low number of women apply to audio programs (Mathew et al., 2016).

In Richard Burgess' 2013 book *The Art of Music Production*, a chapter is devoted to the subject, "Why Are There So Few Women Producers?" Burgess and **Katia Isakoff** (founder of In-Kolab) found that at the time, only five women in the UK belonged to the Music Producers Guild and the Association of Professional Recording Services (APRS) . . . *combined.* That year, out of 6,500 members in the Recording Academy (USA), only 13 percent were women. Burgess goes on to list successful producers and also engages in discussion with one woman about "long hours" being unappealing and prohibitive to child-rearing. He does not come to any conclusions about the subject, instead lamenting the fact that there "seemed to be little interest in the United States in tracking this indicator of equal opportunity" (Burgess, 2013). It wouldn't be until 2018 when the results of the Annenberg study were released that we would see usable data acquired for the music industry.

In 2013, *Sound on Sound* published Rosina Ncube's personal essay, "Sounding Off: Why So Few Women in Audio?" "The worst experience," she writes, is "suggesting an idea, having it rejected; then when a guy suggests exactly the same thing 20 minutes later, everyone loves it. Frustrating? You bet, and it's just one in a string of similar experiences that I have had as a woman starting out in the world of music production" (Ncube, 2013).

On the flip side, in the course of doing this research, I have found myself overwhelmed with the numbers of women being recommended to me for inclusion in the book. (And as I stated in the preface, it's *actually impossible* to include everyone). The numbers from the Annenberg study (provided later in this chapter) are shocking. But after interviewing dozens of women for this book, I think other equally important questions would be, "Why aren't we *featuring* more women in audio? Why do women who begin in audio *leave* the major or profession? How are we *engaging young girls* who love music, math, and science?" And more importantly, "How can we help *each other*?"

Sexism and Gender Bias in the Audio Industry

A study led by Becky Prior, **Erin Barra**, and Sharon Kramer, PhD, and sponsored by the Berklee College of Music uses the term "gender bias" to explore the experience of women in the music industry at large with regard to working climate. Two questions asked related to gender bias were 1) "whether the respondent was treated differently in the music industry because of their gender" and 2) "whether the respondent felt that their gender affected their employment in the music industry." According to the study, "women working in **music production and recording (70 percent)** [emphasis added], performance (68 percent), and music media and journalism (67 percent) had the highest levels of experience with gender bias affecting their employment." Interestingly, only 7 percent of those participating in the survey listed their occupation as "music production and recording." Also of note is that the same study showed that overall, respondents had a high level of job satisfaction (72 percent), but that those working in music production and recording reported a higher level of job satisfaction than the average (82 percent) (Prior et al.).

Whereas "gender bias" might imply more a more subtle inequality (for example, assuming the woman in a mixed-gender group will handle certain tasks like documenting sessions), "sexism" can be either subtle or overt, such as remarking "you probably got this job because you are a woman." Chelsea Land states, "Sexism is the belief that specific characteristics accompany a person's apparent biological sex, while gender bias is the tendency of a person's apparent gender identity to produce certain outcomes due to systemic factors that operate with and without any person's express intent" (Land, 2019). Both of these concepts are differentiated from discrimination, which would be refusing to hire someone based on their sex.

As I was conducting interviews for the book, women recounted to me their experiences with sexism (the word they most commonly used). Even when I contextualized the goal of the book to be an uplifting, coming-together of successful women's profiles rather than tales of struggle, the stories nonetheless emerged.

Even when women felt fortunate enough *not* to have experienced some of the more rude and blatant forms of sexism, they were quick to acknowledge those who *have*, being certain not to discount the experiences of other women, which we all (as women) know exist.

Therefore, I will include some testimonials in the book, but I will leave them anonymous. I think it is important to know what some women have faced and continue to experience. Some readers may want to have their experiences validated; others may be skeptical of the degree to which sexist experiences occur and can refer to these statements as a "reality check."

Here are some examples of what women recounted to me as they made their way through the industry:

> "It is hard for you as a female to find a safe space to create with others. I am finding myself in that situation again, and even worse in New York City. This is supposed to be a very liberal city but it is where I've faced more sexism than anybody anywhere else."
>
> "I was recently upset by a comment that my male music-producing friend had made: that it was 'unfair' that women were getting opportunities in music that men weren't, and getting recognized just because of our gender rather than producing 'good music.'"
>
> "I've heard someone mention that women are less predisposed to working in sound because of our evolutionary characteristics. I've also heard the claim the women in general pay less attention to music. Situations like these make me more motivated to prove that women are just as capable as men in whichever career they choose to pursue."
>
> "People tell me, 'Oh you are so lucky because you are a girl.' . . . Nooooo! I'm *driven* to get what I want."
>
> "I was fired for refusing to [expletive] someone at [male producer's] command. He slandered me all over New York saying people shouldn't hire me because I was

a lesbian. And there was [a different male producer] who wanted me to show him 'how to clean heads.' And [a large commercial studio] said, 'We don't hire women.' "

"When I first started going out on sessions I was in my early 30s. I've always looked a little younger than I am, and I have always been mortified by how many people – even women – would ask me how old I was. We would be having meetings and people would ask, 'how old are you?' Like, 'Do you have any experience?' And I thought, 'Would you ask a man this question?' I think [my male colleague] produced his first album when he was 27 or 28 and I doubt anyone ever asked him how old he was."

"You *never* used to see women at [a large industry conference] who weren't shilling product. They had women in French maid outfits dusting consoles and bikini models posed on meter bridges."

"During a job interview, I was wearing jeans, but was told I should have been wearing a skirt. You're not hiring my [expletive] legs, you're hiring my *ears*!"

"I wonder if there are some jobs I got because I was a female, but I *know* there are some jobs I *didn't get* because I was female."

"I have been greeted with, 'Are you the secretary?' I've had my rear end pinched. I've been told, 'there needs to be some hotness in the room'. Sexism exists and it is not a reflection of your value as a person if you have to struggle. It is valid to know that you will find your networks and people will help you. And to not give up!"

In 2015 **Karrie Keyes** wrote a reflection piece called "Identifying the Battles Worth Fighting." In it, she reflects on the differing views of sexism in the audio industry.

"I have discovered two opposing views from women. The first view that is echoed by women that are established in their careers is that – Yes, there is sexism but it has not hindered or held them back. They feel that they worked hard, often harder than their male co-workers, did not give up and have confidence in their craft. They state that they rarely see sexism and believe that women have made huge strides and the industry is very accepting.

"The second view tends to come from younger women, who are either entering the industry or are in the process of establishing themselves. Their view is that they face sexism on a regular basis, if not daily. They feel they do not start out on equal footing, are passed over for work, not given a chance, and often feel they are set up to fail. So which one is it? Is one the reality or is it somewhere in the middle? Or do they both exist? I think and believe they both exist and are very real."

(Keyes, 2015)

Keyes goes on to cite a survey done in the UK that finds "four in 10 (42%) women aged between 18–34 say they have personally faced a gender barrier, followed by 34% of those aged between 35–54 and 26% of women aged 55 and over" (Half, 2014). She wonders if the experience of the respective generations, who came through the "grin and bear it" age of baby boomers, to the #metoo enlightenment of third-wave feminism reveals the lens through which each respective generation interprets sexism.

Another term, "gendering," is the assumption that an item or concept that, in itself, is ascribed no gender, is made to be associated with "masculine" or "feminine." It is related to sexism because owners or inhabiters of the masculine space or object – which, again, has been arbitrarily described as such – often push out women who try to occupy or own the same. At least two papers speak to this directly, "Technology and the Gendering of Music Education" by Beverly Diamond and "Technology and the Gendering of Music Education" by Victoria Armstrong. When Armstrong observed male pupils consistently occupying the technology suites, she summarized, "The male-dominated atmosphere of the technology suite made the space feel 'off-limits' to Armstrong's young female interlocutors. Such practices extend beyond the classroom, for the gendered discursive and spatial segregation and discrimination noted by Armstrong in the school technology suite has strong parallels in professional recording studios, in music retail, and even in the use of consumer audio in the domestic sphere" (Bennett, 2012).

"Opting Out" of the Conversation

"I don't want to be the spokesperson about the woman in audio who this happened to."

Perhaps more than any other, this statement, shared with me by one of my interview subjects, illustrates one reason why some women are reluctant to talk about 1) sexism and 2) the topic of "women in audio" in general. Additionally, not all women are on the "diversity and inclusion" bandwagon.

The same woman to whom the above quote is attributed continued to explain to me why she avoids talking about sexism.

> "I don't think it helps, because it elevates that stuff. It just gives them more power. I could stand up here and call out people and ruin their lives with things I saw, things that were said to me, but . . . I have power now over that. They didn't knock me off my block. I kept my head down, and did my work, and kept it together. And while I know that people found [audio work] too difficult and quit . . . when I look at the brave women who have come forward in #metoo, I unfortunately think of them as being victims, and I don't think of them being the amazing person they are in the field. That person's aura has been taken over . . . like Rose McGowan. She's a great actress; but now when you bring up her name, she's the 'me too woman'. And these [men] end up winning because their disgusting power has taken over. Where did *you* go? You dissolved into this thing that is *not what you are trying to do*."

Still another woman confided in me, "I don't want to be pointed out, I work hard and want to be recognized for my work, not being a woman."

Yet a third woman told me she thought the term "diversity and inclusion" was a kind of segregating code. "You can't come through *this* door," she said, referring to traditional ways people find opportunities in the industry. "We want you to come through *that* door," she asserted, referring to inclusion initiatives.

A fourth woman explained her view about working in a space that should be gender-neutral (meaning the opposite of "gendered"). "I would like for all women, when they walk into a room, to be given the same benefit of the doubt as a man would be: that they can handle the task in front of them. Sometimes I'm asked what's it like to work with so many men. I don't think I have ever worked with a woman engineer. All the engineers I've worked with are men, but I am not directing men, I am directing the *product*. I have always tried to keep gender out of it."

Sexual Harassment

> "There have been one or two sessions that I couldn't do because I didn't feel safe, so I gave it to a male engineer. I had a 6th sense and if things are off, I try to figure it out."

That sixth sense my interview subject refers to is a sense that was likely *developed* as a result of witnessing and/or experiencing sexual harassment.

In 2006, "Me Too" began as a movement founded by civil rights activist Tarana Burke and became a viral social media movement in 2017 with the hashtag phrase #metoo. The movement has been instrumental in bringing awareness of sexual harassment and assault and in bringing together women within networks of solidarity and support. The *Washington Post* quoted Burke as saying in a tweet, "It made my heart swell to see women using this idea – one that we call 'empowerment through empathy' . . . to not only show the world how widespread and pervasive sexual violence is, but also to let other survivors know they are not alone. #metoo" (Ohlheiser, 2017).

The American Association of University Women (AAUW) has performed studies on sexual harassment in schools for children ages 12 to 18, and in 2011 they found that the phenomenon is *prevalent.*

> Nearly half (48 percent) of the students surveyed experienced some form of sexual harassment in the 2010–11 school year, and the majority of those students (87 percent) said it had a negative effect on them. Verbal harassment (unwelcome sexual comments, jokes, or gestures) made up the bulk of the incidents, but physical harassment was far too common. Sexual harassment by text, e-mail, Facebook, or other electronic means affected nearly one-third (30 percent) of students. Interestingly, many of the students who were sexually harassed through cyberspace were also sexually harassed in person.
>
> (Catherine Hill, 2011)

These statistics are not just harmful "child's play." At universities and in the professional arena, women and men experience sexual harassment in class, on campus, and in the workplace. SoundGirls.org, together with Amy Richards (president of Soapbox, Inc., a feminist lecture agency) and the AAUW, have published the legal definitions of sexual harassment and remedies on the page https://soundgirls.org/sexual-harassment-3/.

It is important to know your rights when it comes to sexual harassment, which "describes unwelcome sexual advances, requests for sexual favors, or other verbal or physical conduct. The behavior does not have to be of a sexual nature, however, and can include offensive remarks about a person's sex" (Catherine Hill, 2011).

In the United States, "Title VII" is a federal law that prohibits discrimination on the basis of sex, race, color, national origin, and religion. Companies with 15 or more employees are subject to the law. The experience of sexual harassment is emotionally overwhelming, and victims are often afraid to report it. "The important thing to remember," the AAUW advises, "is that you are not alone and that you do have options when coming forward." The first thing to remember is to *document everything.*

- "Your experience with the harasser – time, location, details, and witnesses
- Your experience reporting the harassment – time, location, details, and witnesses

- Your productivity – safeguarding and documenting your productivity at work can be essential during and after reporting" (Catherine Hill, 2011)

On Motherhood

As a mother, I find myself giving advice to audio women who face the dilemma of whether to have children. For this book, I have talked to a lot of "audio moms." They are in every field in audio.

Nonetheless, there exists the sentiment that a career in sound is impossible with kids . . . and with good reason, at least in the United States, which is the only industrialized nation that does not offer paid leave for new mothers (Ingraham, 2018). The Family and Medical

Robin Coxe-Yeldham and her daughter, Dakota

Leave Act of 2012 grants women 12 weeks of *unpaid* leave, which allows a new mother time to bond with her baby but no income with which to do so, leaving her with the dilemma of how to return to work so she can provide for her child.

In the book *She Bop II*, producer Ann Dudley (Art of Noise) is quoted as saying, "You work longer hours than a hospital doctor. No woman in her right mind would take that work for those hours and that money, especially if she has children" (O'Brien, 2004).

"But women who want kids sound defeated before they even start," remarks **Fela Davis**. "Kids don't kill your dreams" (Davis). Similarly, **Eiko Ishiwata** says, "I know many women who juggle both family life and career life and excel in both areas. The situation is not black and white" (Ishiwata, 2019). Yes, being a mom is hard work and being a single parent even more so. Still, it is possible to be a mom in the audio business.

Here are a few testimonials from women of various means:

- **Leya Soraide** remembers working on the road with her mother at church gigs (Gaston-Bird, 2017).
- Hollywood producer Irwin Winkler offered to set up a room with a crib and pay for a nurse for **Vickie Rose Sampson** so she could complete her studio training. (She had her husband bring the baby over twice a day to be fed.) (Tucker, 2018)
- An NPR feature tells the story of when **Karrie Keyes'** twin daughters were born. "By the early 90s when Pearl Jam exploded, Keyes was on the road a lot. While she was touring, their dad, aunts and sometimes a nanny took care of them." . . . "It was challenging," Keyes remarked, "It took me probably till they were three or four to actually come to terms with, 'You know what, I'm actually a better mother if I'm doing what I love doing.' So that when I'm here, I'm completely here" (Losneck, 2016).
- **Jamie Angus** recalls, "When I worked at Standard Telecom we had a software engineer who had gone off to have a baby and came back, so they gave her a private telephone line so she could use a modem, a dumb terminal at home, and fully flexible work patterns so they could keep her on and so that she could continue to work. I can't think of any other business that could have been that flexible. They wanted her skills and they were willing to do the extra bit to make it happen" (Angus, 2019).
- **Suraya Mohammed** entered a time-share arrangement with her co-worker and was able to set up a DAW at home so she could work from there (Mohamed, 2019).
- **Carol Bird McLellan** recounts, "I didn't take off much time from work after I had my baby – maybe a month and then I started back to work part-time. [My managers] were very amenable to whatever arrangement I wanted to make. Sometimes I brought my daughter in to the Labs. I would pull out the bottom drawer of my desk, and this little being would sleep down there. After a while I got childcare. She turned out well – she's now a history professor at a small honors college in Florida" (Noll, 2019).
- **Lisa Nigris,** the director of Recording and Performance Technology Services at the New England Conservatory says of her job, "This career will allow somebody to have a family. There are plenty of women who can juggle these things but this is something I was able to grab onto" (Nigris, 2019).

Finally, **Martha De Francisco** made it a point in her interview with me to get the word out about the rewards of motherhood. "I think it's important to mention that because many women think, 'either I am a recording engineer or I am a mother but I cannot be both'. I got married and I have a daughter," she told me. "That was *possible*. I was *able* and *very happy* to – in spite of my career – and she came out really wonderfully. So I was able to combine this tight time schedule and demanding career with being a mother, and that is the best thing I have ever done" (De Francisco, 2019).

On Age

Women are developing their counterpart to the "old boys' club": an "old girls' club," where women who have established themselves are pulling up their younger counterparts through mentoring and networking. But as **Liz Dobson** puts it, "Older women are already doing a lot and have had to work very hard. We should be looking to the industry and saying, 'What can *everybody* do more and better, without putting further pressure on older women?'" (Dobson).

When you are experienced in your field, it's rewarding to help others, but it is also draining and perhaps a bit unfair to expect older women – for example, in higher education – to do it all. **Elizabeth Cohen** states, "I would spend a quarter of my time truly mentoring, which nobody else had to do. Instead of doing research or writing – and yes, I love doing that and I'm good at doing that – but it's that *expectation* or that *need* . . . you know, I had to satisfy the needs for diversity and women, and it's ridiculous that any women in this day and age should not have the privilege of focusing on their art or their skill" (Cohen).

A few women have moved out of the industry to pursue other interests. **Caryl Owen** is enjoying retirement at her home in Wisconsin. "I don't miss the faders," she says, which she admits surprised her. She paints, gardens, and plays the bass. **Carol Bousquet** left the industry early and for years worked in real estate. She still enjoys making connections between people. The late **Joan Lowe** continued to enjoy restoring cars and volunteered with the McKenzie Fire & Rescue District.

The Annenberg Study

In January 2018, the Annenberg Inclusion Initiative released its report, "Inclusion in the Recording Studio? Gender and Race/Ethnicity of Artists, Songwriters & Producers across 600 Popular Songs from 2012–2017." The survey does not directly address the question of women in technical roles, stating that "the goal here is to create measurable change in hiring practices for women and underrepresented racial/ethnic groups across all facets of the music industry – on both the artistic and business sides." However, the statistics for women who work as producers are very bleak – even more so for women of color (Dr. Stacy L. Smith, 2018).

The report finds that out of 300 popular songs (according to the Billboard Hot 100 end-of-year charts), "98% were male and only 2% were female. The gender ratio of male producers to female producers is 49.1 to 1! . . . Of the female producers, only 2 were underrepresented. In terms of song credits, 9 [women] were delineated as producers, 1 was a co-producer, and 3 were vocal producers. Six of the 13 female producers were also artists, reducing the total number of producer-only credits to 7 out of 300 songs" (Ibid.).

It would be interesting to see what the results would be for the industry at large rather than songs that achieved commercial success, but this is a good starting place, and a new report about the audio industry could become the foundation for changes to begin within the larger industry.

FIGURE 1.1 Percentage of women across three creative roles in the music industry (Dr. Stacy L. Smith, 2018).

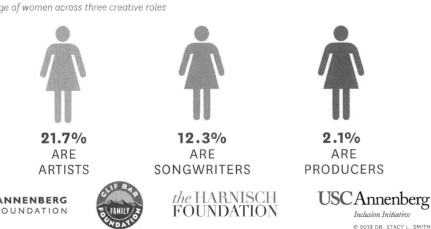

WOMEN ARE MISSING IN THE MUSIC INDUSTRY

Percentage of women across three creative roles

21.7% ARE ARTISTS

12.3% ARE SONGWRITERS

2.1% ARE PRODUCERS

ANNENBERG FOUNDATION   CLIF BAR FAMILY FOUNDATION   *the* HARNISCH FOUNDATION   USC Annenberg Inclusion Initiative © 2019 DR. STACY L. SMITH

FIGURE 1.2 Number of women of color among women in creative music industry roles (Ibid.).

WOMEN OF COLOR ARE INVISIBLE AS PRODUCERS

4 OUT OF **871** PRODUCERS WERE WOMEN OF COLOR

ANNENBERG FOUNDATION   CLIF BAR FAMILY FOUNDATION   *the* HARNISCH FOUNDATION   USC Annenberg Inclusion Initiative © 2019 DR. STACY L. SMITH

Recording Academy Task Force

In 2018, the Recording Academy established a task force, chaired by Tina Tchen, to "effect meaningful change in the organization" and build a "more vibrant membership base" (Recording Academy (TM) Implements Community-Driven Membership Model, 2018). Tchen was assistant to President Barack Obama; chief of staff to First Lady Michelle Obama; and executive director of the White House Council on Women and Girls (Lester, 2019; House, 2011a, 2011b). They announced that industry recommendations and peer review would become part of building its community. Now, each new member must have two professional recommendations in order to apply. A peer review panel then meets to evaluate a range of criteria that include "craft, genre, and overall diversity" in order to determine who will be invited to join (Recording Academy (TM) Implements Community-Driven Membership Model, 2018).

Members of the task force included:

- Stephanie Alexa, vice president of finance and licensing administration, ATO Records
- Michele Anthony, executive vice president and executive management board member, Universal Music Group
- Cam, GRAMMY-nominated artist
- Common, GRAMMY-winning artist
- Sheryl Crow, GRAMMY-winning artist
- Andra Day, GRAMMY-nominated artist
- Giselle Fernandez, award-winning television journalist
- Jimmy Jam, GRAMMY-winning artist
- Beth Laird, CEO and co-owner, Creative Nation
- Debra Lee, chairman and CEO, BET Networks
- Rebeca Leon, cofounder and CEO, Lionfish Entertainment
- Elizabeth Matthews, CEO, ASCAP
- Dr. Stacy L. Smith, founder and director, USC Annenberg Inclusion Initiative
- Trakgirl, music producer, songwriter and entrepreneur
- Ty Stiklorius, founder and CEO, Friends At Work
- Julie Swidler, executive vice president of business affairs and general counsel, Sony Music
- Dean Wilson, CEO, SEVEN20
- **Terri Winston**, founder and executive director of Women's Audio Mission (GRAMMYS, 2018)

Laura Segura Mueller, vice president of Membership & Industry Relations for the Recording Academy stated, "The GRAMMY Awards® are already renowned for being a peer-awarded honor, and our new membership model further reinforces that peer-driven commitment to excellence. . . . Membership is the lifeblood of the Recording Academy and a privilege we strive to uphold. Our new membership model puts the power in the hands of the music community and is designed to build an active, representative membership base that reflects our broader culture. By changing the process to Recording Academy membership, we remain committed to setting a positive example for the music industry as a whole" (Recording Academy (TM) Implements Community-Driven Membership Model, 2018).

The Grammys and the Recording Academy Class of 2019

The Grammys have implemented changes based on the Annenberg study and the work of the Recording Academy task force. Before the Recording Academy welcomed its 2019 "incoming class," the community was 26 percent women, 24 percent underrepresented groups, and 29 percent under the age of 39. In 2019, those numbers shifted: 49 percent are women, 41 percent are from underrepresented groups, and 51 percent are under the age of 39. "While the improvement is clear, the work is just beginning," the article states. To illustrate the work ahead, Segura Miller, vice president of membership for the Recording Academy, states, "With such dire statistics industry-wide, we will face challenges with future new member classes if not enough women and people of color are being hired, mentored, and have access to opportunities to lead and excel. The industry-wide change we need will only be achieved when new voices are encouraged to rise through the ranks" (Hertweck, 2019).

Audio Engineering Society 2019

In 2018, **Colleen Harper** joined the AES as its first woman to become its executive director. Along with the election of **Nadja Wallaszkovits** (president) and **Agnieszka Roginska** (president-elect), four women served as governors-at-large in 2019: **Leslie Gaston-Bird**, **Piper Payne**, **Martha de Francisco**, and **Jessica Livingston**. **Valerie Tyler** was reelected as secretary. With such a large presence of women on the board of governors, the organization has come a long way since its founding 70 years ago.

There is still work to be done: a 2015 paper by Mathew et al. revealed only 7 percent of AES members are women (out of members opting to report their gender) (Mathew et al., 2016). In 2016, Gaston-Bird, the first African American elected to the AES Board of Governors, worked with the AES Board of Governors to establish the Diversity and Inclusion Committee, co-chaired by Payne, whose mission is "to ensure diversity in the AES worldwide and the audio industry as a whole by improving accessibility, welcoming diverse genres, embracing emergent audio fields and research, and radiating inclusiveness to all races, gender and gender identities, physical abilities, ages, and nationalities." The committee is comprised of over 50 members from industry and academia and has published guidelines for conventions and conferences that outlines several goals related to the committee's mission, one of which is to increase the visibility of women and underrepresented groups at these events.

You can find out more about the AES Diversity and Inclusion Committee at http://aes.org/community/diversity.

Note

1.   Established in 1967, the Trustees Award Special Merit Award is presented by vote of the Recording Academy's National Trustees to individuals who, during their careers in music, have made contributions, other than performance, to the field of recording.

Bibliography

Albertoni, Rich. 2011. *Rock Pioneers the Millingtons Still Fight Stereotypes*. Available: https://isthmus.com/music/rock-pioneers-the-millingtons-still-fight-stereotypes/.

Angus, Jamie. 2019. Interview with Jamie Angus. *In*: Gaston-Bird, L. (ed.).

Baguley, Richard. 2018. *Sophie Germain: The Mathematics of Elasticity*. Available: https://hackaday.com/2018/03/20/sophie-germain-the-mathematics-of-elasticity/ [Accessed February 26, 2019].

Barrett, Sue. 2001. Revelation in the Studio: Women Producers and Engineers. *Femmusic* [Online]. Available: http://femmusic.com/wp/index.php/tag/revelation-in-the-studio-women-producers-and-engineers/.

Barrett, Sue. 2008. Making Music: Virgo Rising – The Once and Future Woman (1973). *FolkBlog* [Online]. Available: https://web.archive.org/web/20081013091823/www.onlinefolkfestival.com/blog/2008/08/05/virgo-rising/#more-300 [Accessed August 5, 2019].

BBC. 2018. *Daphne Oram: Still Point – Excerpt (Prom 13)* [Online]. Available: www.bbc.co.uk/programmes/p06g8d0f.

Bell, Melanie. 2017. Learning to Listen: Histories of Women's Soundwork in the British Film Industry. *Screen*, 58, 437–457.

Bennett, Marie. 2012. Technology and the Gendering of Music Education – By Victoria Armstrong. *British Journal of Educational Technology*, 43, E93–E94.

Berrett, Jesse. 1999. David Bowie. *Rolling Stone*. New York.

Biography of Grace Murray Hopper [Online]. Yale University. Available: http://www.computinghistory.org.uk/det/1791/Grace-Murray-Hopper/ [Accessed May 23, 2019].

Bletchley Park Names 'Secret' World War II Codebreakers. *BBC News* [Online]. Available: www.bbc.co.uk/news/uk-england-beds-bucks-herts-24378091.

Boden Sandstrom Papers, 1975–2015. 1975–2015. Five College Archives and Manuscript Collections.

Bousquet, Carol. 2019. Interview with Carol Bousquet. *In:* Gaston-Bird, L. (ed.).

Branstetter, Leah. Cordell Jackson. *Women in Rock n' Roll's First Wave* [Online]. Available: www.womeninrockproject.org/reference/jackson-CORDELL/#1546233107155-4a3ee8fd-2a18092e-99db09e7-7ef7.

Brooks, C.N. 2003. *Architectural Acoustics*. Jefferson, NC: McFarland & Co.

Burgess, Richard. 2013. *The Art of Music Production Theory and Practice*. Oxford University Press. New York.

Busch-Vishniac, Ilene. 1998. A Sound Specialist Takes Charge at Engineering. *Johns Hopkins Magazine. In:* Hendricks, M. (ed.).

Callahan, Mike. 2006. The Vee-Jay Story [Online]. Available: https://www.bsnpubs.com/veejay/veejay story1.html

Campbell, Madeline. 2018. Susan Rogers. *Women in Sound*.

Charman-Anderson, Suw. 2009. Pledge 'AdaLovelaceDay'. *Pledge Bank* [Online]. Available: www.pledgebank.com/AdaLovelaceDay [Accessed June 30, 2019].

Chojnacki, Bartłomiej. 2018. Prof. Marianna Sankiewicz-Budzyńska. *Audio Engineering Society Polish Section* [Online]. Available: http://aes.org.pl/prof-marianna-sankiewicz-budzynska/.

Clance, Pauline and Imes, Suzanne. 1978. The Imposter Phenomenon in High Achieving Women: Dynamics and Therapeutic Intervention. *Psychotherapy*, 15, 241–247.

Clayman. Vee-Jay Records, est. 1953. Available: www.madeinchicagomuseum.com/single-post/vee-jay-records.

Cohen. Interview with Elizabeth Cohen. 2019. *In:* Gaston-Bird, L. (ed.).

Collier. Interview with Cossette Collier. 2019. *In:* Gaston-Bird, L. (ed.).

Sun Records Company. *Marion Keisker* [Online]. Available: www.sunrecordcompany.com/Marion_Keisker.html.

Conover, Kirstin. 1991. Before Elvis, Cordell Jackson Rock-and-Rolled 'Em in Memphis. *The Christian Science Monitor* [Online]. Available: www.csmonitor.com/1991/0820/20131.html [Accessed June 29, 2019].

Council on Women and Girls Leadership [Online]. Washington, DC: The White House. Available: https://web.archive.org/web/20111001194414/www.whitehouse.gov/administration/eop/cwg/who/leadership.

Cymascope. A History of Cymatics. *Cymoscope.com* [Online]. Available: www.cymascope.com/cyma_research/history.html [Accessed June 6, 2019].

Dance, Helen. 1998. Helen Oakley Dance Interview by Monk Rowe, February 12. San Diego, CA. *In:* Rowe, M. (ed.).

Dance, Francis. 2001. Helen Oakley Dance 1913–2001. *The Last Post* [Online]. Available: https://web.archive.org/web/20020109195943/http://www.jazzhouse.org/gone/index.php3?view=991165404 [Accessed September 27, 2019].

Davis, Fela. 2019. Interview with Fela Davis. *In:* Gaston-Bird, L. (ed.).

The Delian Mode. 2009. Directed by Kara Blake. Videographe.

Densmore, Francis. 1950. Songs of the Chippewa. *In:* Congress, L.O. (ed.), *Archive of Folk Song*. Washington, DC: Library of Congress.

De Francisco, Martha. 2019. Interview with Martha De Francisco. *In:* Gaston-Bird, L. (ed.).

Dobson, Liz. 2019. Interview with Liz Dobson. *In:* Gaston-Bird, L. (ed.).

Dileo, Emily (FERRIGNO). 2019. *RE: RE: Permission Request.* Type to Gaston-Bird, L., April 30.

Dodero, Amy. 2016. *In Conversation: Shiva Feshareki on Daphne Oram's 'Still Point'.* Available: www.lcorchestra.co.uk/conversation-shiva-feshareki-still-point/.

Essinger, James. 2014. *Ada's Algorithm: How Lord Byron's Daughter Ada Lovelace Launched the Digital Age.* Brooklyn, NY: Melville House.

Fanny: Chart History. *Billboard Magazine* [Online]. Available: www.billboard.com/music/fanny [Accessed July 3, 2019].

Female Future – Nonprofit Programs Train Women and Girls for Audio Careers. 2017. *Mix* [Online], 41. Available: www.mixonline.com/news/female-future-nonprofit-programs-train-women-and-girls-audio-careers-430657 [Accessed September 27, 2019].

Female Pioneers in Audio Engineering [Online]. 2017. Intelligent Sound Engineering. Available: https://intelligentsoundengineering.wordpress.com/2017/08/07/female-pioneers-in-audio-engineering/ [Accessed June 6, 2019].

Feshareki, Shiva. 2019. *A Quest to Find 'Real'.* Doctoral thesis, Royal College of Music.

Findingada.COM. *Who Was Ada?* Available: https://findingada.com/about/who-was-ada/ [Accessed February 6, 2019].

Gaston-Bird. 2017. 31 Women in Audio: Leya Soraide. *31 Women in Audio* [Online]. Available: http://mixmessiahproductions.blogspot.com/2017/03/31-women-in-audio-leya-soraide.html [Accessed March 9, 2017].

Germain, Sophie. 1821. *Récherches sur la théorie des surfaces élastiques.* Paris.

Gilbert, Andrew. 2015. *Feminist Icon Celebrates Album's 40th Anniversary.*

Grace Murray Hopper [Online]. The Centre for Computing History. Available: www.computinghistory.org.uk/det/1791/Grace-Murray-Hopper/.

Grammys. 2018. *Recording Academy Names Diversity and Inclusion Task Force Members* [Online]. Recording Academy Grammy Awards. Available: www.grammy.com/grammys/news/recording-academy-names-diversity-and-inclusion-task-force-members.

Guitar Granny // Cordell Jackson Was Playing Rockabilly When Elvis Was a Babe. 1992. *Tulsa World.* BH Media Group, Inc.

Half, Robert. 2014. *More Than a Third of UK Female Employees Have Faced Barriers During Their Career, While Nearly Half of HR Directors Believe Progress Is Being Made* [Online]. Robert Half, Inc. Available: www.roberthalf.co.uk/press/more-third-uk-female-employees-have-faced-barriers-during-their-career-while-nearly-half-hr.

Hatch, Kristen. 2013. Cutting Women: Margaret Booth and Hollywood's Pioneering Female Film Editors. *Women Film Pioneers Project* [Online]. Available: https://wfpp.cdrs.columbia.edu/essay/cutting-women/.

Herrick, Sophie. 1891. Visible Sound. *Century Illustrated Magazine.* American Periodicals.

Hertweck, Nate. 2019. Recording Academy Advances New Membership Model, Inviting This Year's Class. *Recording Academy Grammy Awards* [Online]. Available: www.grammy.com/grammys/news/recording-academy-advances-new-membership-model-inviting-years-class?utm_source=email&utm_campaign=member%20invite&utm_term=social.

Hill, Catherine. 2011. *Crossing the Line: Sexual Harassment at School* (Executive Summary).

Hillery, Louise. Vivian Carter: From Gary Roosevelt High School to Introducing the Beatles. *The Indiana History Blog* [Online]. Available: https://blog.history.in.gov/tag/livin-with-vivian/ [Accessed July 5, 2018].

Hoffmann, Frank. 2004. *Encyclopedia of Recorded Sound.* London: Routledge.

Honey, Maureen. 1999. *Bitter Fruit: African American Women in World War II.* Columbia: University of Missouri Press.

Hutton, Jo. 2003. Daphne Oram: Innovator, Writer and Composer. *Organised Sound*, 8, 49–56.

In 1987, June Millington Cofounded the Institute of Musical Arts (ima) in Bodega. 1994. *The Word Is Out!* 4(1).

Ingraham, Christopher. 2018. The World's Richest Countries Guarantee Mothers More Than a Year of Paid Maternity Leave: The U.S. Guarantees Them Nothing. *The Washington Post*, February 5.

Ishiwata Eiko. 2019. Interview with Eiko Ishiwata. *In:* Gaston-Bird, L. (ed.).

Jones, Leslie Ann. 2017. 'Paying Attention' Heyser Lecture. Audio Engineering Society, October 18, New York.

Jrobinson. 2017. *Helen Oakley Dance: The Dances of Bittersweet Hill* [Online]. Aquarium Drunkard. Available: https://aquariumdrunkard.com/2017/03/20/helen-oakley-dance-the-dances-of-bittersweet-hill.

Karaś, Dorota. 2018. Zmarła znana polska elektronik związana z Politechniką Gdańską. *Trójmiasto* [Online].

Keyes, Karrie. 2015. Identifying the Battles Worth Fighting. *Soundgirls.org* [Online]. Available: https://soundgirls.org/identifying-the-battles-worth-fighting/ [Accessed 2015].

Killick, Cynthia. 2018. A Sound Revolution. *Lulu.com*.

Kolappan, B. "India's first woman sound engineer, unsung yet." *The Hindu*, September 18, 2019 Wednesday [Online]. Available: http://ct.moreover.com/?a=40222142566&p=2a4&v=1&x=o-vTiJwLukE1rDRPkBVCZw [Accessed October 4, 2019].

Koomar, Susan. 2007. Working with the Stars. *Pocono Record*, January 21, 2007 [Online]. Available: www.poconorecord.com/article/20070121/news/701210314 [Accessed September 27, 2019].

Kozinn, Allan. 2009. Wilma Cozart Fine, Classical Music Record Producer, Dies at 82. *The New York Times*, September 24.

Land, Chelsea. 2019. *RE: Sexism vs Gender Bias*. Type to Gaston-Bird, L., July 12.

Lane, James. 2011. Vivian Carter and Vee Jay Records. *Traces of Indiana and Midwestern History*, 23(1), 48–55, Winter.

Leonard, Theresa. 2019. Interview with Theresa Leonard. *In:* Gaston-Bird, L. (ed.).

Lester, Melia. 2019. Tina Tchen on Time's Up and Working for the Obamas. *The Sydney Morning Herald* [Online]. Available: www.smh.com.au/world/north-america/tina-tchen-on-time-s-up-and-working-for-the-obamas-20190205-p50vq2.html [Accessed September 27, 2019].

Losneck, Caroline. 2016. Meet the Woman Who's Been Pearl Jam's Sound Engineer for 24 Years. *Weekend Edition Sunday* [Online]. Available: www.npr.org/2016/09/04/492433224/meet-the-woman-whos-been-pearl-jams-sound-engineer-for-24-years [Accessed June 26, 2019].

Mathew, et al. 2016. *Women in Audio: Contributions and Challenges in Music Technology and Production*. Audio Engineering Society Convention 141.

McFadden, Kristin. Frances Densmore: A New Woman, but Not Without Complication. *Digitizing American Feminisms*. Projects from Oberlin College.

Millington, June. 2019.Interview with June Millington. *In:* Gaston-Bird, L. (ed.).

Mohamed, Suraya. 2019. Interview with Suraya Mohamed. *In:* Gaston-Bird, L. (ed.).

Morris, Bonnie. 2015. Olivia Records: The Production of a Movement. *Journal of Lesbian Studies*, 19, 290–304.

Mozans, H.J. 1913. *Woman in Science*. New York: Appleton.

National Park Service. *Charles Cros: The Paléophone Process* [Online]. National Park Service, U.S. Department of the Interior. Available: www.nps.gov/edis/learn/historyculture/origins-of-sound-recording-charles-cros.htm [Accessed May 23, 2019].

Ncube, Rosina. 2013. Sounding O!: Why so Few Women in Audio? *In: Sound on Sound*. London: SOS Publications Group.

Nigris, Lisa. 2019. Interview with Lisa Nigris. *In:* Gaston-Bird, L. (ed.).

Noll, Michael. 2019. *RE: Bell Labs: Women in History*. Type to Gaston-Bird, L.

O'Brien, Lucy. 2004. *She Bop II: The Definitive History of Women in Rock, Pop and Soul*. New York: Continuum.

Ohlheiser, Abby. 2017. The Woman Behind 'Me Too' Knew the Power of the Phrase When She Created It – 10 Years Ago. *The Washington Post*, October 19.

Oliver, Myrna. 2002. Kay Rose, 80; First Female Sound Editor to Win an Academy Award. *LA Times*, December 16.

Pearson, Harry. 2014. *Trustees Award: Wilma Cozart Fine*. Available: www.grammy.com/grammys/news/trustees-award-wilma-cozart-fine [Accessed September 27, 2019].

Phillips, Anna. 2011. Crowdsourcing Gender Equity: Ada Lovelace Day, and Its Companion Website, Aims to Raise the Profile of Women in Science and Technology. *American Scientist*, 99, 463–464.

Plunkett, Donald. 1998. Reminiscences on the Founding and Development of the Society. *Journal of the Audio Engineering Society*, 46, 5–6.

Post, L. 1989. The Institute for the Musical Arts. *Hot Wire, Chicago*, 5, 46.

Powers, Ann. 2015. You've Got a Home: June Millington's Lifelong Journey in Rock. *The Record, Music News from NPR* [Online]. Available: www.npr.org/sections/therecord/2015/11/19/456581427/youve-got-a-home-june-millingtons-lifelong-journey-in-rock.

Pożegnanie doc. dr inż. Marianny Sankiewicz-Budzyńskiej. 2018. *Politechnika Gdańska* [Online]. Available: pg.edu.pl/aktualnosci/-/asset_publisher/hWGncmoQv7K0/content/pozegnanie-doc-dr-inz-marianny-sankiewicz-budzynskiej [Accessed September 27, 2019].

Prior, et al. Women in the US mUSic indUStry. Berklee Institute for Creative Entrepreneurship Available: www.berklee.edu/sites/default/files/Women%20in%20the%20U.S.%20Music%20Industry%20Report.pdf [Accessed September 27, 2019].

Recording Academy (TM) Implements Community-Driven Membership Model. 2018. Santa Monica, CA: The Recording Academy.

Riverwalk Jazz. 2006. *Hot Chamber Jazz: The Benny Goodman Trio 1935–1954* [Online]. Stanford University Press. Available: http://riverwalkjazz.stanford.edu/program/hot-chamber-jazz-benny-goodman-trio-1935-1954 [Accessed June 30, 2019].

Rose, Kay. 2001. The First Independent: An Interview with Legendary Sound Editor Kay Rose Part 1. *In:* Shatz, L. (ed.). *Cinemontage.*

Sandstrom, Boden, and Vetter, Charlene. 2001. The Sound of Activism. *Journal of Lesbian Studies,* 5, 161–167.

Schroeder, Manfred. 1961. *Natural Sounding Artificial Reverberation.* Audio Engineering Society Convention 13.

Spevak, Jeff. 2013. Happy Days Here Again for Record Producer. *Rochester Democrat and Chronicle* [Online]. Available: www.democratandchronicle.com/story/lifestyle/2013/12/12/happy-days-here-again-for-record-producer/4005681/ [Accessed June 29, 2019].

Smith, Stacey. 2018. Inclusion in the Recording Studio? Gender and Race/Ethnicity of Artists, Songwriters & Producers Across 600 Popular Songs from 2012–2017. *Annenberg Inclusion Initiative.* USC Annenberg.

Steiner-Adair, Catherine. 1995. *Women in Audio: Project 2000.* 99th Convention of the Audio Engineering Society, October. New York, Audio Engineering Society.

Sterling, Bruce. 2017. Luigi Federico Menabrea Paying Tribute to Ada Lovelace. *Wired* [Online]. Available: www.wired.com/beyond-the-beyond/2017/05/luigi-federico-menabrea-paying-tribute-ada-lovelace/ [Accessed February 6, 2019].

Stone, W.H. 1879. *Elementary Lessons on Sound.* Macmillan & Company.

Sutton, Allan. 2018. *American Record Companies and Producers, 1888–1950.* Denver: Mainspring Press.

Thomas, Michael. *Octacros* [Online]. Available: www.mgthomas.co.uk/Records/LabelPages/Octacros.htm [Accessed May 23, 2019].

Tina Tchen to Join Office of the First Lady as Chief of Staff [Online]. Washington, DC: The White House [Accessed January 5, 2011].

Tsui, Kitty. 1991. The Institute for the Musical Arts. *Outlines: The Voice of the Gay and Lesbian Community,* 4, 34.

Tucker, April. 2018. Kay Rose Profile.

Vaillant, Derek. 2002. Sounds of Whiteness: Local Radio, Racial Formation, and Public Culture in Chicago, 1921–1935. *American Quarterly,* 54, 25–66.

Van der Tuuk, Alex. c. 2017. Aletha Dickerson: Paramount's Reluctant Recording Manager. *Vintage Jazz Mart: The Magazine for Collectors of Jazz and Blues 78s and LPs* [Online]. Available: www.vjm.biz/new_page_18.htm.

Various. 1973. *Virgo Rising: The Once and Future Woman.* Reno, NV: Thunderbird Records.

Various. Introducing . . . the Beatles. *Wikipedia* [Online]. Available: https://en.wikipedia.org/wiki/Introducing . . . _The_Beatles [Accessed September 27, 2019].

Voice, A.C. 2018. History of the Record Industry, 1877 – 1920s: Part One: From Invention to Industry. *Medium* [Online]. Available: https://medium.com/@Vinylmint/history-of-the-record-industry-1877-1920s-48deacb4c4c3 [Accessed May 23, 2019].

Waller, Mary Désireé. 1962. Chladni Figures – A Study in Symmetry. *American Journal of Physics,* 30, 935.

The War Gave Mary Howard Her Big Chance to Make Good in Recording; She Did – and How! 1948. *Audio Record.* New York: Audio Devices, Inc.

We Have Also Sound-Houses. 1979. Directed by Ian Chambers. *BBC Radio,* 4.

Williamson, Lori. 2018. Frances Densmore. *Collections Up Close Blog* [Online]. Available: www.mnhs.org/blog/collectionsupclose/9868 [Accessed March 22, 2018].

Wilma Cozart Fine: Trustees Award Acceptance. 2011. Directed by Grammy.com.

Wolk, Douglas. 2013. The First All-Female Rock Band Signed TO A Major Label, and Other Pioneer Girl Groups. *MTV.*

Women in Audio: Project 2000 [Online]. 1998. Available: https://web.archive.org/web/19980525212416/www.hyperbeat.com/wia/ [Accessed September 27, 2019].

Young Ladies' Radio League. *Bletchley Park – Enigma – in Remembrance.* Available: https://ylrl.org/wp/bletchley-park/ [Accessed May 23, 2019].

2 Outreach: Organizations and Current Initiatives

Pull

In her book, *Pull*, Dr. Pamela Laird describes how throughout history, the concept of "social capital" has determined the upward mobility of American citizens. She dispels the notion of the "self-made man," showing that famous figures such as Ben Franklin and Andrew Carnegie each had help from networks that enabled them. "By the late 1980s," she writes, "it became clear that nobody can penetrate a workplace ceiling without pull from above. That pull requires social capital" (Laird).

"Terms such as 'glass ceiling,' 'networking,' 'mentor,' 'gatekeeper,' 'keeper,' and 'role model' have become buzz words in the social capital lexicon over the past four decades," she writes. As mentioned in the preface to this book, there seems to be a "women's version" of every modern professional society. Characteristic of these groups is the mission of "fostering their own members' business success," as distinct from the goal of merely furthering study in a given occupation (Ibid.).

For example, in 1915 the Medical Women's National Association (now known as the American Medical Women's Association) stated its purpose was "to bring Medical Women into communication with each other for their mutual advantage" and engaged making post-graduate training opportunities available as well as publicizing internships (Ibid.). Laird gives several such examples from the early twentieth century.

In the pages that follow, we see that the trend has continued today. The advent of social media has made possible all of the essential ingredients for "Pull": networks, mentoring, and role models (Ibid.). Just a few of the organizations engaged in this work are profiled in this book; altogether there are now more than 60 groups working to elevate the visibility of and resources available to women in audio (see Appendix 2: Women's Audio Organizations).

Digital Audio Eco Feminism

Researcher **Dr. Liz Dobson** contributed a chapter to the book, *Creativities in Arts Education, Research and Practice* called "Digital Audio Eco Feminism (DA'EF): The Glocal[1] Impact of All-Female Communities on Learning and Sound Creativities" (Leon et al., 2018).

The goal of the study was to examine the emergence of woman-centered, music technology-focused organizations and how they were engaging with and creating change within this traditionally male-dominated space.

Dobson traveled to the USA and engaged in a qualitative research study with Beats By Girlz, Women's Audio Mission, The Seraphine Collective, The Girls Rock Camp Alliance, and the Society of Women in TeCHnology. The conversations led to three observed "themes" that emerged:

- how the minority centers the marginalized,
- DA'EF learning and creativities, and
- the challenging economics of DA'EF organizing (Dobson, 2018).

Dobson heard that the "identity and presence a few women can draw attention to the absence of many (the marginalized)." In response to this marginalization, women's music technology groups have created social capital (see: "Pull" earlier in this chapter) as well as cultural, economic, and social capital. They have done this by creating spaces, programs, and social activism (Ibid.).

With respect to "learning and creativities," Dobson documented the natural emergence of "interthinking situations, where people with diverse experiences and backgrounds are thinking together, solving problems and developing knowledge jointly through dialogue" (Ibid.). Another name for this might be peer-learning.

Finally, the economics of organizing: **Terri Winston** confides that "[people] don't realize . . . what it's taking for us to do this. It's taking the concentrated efforts of five staff members and 18 board members and all of the people that support Women's Audio Mission which is lot of people. So, it's 1300–1500 pages of grants every year. It's an incredible amount of energy" (Ibid.).

Dobson concludes, in part, that women bear the great burden of trying to increase the participation of women in the field, but that "as a global society we need to find better ways of taking collective responsibility for gender inequalities, and to develop strategies that do not further deplete potential contributions of women who are sound, digital audio and music experts."

Erin Barra and Beats By Girlz (USA)

Erin Barra didn't like being called "The Ableton Lady," but after talking with her publicist, decided to surrender to it. "If that's what people want and that's where the traction is . . . I'll go for it" (Barra). Since starting her work with Ableton in 2012, she still freelances with them. She is currently an associate professor at Berklee College of Music and Berklee Online, and one of the primary authors of the 2018 study, "Women in the U.S. Music Industry" (see Chapter 1).

Barra is one of the pioneering electronic singer/songwriters using digital technology on stage. "There was always Imogen Heap and Björk . . . but it was almost esoteric in a way, especially the way the technologies were being presented, and I feel like what I ended up doing was extremely digestible to a lot of people. It was pretty stripped-down live looping, live sampling, playing instruments, singing songs, but in a context that people gravitated towards. And I made a lot of videos. It's interesting, because now you see lots of women creating videos in their bedrooms of their songs using laptops or digital technologies. When I started, there really wasn't anyone else doing it. So, I think I had a lot of impact on other female singer/songwriters getting into technology. I think the massive impact was 2010–2013; I was hitting it hard as an artist, and now I'm on the backend of that, helping other artists to do what I was doing" (Barra).

Erin Barra, founder of Beats by Girlz

"I use a lot of MIDI controllers and synthesizers. Things have changed so much. At first, I used a Wurlitzer A200 that would break all the time," she recalls. "I also used the Virus T1, the APC 40, and a Melodica, then shifted to working more 'in the box.'" Now she uses Ableton Push, a lot of Akai products (like the LPK 25 and LPD 8), the Novation launch pad, Launch Key, Roli Seaboard, and Roli Blocks. "That's just the hardware," she says (Ibid.).

In her previous work, she was using digital tech and doing well, but people were more interested in how she was making music as opposed to the music itself. "That's when I started to let go of the artist identity and started to become who I am" (Ibid.).

Beats by Girlz was founded in 2013. "We are making a huge difference," she says. "You can tell in the response we get: there are three people a month wanting to start chapters in their communities. We license everything for free as long as they are offering seats in the classroom" (Ibid.).

The curriculum, designed by Barra, is Ableton-based with options to study the Ableton Push or not. "It is written so people have fun right away," she explains. "It is application-based, rather than describing how mechanisms work. We get people excited" (Ibid.).

There are different Beats By Girlz communities based on locale. "Each community is different and they provide the community with what they need: support, hardware, curriculum, resources, etc. Minneapolis is extremely vibrant! They have really run with it and they have a great group of women spearheading it. And we help" (Ibid.).

Originally the concept was funded by the Lower East Side Girls Club in New York. Now Beats by Girlz have fundraisers, donations, silent auctions, and raffles, which keep the overhead very low. But Barra says, "No matter how you do it, it's hard" (Ibid.).

The students share their content, and the student profiles are online at www.beatsbygirlz.com/student-features. "It's less about the tools and more about letting women have an experience. It informs who they believe themselves to be. Maybe they become coders or game programmers. It is a gateway" (Ibid.).

"The takeaway is *role support*. The sheer fact of having someone to look at who is representative of people who aren't represented enough makes more of an impact. When I came to Berklee, I was the only woman teaching this class (Ableton). But now 14 out of 15 people in my class are women and there is role support and they are excited. That is the latest pathway to raising visibility. Perception is more important than skills" (Ibid.) (Barra).

To learn more, visit BeatsByGirlz.org.

Fun Facts: Ableton Live and Ableton Push

Simply put, Ableton Live (released in 2001) is software that allows users to create music. Its interface has revolutionized music-making, production, and live performance. Ableton's interface is relatively intuitive. A user can begin making music by watching videos and completing tutorials in the program. However, there is an opportunity to go much, much deeper by linking controls and parameters together within the program. Ableton Live for Max also allows users to interface with Cycling 74's Max/MSP and use MIDI commands to control just about anything you can imagine. The Ableton Push is a hardware controller that allows users to create sequences and interact with Ableton on the fly.

Ableton has become an industry giant, hosting the annual Loop conference and offering the highly sought after "Ableton Certified Trainer" title. They even offer refurbished units of the Ableton Push to high school music programs, offering children – who in some cases have no access to acoustic instruments – an opportunity to learn music and technology (Slater, 2017).

Phebean Adedamola Oluwagbemi (Nigeria)

Phebean Adedamola Oluwagbemi is one of the few women doing audio engineering in Nigeria, specifically live sound. She would guess there are less than 10 in the entire country.

Febe first got an internship to train with Edward Sunday at Azusa Productions and has done many large- and small-scale productions. "When it comes to concert production, that is my strength," she says. "I also do studio recording and mastering but mostly handle rehearsals and recording sessions at Azusa."

She currently plans to further her studies and pursue a degree in sound design for film. "I'm looking at traveling to do that," she explains. "Traveling would afford more opportunities, and I could network with people who are already doing this. We do not have enough people in Nigeria who do sound design for film on a professional level, but we do have music producers who do sound design on the side. This is one area that is really lacking" (Gaston).

Audio Girl Africa (Nigeria)

"I started Audio Girl Africa when I discovered the gender gap/disparity in the field," recalls **Phebean Adedamola Oluwagbemi**. "I started reaching out and discovered that we didn't even make up one percent in any field to do with audio production. That's for Africa in general." She tried Ghana and South Africa and found there were maybe one in a 100 or one in 50. "And they didn't know each other," she continues. "So, Audio Girl Africa empowers

Phebean Adedamola Oluwagbemi (Febe), founder of Audio Girl Africa

Phebean Adedamola at work

African women who would like to take up careers in audio production and engineering. We are committed to increasing the level of involvement of African women in audio technology by providing the necessary resources and right support system, regardless of your gender, as long as you have a passion for it" (Adedamola).

(L-R): Audrey Aramide Bada, Phebean (Febe) Oluwagbemi, Mercy Kisusa, Joy Elumezie, Folusho Ogunjob

Audio Girl Africa has so far run two successful workshops in Lagos and Ibadan for 60 girls in total. At the workshop, they used a Behringer X32 mixer and taught the girls audio signal flow. Girls as young as 12 attended the one-day workshop. They also offered a music production class with FL Studio and PreSonus Studio One software (Ibid.).

Just as importantly, the experience continued after the workshop. Participants were given access to an online platform to do more audio work and to have weekly chats with girls in the field. They are developing a mentoring program where mentors can be anywhere in the world, but it is also important for the girls to have Black women in Africa as role models (Ibid.).

Febe is now looking at how to support the girls with school and internship opportunities and help them become certified. Now they are looking at a site for girls from different countries to come together, creating opportunities to study, get placements, and get to know the industry better. "In four to five years we want to create an institute where girls can get certified" (Ibid.).

You can learn more about Audio Girl Africa at www.audiogirlafrica.com.

■ Female Frequency (USA)

According to their website, "Female Frequency is a community dedicated to empowering female, transgender & non-binary artists through the creation of music that is entirely female generated" (Frequency, c. 2019).

"We keep everything open to everybody no matter how you identify. We try to keep everything that we put out there friendly in that way . . . so nobody feels like they're not welcome," says founder **Dani Mari**. "I realize we're in a new time where we're all trying to figure out what pronoun to use and how to properly communicate that. I think it's still a little bit gray. I've also noticed some of the pronouns that are used on the East Coast are different than the West Coast, and I'm sure it's different when you go to different countries as well. If we ever had a situation where somebody felt like the direction of the workshop or anything we said made them feel like they weren't included, we would definitely be open to any kind of feedback to make sure that we're be making everybody feel comfortable" (Mari).

Female Frequency is based in New York and Los Angeles and hosts workshops for women in audio. For example, one of their early events featured Psychic Twin (Erin Fein and Rosana Cabán). "They came in and talked about their live setup and recording process and then they played a live set. Afterwards, they let people come up and try out the different instruments and vocal pedals. They also had a SoundGirl running sound for the workshop. We found that model [for workshops] to be really great, because no matter what their level or experience, the participants can get something out of it: whether they're a musician and they want to know more about how to use the equipment live on stage, or if it's somebody that's working on a recording and wants to know more about a certain DAW, or someone that is a fan of the music and wants to hear a live show. We also have been hosting workshops about how to do a live sound check. We did that recently here in Brooklyn with SoundGirls and Tom Tom Magazine" (Ibid.).

Female Frequency has workshops in Brooklyn and in Los Angeles. For more information, visit www.femalefrequency.com.

Dani Mari (USA)

In a project reminiscent of the 1970s project *The Changer and the Changed* (see Chapter 1), **Dani Mari** wanted to make an album featuring all women. "I reached out to Women In Music back in the day to find a female producer to work with. **Julie Kathryn** (I Am Snow Angel) reached back out to me. We met up for coffee and when we talked about making an album together we said, 'Hey, why don't we make an album made entirely by women?' So from that we invited Claire London to help out, and the three of us created Female Frequency – initially with the mission of creating an album made entirely by women. Through that experience, we met **Karrie Keyes** from SoundGirls who introduced us to some of the engineers that we worked with on the album." **Jett Galindo** is one of the engineers, along with Kerry Pompeo, Dara Hirsch, Steph Durwin, and Kimberly Thompson. Maria Rice was the mastering engineer.

"I always kind of saw myself as a singer-songwriter like Carole King," says Mari, "but I never thought of myself as being a producer or running sound or being behind the soundboard because I wasn't exposed to seeing women in that position at that time. After the experience of working with Female Frequency it opened my eyes to all different women in audio that have been doing it for a long time that I really respect and admire. From that experience

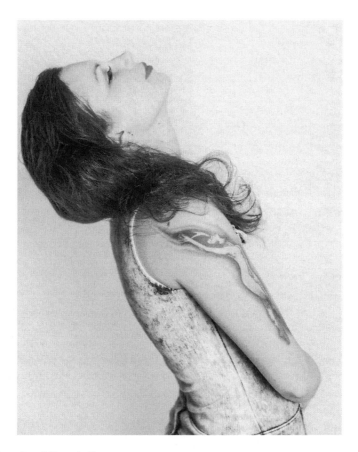

Dani Mari, cofounder of Female Frequency
Source: Photo credit Angelys Ocana

I co-produced an album with Rosana Cabán who was the drummer for Psychic Twin, and have gained the confidence to produce music on my own. My latest project is called 'Primitive Heart'" (Mari).

Female Frequency has been influential not only for other women, but for Mari as well.

"I've held myself back in the past with not doing things that I didn't think I could, or I didn't see other people like me doing that." When asked what young and novice readers should know, she replies that you should not be afraid to try things. "There is a strong network of women out there that can help you with finding resources, or education and mentoring, or whatever you want to be able to reach your goals – whatever you want to be involved in audio."

As an example, she points again to **Karrie Keyes** as an amazing resource. "I'm so inspired by her, she's been really helpful. When we first started Female Frequency we didn't really know what direction we wanted to go, and we were still figuring things out. She was just really helpful with that. She has been a great supporter of us and a lot of other women's organizations" (Ibid.).

I Am Snow Angel (Julie Kathryn), cofounder of Female Frequency
Source: Photo credit Julia Drummond

I Am Snow Angel (USA)

Julie Kathryn from northern New York goes by the stage name I Am Snow Angel. She received a BA in Psychology from Cornell University and an MS in social work. She wasn't trained as an engineer; rather she "stumbled into production," she says. "I was always musical but also nerdy," she says, mentioning her love of AP (Advanced Placement) Calculus. "So, it was creativity plus OCD (obsessive compulsive disorder) mathematical precision." She played gigs and open mics but admits she struggled with the vocabulary and with speaking up. She took songwriting lessons with Tony Conniff, who taught her how to use GarageBand. She started using the synths within the software to make demos at home and fell in love with it. "I was *activated*," she recalls. "I stayed up all night" (Kathryn).

She estimates that her progression from experimentation to proficiency took about four years, at which time she became a full-service producer and sound designer. During that time Conniff taught her Logic Pro, and she met and worked with producer-engineer Jason Cummings. She also worked with **Erin Barra**, learning how to use Ableton Live in the fall of 2013. "It felt like an even playing field," she says of meeting Barra. "I thought, 'Hey, I can do this.' " Mike Doughty (of the avant-garde group Soul Coughing) had a remix contest, which she won. Her brother offered her some inspiring words. "I have faith in you. I think you're going to claw your way in," recalls Kathryn (Ibid.).

She cofounded **Female Frequency** with **Dani Mari**, who was looking for a female producer to create an album with female artwork, songwriting, engineering, and producing.

"I was good at math and computers, but when I had to learn GarageBand, I thought, 'Nah, that's not for me.' Why did I think that? It's a combination of not seeing women doing it. It feeds on itself. So many subtle cues come from outside and reinforce themselves. So when I show up with my gear people ask, 'who programs your set for you?' It gets patronizing. There's an assumption that femaleness is not attached to engineering" (Ibid.).

The mission of Female Frequency is to raise the visibility of female, transgender, and nonbinary artists and engineers. Producers create their own music. "I now have more female clients than I can take. We make all the music ourselves," says Kathryn.

Women in Sound Zine and Madeleine Campbell (USA)

"Music has always been part of my life," says **Madeleine Campbell**, author and publisher of Women in Sound, a magazine featuring interviews with women in the field. "I grew up always playing in school orchestras . . . I studied piano and cello. I auditioned at Duquesne University in Pittsburgh, and took cello all 4 years and took audio classes, and that got the ball rolling. Between that and recording friends for fun, once I left school I continued to do studio work" (Campbell).

Eventually she found a venue called Brillobox. "The bartender tapped me on the shoulder and said, 'Hey, you work in sound, don't you? We need another sound tech.' I told him I'd done recording but not live, but the engineer was so gracious and showed me how to use a digital board, since I had only used analog boards to that point." Then she got a call from Hebra Kadry, who had heard she was doing front of house and introduced her to a band that needed an engineer (Ibid.).

"I didn't set out to make a zine. I just wanted to connect with more women in the field," she says. "And where I was, I was the only woman in the studio where I interned and didn't

Women in Sound, Issue 1
Source: Artwork by Elly Dallas

Women in Sound, Issue 6
Source: Artwork by Maggie Negrete

know women in the field. I started scouring liner notes and found quite an incredible community." Campbell was interning at Treelady Studios in Pittsburgh, Pennsylvania. She started working on the zine in 2014 and released the first issue in October 2015 (Ibid.).

Her latest issue, number six, has already sold out. "I'll do a reprint soon. It's hard to keep up, especially when I leave for a couple of months at a time" (Gaston). Issue number seven is in pre-production. Over the course of the zine's five-year history, she has done interviews with over 72 studio and mastering engineers, live engineers, and artists/producers (Ibid.).

You can learn more about the Women in Sound zine at www.womeninsound.com.

Gender Amplified (USA)

In 2006, **Ebonie Smith** proposed a thesis focused on hip-hop music production for her undergraduate degree in Africana Studies at Barnard College, Columbia University. Upon doing research about the field and finding a dearth of female producers, she talked to her thesis advisor, Kim Hall, about setting up a conference called "Gender Amplified: Women and Technological Innovation in Hip Hop" (Smith, 2013).

With support from the Barnard Center for Research on Women, the Africana Studies Program, and Dean Vivian Taylor, she had a support system that made the conference a success. On April 14, 2007, the event launched. Highlights of the day included a networking brunch, keynote, a screening of "Lady Beat Makers, Volume 1" by Feminixx.com founder Tachelle Wilks, a conversation with DJ Spinderella (of Salt N' Pepa), and a panel discussion. The evening ended with performances by DJ Rheka, DJ Sparkles, and DJ Ayana Soyini (Ibid.).

Energized by the event, Smith established Gender Amplified as "a non-profit organization that aims to celebrate women in music production, raise their visibility and develop a pipeline for girls and young women to get involved behind the scenes as music producers."

In 2017, Gender Amplified joined Art Girl Army to get girls into music production and audio engineering, co-hosting a hands-on workshop followed by a conversion with artist/activist Genesis Be on "how to combine production techniques and political activism in powerful ways" (Gaston-Bird, 2017a).

More recently, Gender Amplified had a landmark year in 2019 with the creation of a music production camp initiative called the CTRL ROOM series in partnership with Atlantic Records and the Clive Davis Institute of Recorded Music at NYU. They brought together top artists, aspiring student musicians, and established music producers from the Gender Amplified community. Prominent artists who participated included Elle Varner, Spencer Ludwig, and Coco & Breezy (Mengesha, 2019).

You can learn more about Gender Amplified at www.genderamplified.org.

Ebonie Smith, founder and president of Gender Amplified, in 2019

Ebonie Smith (USA)

When **Ebonie Smith** was four years old, her mother bought her an album, *Blacks' Magic* by Salt-N-Pepa. "That album set the tone for who I would become as a woman and as a human being," she says. "Also, I adored listening to the radio in the car. It always had to be on. As a kid I developed a very emotional relationship to music and to sound" (Gaston-Bird, 2017a).

She also played basketball and wanted to go to the WNBA, but instead she chose to go to Barnard College and got a campus job as an audio/visual technician. "It exposed me to the world of audio recording, and I have never looked back." She went on to earn a master's degree in Music Technology from New York University and now works as an audio engineer and producer for Atlantic Records (Ibid.).

In 2016, Smith received a Grammy certificate for her work as assistant engineer on *Hamilton, the Original Broadway Cast Album*. When anyone asks her about the obstacles she encountered, she says she doesn't know how to answer them. "There is always this assumption that I have struggled because I am a woman. Music is the most enjoyable thing I do in life, and my career has been pretty smooth. Nevertheless, I would never negate the fact that there is gender-based inequality in the world or deny that women face challenges in all professional areas. However, questions about challenges and obstacles shouldn't overshadow the other myriad points I could address about audio and music production" (Ibid.).

Her commitment to supporting women and girls in music production led her to found **Gender Amplified, Inc**.

Ebonie Smith, founder and president of Gender Amplified

Source: Photo credit Xavier Li

▓ The Creator's Suite (USA)

"I want **The Creator's Suite** to be a prominent force in the movement to bring gender equity to the music industry." **Abhita Austin**, founder of The Creator's Suite, has found her calling, to play her role in being a catalyst for change for women in the industry. "I'm not about sitting back and talking about it. To me it's about creating opportunities for us to flourish." On The Creator's Suite's Instagram page, she doesn't use the hashtag #women in her posts. "I don't hashtag 'women' anything. I want to normalize our presence so when you think 'music production' you think 'women and men.' You can find everything I'm doing with women in music production under #thecreatorssuite" (Austin).

But something interesting happened with the audience of her newly created Instagram page. "It was 65% male and 35% female," she recalls. "And it was interesting to watch how men just thought it looked cool and followed the page, but they slowly started to realize this is a space for women. Now it's 35% male. What I'm trying to do is bring more visibility to the musical and technical excellence of women in music production. These women have been here. I'm just putting a spotlight on them" (Ibid.).

The Creator's Suite's social media presence features profiles and live Instagram chats with female producers. Their signature event, "Input < > Output," takes place in New York and is an opportunity to watch some of the most talented women producers perform their music live and then break down their creative process (Ibid.).

To support the endeavor, Austin has moved into the corporate audio-visual sector and closed her boutique recording studio Hidden Chapel Studios. Instead she says, "the goal now is to put my energy into The Creator's Suite and build this much-needed platform for women in music production. And the feedback that I've been getting is crazy! There were 200-plus

An event held by The Creator's Suite
Source: Photo by Kevin Vallejos, courtesy of The Digilogue

women at our annual Celebration of Women In Music Production this past March. One woman who attended said that she was crying during the beat cypher portion of the program. Another woman said she was on our page for 30 minutes and told me, 'this is the most amazing experience, I can't believe this' . . . I mean this is incredible. It's really fulfilling, but it screams that this type of space for women to see themselves is desperately needed. The end goal is to create an online education platform for women in music production, by women in music production." You can find out more about The Creator's Suite at www.thecreatorssuite. com (Ibid.).

Abhita Austin (USA)

During high school, **Abhita Austin** was a drummer and a visual artist and enjoyed running track. Gradually she became interested in music production. She started at New York University with an undeclared major. "It was kind of a blur because these concepts of audio production were new to me. Even when I first started assisting, I was still figuring things out, because it's different when you get in a room with a client" (Austin).

Austin received a bachelor of music in music technology at NYU. While still attending NYU, she got her opportunity to work in the industry when she scored an internship at Quad Recording Studios. "They tested you to see if you were ready to be an audio assistant. There was such an influx of interns at the time. So I passed that test, but you're still figuring out a lot of things."

After interning and graduating to an assistant engineer at Quad, she went to the Cutting Room. "They paid me, like, a dollar more," recalls Austin. "And I was just hopping

Abhita Austin, founder of The Creator's Suite
Source: Photo by Claudia Hayden

Austin during her time at NYU

around, assisting; I worked at Kampo which was across the street. During this time, I was drumming in an all female band, and someone in the band got connected with the VP of the A&R department at Warner Brothers Records, and he was working at Chung King. He was so interested in our band, that he let us use the room that he had a long-term rental on – 'the Purple Room'. That VP found out I was an engineer and started putting me on sessions. That was around 2000 when tape machines were going out of the studios and they were bringing in Pro Tools rigs."

Austin began engineering on bigger name sessions, with popular celebrities like Da Brat, Anthony Hamilton, and Common. Over time, however, freelancing and the audio engineer's lifestyle became draining. "I got burned out. I didn't know that becoming an audio engineer meant you were signing up to be an entrepreneur. I was tired of being paid after 90 days and chasing record companies for my checks. I got spiritually drained, so I stopped" (Ibid.).

She started working a nine-to-five job for a few years, but she began to miss being in the studio. Eventually she moved back to Long Island, New York, and worked for Sabella Recording Studios. "It was cool because it was all live instruments and rock bands. I was used to drum programming with MPC's and vocal miking, so I had to start as an assistant again.

"But it was great because I learned how to properly tune and track drums. The studio's clientele was mostly rock, but had a strong jazz and classical community of musicians that booked sessions. There were also a lot of Haitian artists recording konpa and zook records."

During her time at Sabella Recording Studios, from 2004 to 2009, Austin learned how to run a recording studio. "I was calling record companies, creating studio tours; I learned Photoshop, I was promoting the studio. That's where I started to put together advertisements and learn about marketing" (Ibid.).

Before long, she decided to go into business for herself and created Hidden Chapel Studios. "My dad planted that in my head. He said 'there's a cap on your success if you're working for somebody else,' she recalls (Ibid.).

Austin would soon discover that diversifying and putting all of her talents to use would have tremendous payoff. "I had a client I was recording audio for, and I happened to have a USB flip camera," she says. "I put it on a tripod and recorded the session to give them an extra perk. I edited the video and added 'lower third' titles in iMovie, and the client said, 'Oh my God, this is amazing! So they hired me to do 32 more videos. And I thought, 'This could be a thing!'" That's when she fell in love with the fusion of audio and video (Ibid.).

Around 2013, she met **Ebonie Smith**. She began volunteering with **Gender Amplified**, first as an instructor and then doing videography and social media work for them. Austin also became involved with the Pushing Buttons Collective, an inclusive space for men and women interested in beatmaking. Recently in 2018 Austin was accepted to NEW INC – The New Museum's incubator for creatives working at the intersection of art, design, and technology. She currently freelances with American Express and Simon & Schuster Audio as she continues to support a global community of women in music production through The Creator's Suite.

Yorkshire Sound Women Network (England)

"In order to learn," says **Dr. Liz Dobson**, "We have to be in an environment which is lower risk, which – in this case – means not being the only woman in a world which associates masculinity and technology; to remove that risk to be in a community of peers; and to have a chance to put your hands on and use equipment. To make mistakes and learn from that" (Dobson).

Dobson studied child psychologist Lev Vygotsky, who showed "how language is a tool for higher mental development" (Ibid.). Together with the study of social/cultural capital and her previous research in the social psychology of collaboration and creative output in music technology situations, Dobson was inspired to invite women to join her for the first meeting of the **Yorkshire Sound Women Network** in Huddersfield, England, in 2015.

Further inspiration for the project came from The Women's Institute, the largest voluntary women's organization in the UK, where women come together and empower each other (Institutes).

The first challenge in defining their long-term goals was to reconcile the range of articulated desires of a diverse group. Some were good project managers who wanted to shape long-term change, others with a more immediate wish to *do* and *create*. "There's a feeling that it's sort of an emergency," she says, and a fear of not getting things right the first time. "I believe that all of this is important, in a sense, but it's not an emergency because if we think of it that way we are going to burn out. We have to pace ourselves and we have to be sustainable." She also reiterates that there are multiple organizations dedicated to the cause of encouraging women in music technology. "We don't have to do things the same way, all of these things are complementary" (Dobson).

One key development for YSWN was obtaining funding from the University of Huddersfield for a project manager, who has since brought in multiple grants. They now deliver a massive portfolio according to the agreed-upon YSWN business plan, which includes strategic priorities around education, community, and advocacy to industry. "For example, with our Youth Music fund we have been funded to deliver 20 workshops in schools, create two new music tech clubs for girls and some new industry internships. Also, when organizations with little or no funding approach us looking for women to hire, we reiterate that [budget-less] approach is detrimental to women's success and encourage the organizations work harder at finding the funds, which they ultimately do. It means we are making a difference," says Dobson.

In their first three years, they held 80 workshops. Their activities also involved mentoring, skill-sharing, musical performances, and creative events throughout the region of Yorkshire (where Huddersfield is located). You can find out more about Yorkshire Sound Women at yorkshiresoundwomen.com.

International Women Working in Film

"The idea for International Women Working in Film began 3 days before the 2018 Oscars. That was the year of Francis McDormand's now-famous **inclusion rider** speech.

"I was running out of patience with male cries of 'reverse discrimination' whenever someone posted on social media asking specifically for a woman production sound mixer for a job. I felt it was a knee-jerk reaction from men. They weren't asking *why* a woman might be wanted for that job and it was absurd that we ended up being labeled as difficult and discriminatory" (Dovi).

Lori Dovi is a union production sound mixer based in Los Angeles who has seen much discrimination in her field for the past 25 years. "The backlash was unbelievable," she recalls. But there are legitimate reasons for seeking out a woman. "You might be working with kids, women who are rape and incest survivors might need to be miked by a sound woman because of wardrobe and privacy issues. I'm in a union with 500 sound mixers and there are only 5 women at the top. And this was during the #metoo movement" (Ibid.).

Dovi set out to create a networking tool for women. "We need our own space where we can find jobs as well as a safe space where we support each other not just nationally but internationally, too. That was how the IWWF Facebook page started. The international scope came to me because the issue is global. There are so many 'above the line' initiatives now, that I felt that also including women in a 'below the line' initiative made the most equitable sense.

"We have to create mentorship programs so we can get women in higher end shows for a proper standard of living," Dovi says emphatically. But she also is concerned that some well-known women directors and producers don't seem to be engaged. "Are they really paying it forward? Are they hiring women? I'm not always convinced" (Ibid.).

Since its formation, Dovi has welcomed over 5,000 women to her network on Facebook, which is for women working in all occupations of film around the world. There is also a directory of professional women where you can choose to join for free or pay for various tiers of increased visibility. You can find more about International Women Working in Film at womenworkinginfilm.com.

Sound Women (United Kingdom)

Maria Williams established Sound Women in 2011 to "encourage, promote and support women in UK radio." Although it is now defunct, the website still offers its podcast series as a legacy item. All of the assets can still be listened to at audioboom.com/channel/soundwomen.

Journalist Miranda Sawyer remembers the impetus for the group's formation. "The trigger, as I recall, was the 2011 Sony awards, a celebration of radio where almost every woman who got on stage was there to present, rather than receive an award; where each of those women's looks were commented upon by the host; where TalkSport won Station of the Year – a station which, at the time, featured an online quiz that rewarded correct answers with a video of a 'lovely' removing her clothes. Gah" (Sawyer, 2016).

Williams put a more positive spin on the group's accomplishments during its active phase. "It runs an annual mentoring scheme in partnership with the BBC Academy and Skillset, has many active regional groups, and releases original research. It also holds regular networking events and training workshops, and 2013 saw the first Sound Women festival" (Williams, 2013).

In the report, Sound Women discovered:

- One in five solo voices on the radio is female.
- That figure is one in eight during peak-time breakfast and drive hours.
- In a cohosted show, you are nearly ten times as likely to hear two-plus male presenters as you are to hear two-plus female presenters.
- Solo women are more likely to be on air at weekends than during the week (Women, 2013).

Ann Charles, who worked with Williams and now organizes RadioTechCon, continues to seek opportunities to get women training in radio. She comments that although Sound Women is no longer operating, there are still efforts to do training for women, but it's difficult to find funding. "We need to stop talking about diversity and start paying for it, and not leaving it to women on the ground. . . . It's this seesaw: running a small business and wanting accessible pricing, but [we] have to be paid or [we] can't do this at all" (Charles, 2019).

SOUND WOMEN 2011-2016
FIVE AMAZING YEARS
ONE EXTRAORDINARY LEGACY

Sound Women was set up to help women get more out of working in radio and to help the radio industry get more from the talents of women. Gender equality in the radio and audio industry is no longer just Sound Women's business, it's everyone's business!

INCOME

- ● Volunteer time ● BBC (Diversity & local radio)
- ● Member Subs ● Creative Skillset ● Event income

EXPENDITURE

- ● Trainers ● Events ● MD ● Running costs ● Admin
- ● Website, social, marketing ● Accounting

RESEARCH

Sound Women commissioned research to explore the role and influence of women in the radio and audio industry.

- It's no longer acceptable to have all male panels at events
- There are more women on air than ever before
- Sound Women has changed how the radio industry thinks

> **Having women's voices on-air is hugely important, both to female radio audiences and to aspiring female presenters.**
> Maria Williams, Sound Women Founder

> **Sound Women leaves a proud legacy which the whole industry must build on.**
> Helen Boaden, BBC Radio

> **Sound Women has been a force for positive change and had a huge influence in shaping the debate around diversity.**
> Steve Ackerman, Somethin' Else

RUN BY VOLUNTEERS

Total Hours 12,040

- Social Media
- Podcasts
- Board
- Events
- That's equivalent to one and a half full time employees over five years
- Regional Groups
- Mentoring
- Research

TRAINING AND EVENTS

2,500 people attended training, events and regional meetings

SOUND WOMEN 2011-2016

- Brand You
- SW in Scotland
- Negotiating techniques
- Aim High
- SW On Tour
- SW On Tour
- Voiceover training
- Get That Job
- SW On Tour
- Return To Work
- Regional Meetings
- Freelance pitching
- Tech Train
- Future Festival
- SW On Tour
- **Tech Train**
- Meet the Patrons
- Inspiration Festival
- Confidence training

MENTORING

Building confidence and opening doors, six Sound Women mentoring schemes have supported a total of 115 women.

Got a new job	Was promoted	Got my own show	Set up a station	My idea was commissioned

PARTNERS

THE RADIO ACADEMY

BBC

RADIOCENTRE

creative skillset

somethinelse

PODCASTS

MORE THAN **30** PODCASTS
TENS OF THOUSANDS OF DOWNLOADS

SOCIAL MEDIA

More than **8,000** Twitter followers

PATRONS

Jane Garvey **Annie Nightingale** **Angie Greaves**

Sound Women's summary infographic

Terri Winston and Women's Audio Mission (USA)

"If I had spent money, time, and energy on research, that would have been 15,000 women and girls who would not have been trained," says **Terri Winston**, founder and executive director of Women's Audio Mission (Winston).

During their 17-year history, Women's Audio Mission has been continuously designing and retooling classes they knew were helping to "change the face of sound," as their motto says. "We didn't need a study, we could see from the convention floor that there were no women in the industry," she explains, referring to the visible lack of women at audio conferences and conventions. "We are not responding to studies, but Silicon Valley is responding to us. Intel funded us to open up and serve a few more schools in East San Jose which is adjacent to Silicon Valley, where poverty exists next to extreme wealth. We've had investments from Google, Cisco, Adobe, Dolby. The tech community is responding to how we're approaching and connecting at-risk girls to STEM" (Ibid.).

From their first AES booth over ten years ago that brought in $10,000 of audio gear, to their arsenal of three studios – two in San Francisco and one in Oakland – 17 years of hard work and dedication have gone into making WAM what it is today, a powerhouse of training, opportunity, and inspiration for young girls and women. They are the "only professional recording studios in the world built and run by women," providing "hands-on training, work experience, career counseling and job placement to over 2,000 women and girls every year in creative technology for music, radio, film, television and the internet" (womensaudio mission.org).

As a result, there have been over 750 placements: students have gone on to paid employment with Skywalker Sound, Andra Day, National Geographic, Alanis Morrissette, Google, Dolby Laboratories, and Pixar, among others (Dobson, 2018).

Placement is only part of the answer; retention is the other. Winston and her group engage in dialog with companies from Dolby to iZotope to festivals like Outside Lands, all of whom have approached her to help improve the working climate, thus taking an active role in creating environments more suitable for women (Ibid.).

Terri Winston, founder and executive director, Women's Audio Mission

The premise of having a space only for women and girls is at the core of their success. Co-ed spaces are simply not the ideal for the workshops and training they do. "The social norms are confining at school. When I went home or to my dad's lab, [it] was freedom; but when I went to school, suddenly 'I'm not supposed to be doing that stuff'" (Gaston-Bird, 2017b).

Winston's father is a research scientist and mechanical engineer, so from a young age, "his lab was my playpen, and he was always fixing things: the car, the television, the radio, so that was all big fun for me. Trips to the hardware store, one of my favorite places in the world, were all big influences on me. I was also a songwriter/musician from early on so audio was a natural way to combine both of these loves of mine. I definitely have my 10,000+ hours with tape recorders of all varieties from my childhood" (Ibid.).

"My earliest experience was recording myself playing guitars and singing, back and forth on two cassette tape boom boxes, probably in middle school," Winston recalls. "Then in college, I was studying electrical engineering, and I started recording the bands I was in on various analog 4 tracks, bouncing a ton of tracks. We did a lot of overdubs. I am a guitar player so I was always tinkering with amplifiers, biasing tubes. We eventually were signed to Polygram, toured a bunch with the Pixies and Throwing Muses, and that's when I ended up working in proper recording studios. My biggest influence during that time was working with Lenny Kaye (Patti Smith Group). That's how I got the engineering bug and learned what it meant to be a producer" (Ibid.).

Trying to pick out landmark achievements for WAM isn't easy. There was never an "*a-ha!*" moment. Instead, there are so many successes to list, but when pressed to highlight one, Winston talks about an experience centered around young, African-American women. "One of the most memorable projects was a project with ESPN. I went to speak at a school in Michigan [at the request of] a student who arranged the speaking engagement. A few years later she had graduated and ended up at ESPN. They were doing a piece on Black women athletes and they wanted an anthem to be created and recorded, and she convinced her boss to have Women's Audio Mission do it. They paid for us to get an all African-American engineering crew from our people, so we picked a local Bay Area artist (Rayana Jay), we had a Disney songwriter come out, we had an entire AA crew and everybody just. . . . It was an emotional recording session, you could see how happy everyone was. They cried. Incredible experience. The single got picked up by Hollywood Records and was seen on *The View, Good Morning America*, the music video was on ABC, NBC; super young engineers got that credit on that project and the traction helped them launch their freelance careers. It was an intentional choice that everybody made and had a big ripple effect" (Winston).

WAM started a conference series called WAMCon in Boston at iZotope in 2017, which has since expanded to include Nashville, Los Angeles, and New York, connecting 500-plus women across the country to training and mentorship with the best engineers and producers in the industry, from Grammy-winning mastering engineer **Emily Lazar** (David Bowie and Foo Fighters) to Grammy-nominated producer/songwriter Linda Perry (P!nk and Dolly Parton). Their internship placements include Alanis Morisette and Tracey Chapman.

"We don't spend money on public relations; we spend it on getting in the classroom. We overhauled two entire school districts, and we are looking at a third. We were the first to be at the table and provide training programs specifically tailored to women and girls. That's the pith of Women's Audio Mission" (Winston).

You can learn more about WAM at womensaudiomission.org.

SoundGirls.org

SoundGirls

In October 2012 at the 133rd AES Convention in San Francisco, **Terri Winston** moderated a panel called "The Women of Professional Concert Sound." The invited speakers included **Karrie Keyes** and **Michelle Sabolchick Pettinato**, as well as Claudia Engelhart, Deanne Franklin, and Jeri Palumbo (AES, 2012).

"We had all been in the business for 20 years or more," states SoundGirls' website, "yet most of us had never met before that day, and within minutes we bonded like long-lost sisters. We were struck by how similar our experiences, work ethics and passions were and wondered why our paths had never crossed and how our careers would have been different had we been there to support each other through the years. We were empowered" (SoundGirls.org, 2019).

In an article published *by Pro Sound News* in 2013, Keyes recalls how she felt after the event. "I left the AES panel with the feeling that none of us wanted the conversation to end. . . . My idea was that we could do something to find the women working in this industry and create a way to communicate with each other" (Welch, 2013).

In 2013, they began operating under the fiscal sponsorship of a nonprofit called the Northern California Women's Music Festival, and in March of that year, they established their website, SoundGirls.org. The site began by highlighting profiles of women working in live sound, and over time their efforts grew to include articles from "Sexism and dealing with it in the work place" to "Understanding RF"; links to instructional articles such as "How to design a stage plot"; invitations to volunteer on live shows and learn new skills; and links to sign up for pro sound workshops.

But perhaps the most exciting goal SoundGirls accomplished was the establishment of live sound camps for young girls in 2016. The group did extensive fundraising to make the one-week experience possible. The funds provided scholarships for any girl who wanted to attend. Ultimately, over 100 girls (70 percent of whom received scholarships) attended four camps across the USA. There were numerous volunteers and sponsors who invested time and money into the course, which was promoted as "a one-week camp for young girls & women ages 12–18 (all genders and non-binary people welcome) who want to learn about live music production." Altogether, there were 16 girls and a few boys who attended the course, which focused on live event safety, terminology, signal flow, microphones, and working a real show.

"By the end of the week [the students] were all able to read a stage plot, make an input list, wire the stage, line check and were able troubleshoot a few problems (phantom power and a monitor amp that was not on), without our assistance," reported Keyes. "We are already getting inquiries from people who wish to bring it to their communities" (Keyes, 2017).

At the camp, author Joy Lanzendorfer asked 17-year-old camper Mary Vogel for some insights. Vogel responded, "You're creating something live right in front of you. . . . You're making it richer. You're taking out the little buzzes and snaps and things you just don't want to hear because it takes away from the performance" (Lanzendorfer, 2017).

I asked Keyes how the atmosphere or vibe changes when women get together in crews or for workshops. "Women working together is powerful," she replies. "For some women, it's the first time to be surrounded by women versus men. We are finally with our peers. No one dominates the conversation. When men attend our workshops, many comment on how refreshing it was that the dynamic was completely different, and everyone is learning and working towards a common goal. They also comment that they have never been the only guy in a room – ever in their lives. They comment that for the first time they understand what women face every day. All people are welcome to attend our workshops and seminars, most of the time there are only a few men who attend" (Keyes, 2019).

SoundGirls has hosted several workshops with manufacturers DiGiCo, SSL, Allen & Heath, d&b audiotechnik, VUE audiotechnik, QSC, Sennheiser, Meyer Sound, Shure, DPA, and Klang, encouraging women to become trained and certified in the operation of live audio equipment (Ibid.).

Most recently, Spotify has sponsored SoundGirls' database of women in audio, the EQL Directory. The goal was to get rid of the excuse that organizers of conferences, programs, events, and recording sessions "can't find any women." Initially the database was on a Word-Press site where women could enter their own information, but with Spotify the resource now has a sleek look and is called "The EQL Directory: A Database for Women and Gender Non-Conforming Audio Professionals" which can be accessed at makeiteql.com. A press release published by Spotify in 2018 reads, "Powered by SoundGirls, made possible by Spotify, this is your resource to make putting together an inclusive team that much easier. Search the directory or add your profile to the community of women changing the face of audio" (Introducing The EQL Directory: A Database for Women and Gender Non-Conforming Audio Professionals, 2018).

The year after Keyes received her She Rocks award, she organized women from Sound-Girls.org to take care of all of the sound for the She Rocks awards ceremony.

The sound crew for the She Rocks Awards ceremony (2018)

The group has since expanded and includes 17 chapters in the United States, as well as chapters in Australia, Canada, Scotland, and England. Some notable alumni include Zionya Nolan, who now works at 8th Day Sound, and Kate Lee, who is now at Rat Sound Systems. Others have gone on to work with Sennheiser.

You can find more information at SoundGirls.org.

Note

1.   DA'EF work is viewed from a **glocal** perspective, in the sense that 'people are neither wholly global or wholly local – they are **glocal**' (Eriksen, 2010, p. 318) cited in Dobson 2018. Digital Audio Eco Feminism (DA'EF): The Glocal Impact of All-Female Communities on Learning and Sound Creativities. *Creativities in Arts Education, Research and Practice.* Leiden, The Netherlands: Brill | Sense.

Bibliography

About the WI [Online]. National Federation of Women's Institutes of England, Wales, Jersey, Guernsey and the Isle of Man. Available: www.thewi.org.uk/about-the-wi [Accessed September 27, 2019].

Adedamola, Phebean. Interview with Phebean Adedamola.

Aes. 2012. *Live Sound Seminar LS7* [Online]. New York: Audio Engineering Society. Available: www.aes.org/events/133/livesoundseminars/?ID=3277 [Accessed June 17, 2019].

Austin, Abhita. 2019. Interview with Abhita Austin. *In:* Gaston-Bird, L. (ed.).

Barra, Erin. 2019. Interview with Erin Barra. *In:* Gaston-Bird, L. (ed.).

Campbell, Madline. 2019. Interview with Madeline Campbell. *In:* Gaston-Bird, L. (ed.).

Charles, Ann. 2019. Interview with Ann Charles. *In:* Gaston-Bird, L. (ed.).

Dobson, Elizabeth. 2018. Digital Audio Eco Feminism (DA'EF): The Glocal Impact of All-Female Communities on Learning and Sound Creativities. *In: Creativities in Arts Education, Research and Practice.* Leiden, The Netherlands: Brill | Sense.

Dobson, Elizabeth. 2019. Interview with Liz Dobson. *In:* Gaston-Bird, L. (ed.).

Dovi, Lori. 2019. Interview with Lori Dovi. *In:* Gaston-Bird, L. (ed.).

Female Frequency. c. 2019. *Female Frequency* [Online]. Available: www.femalefrequency.com [Accessed September 27, 2019].

Gaston-Bird, Leslie. 2017a. 31 Women in Audio: Ebonie Smith. *31 Women in Audio* [Online]. Available: http://mixmessiahproductions.blogspot.com/2017/03/31-women-in-audio-ebonie-smith.html [Accessed March 31, 2017].

Gaston-Bird, Leslie. 2017b. 31 Women in Audio: Terri Winston. *31 Women in Audio* [Online]. Available: http://mixmessiahproductions.blogspot.com/2017/03/31-women-in-audio-terri-winston.html [Accessed March 21, 2017].

Introducing the EQL Directory: A Database for Women and Gender Non-Conforming Audio Professionals. 2018. *Spotify for Artists* [Online]. Available: https://artists.spotify.com/blog/introducing-the-eql-directory [Accessed June 17, 2019].

Kathryn, Julie. 2019. Interview with Julie Kathryn. *In:* Gaston-Bird, L. (ed.).

Keyes, Karrie. 2017. Live Sound Camp for Girls. Available: https://soundgirls.org/live-sound-camp-for-girls/ [Accessed September 27, 2019].

Keyes, Karrie. 2019. Interview with Karrie Keyes. *In:* Gaston-Bird, L. (ed.).

Laird, Pamela. 2006. *Pull: Networking and Success Since Benjamin Franklin* (Harvard Studies in Business History).

Lanzendorfer, Joy. 2017. Why Aren't There More Women Working in Audio? *The Atlantic* [Online]. Available: www.theatlantic.com/entertainment/archive/2017/08/why-arent-there-more-women-working-in-audio/537663/ [Accessed June 17, 2019].

Leon, et al. 2018. *Creativities in Arts Education, Research and Practice.* Leiden, The Netherlands: Brill | Sense.

Mari, Dani. 2019. Interview with Dani Mari. *In:* Gaston-Bird, L. (ed.).

Mengesha, Ethiopia. 2019. The Control Room Series, The First Workshop. *Gender Amplified* [Online]. Available: https://genderamplified.org/the-ctrl-room-series-the-first-workshop/ [Accessed March 12, 2019].

Sawyer, Miranda. 2016. Farewell, Sound Women, You Made a Difference. *The Observer: Radio* [Online] Available: https://www.theguardian.com/tv-and-radio/2016/nov/06/how-sound-women-changed-the-conversation-radio-gender-representation [Accessed 27 September, 2019].

Slater, Maya-Roisin. 2017. The Untold Story of Ableton Live – the Program That Transformed Electronic Music Performance Forever. *Vice* [Online]. Available: https://www.vice.com/en_uk/article/78je3z/ableton-live-untold-story-transformed-electronic-music-performance [Accessed September 27, 2019].

Smith, Ebonie. 2013. The Beginning: Gender Amplified 2007. *Gender Amplified* [Online]. Available: https://genderamplified.org/conference-2007/ [Accessed April 17, 2013].

Soundgirls.ORG. 2019. *About Us* [Online] Available: https://soundgirls.org/about-us/ [Accessed September 27, 2019].

Welch, K. 2013. Veteran Engineers Create SoundGirls.Org. *Pro Sound News. NewBay Media LLC.*

Williams, Maria. 2013. Sound Women. *Sound Women* [Online]. Available: https://mariawilliamsdotorg.wordpress.com/sound-women/ [Accessed September 27, 2019].

Winston, Terri. 2019. Interview with Terri Winston. *In:* Gaston-Bird, L. (ed.).

Only 1 in 5 Solo Voices Is Female. *Sound Women on Air – 2013.* Sound Women. [Online] Available: https://web.archive.org/web/20170323000109/http://www.soundwomen.co.uk/research/ [Accessed September 27, 2019]

3   Radio and Podcasts

Careers in Radio and Podcasts

▶ **Broadcast engineer:** A broadcast engineer oversees the equipment responsible for the transmission of radio programming. This can include broadcast towers, studio-to-transmitter links (STLs), satellite feeds, and remote broadcasting equipment (for example, to cover special events like sports or political conventions). **Ann Charles** elaborates on the difference between broadcast engineers and sound engineers in the UK: "Broadcast engineer is a really broad term," she says. "It's a spectrum of jobs, where there is some overlap between a broadcast engineer and sound engineer. However, *a sound engineer is **not** a broadcast engineer*," she emphasizes. "A sound engineer specializes in sound: they might upgrade the sound desk (console), help with complicated mixes for news features, or place mics on the orchestra. A broadcast engineer looks after the transmission chain: how audio gets from the studio to the transmitter and thus to the audience. Half of their job is to help announcers who have broken something and are live on air" (Charles, 2019).

▶ **Board operator/Broadcast mixer:** The board operator might load music into the automation software, make sure shows download from the network for live broadcast, and for larger stations operate the host's mic and switch between live elements. Their oversight of the feed usually ends where the broadcast engineer's job begins; in other words, the mixer makes sure the audio makes it out of the studio, but what happens after that is generally the domain of the broadcast engineer.

▶ **Radio Announcer/DJ/Host:** A radio announcer (DJ or host) is the person you hear on the radio and will read the news or play music for the broadcast audience. For sports programming, they will call the live action and work with the engineer to read into commercial breaks. In the case of "automated" music programming, the host "voice tracks" their front-announce (introduction) to the next song, or back-announce the previous song (letting listeners know what just played). This can be done hours before the broadcast, and a computer automatically fires the various voice and music elements in sequence.

▶ **Programming director (PD)**: The programming director makes decisions about the content: what types of songs and programs play and when. This is done in consultation with the general manager (GM) and market research teams.

▶ **Producer:** A radio producer might conceive of stories to cover for news. They might contact talent and guests, write scripts, and even record and edit the program.

▶ **Research and Development (R&D) and Innovation:** This includes the development and creative use of technology and radio apps for mobile devices. For example, the BBC is developing "Cloud-Fit" production. In this new era of apps and timeshifting, radio services can be expanded. According to the BBC website, "our current generation of iPlayer makes great use of cloud computing to move files and streams around, but it can't undertake any production operations: content still has to be created using traditional broadcast equipment in a physical production facility. R&D are working with colleagues from around the BBC to join up these two areas, enabling broadcast centre-style production operations to occur within a software-defined cloud environment" (*Cloud-Fit Production Architecture*, 2019).

There are also technologies like "smart speakers" (such as Alexa, Google Home, or Apple Home Pod) that people use to play radio. "It's becoming a radio replacement in many households," says broadcast engineer Ann Charles. "If someone says 'Capital', the system should know which capital. London? Nairobi? If you say the individual letters W-B-E-Z it shouldn't be picked up as 'wubbez', and 'Radio 1–0–2' is not 'Radio one hundred two'" (Charles, 2019).

▶ **Accessibility expert:** For users that are sight- or hearing-impaired, apps and other tools can give these audiences access to programming and features.

▶ **Audiobooks** usually require a single voice talent to read while an engineer records and perhaps edits the various takes. Editing sometimes happens at faster-than-real time speeds.

▶ **Podcasts** can be produced by anyone who can connect a microphone to their computer and learn how to edit, which is a good thing because it provides more people access to a more targeted audience. It benefits women who prefer to learn the trade outside of traditional structures to develop their content to their intended demographic.

▶ **Radio dramas** are an interesting blend of old and new production styles. Examples from early radio history include *The Shadow* or *Masterpiece Theater* in the USA. Today, radio dramas are more sophisticated, and some have been exploring **binaural** reproduction. Such programs are promoted by the BBC in England, including *Ring*, which was mixed by engineer Catherine Robinson.

Tools of the Trade: Radio

This is a list of common tools used in radio.

Handheld mic: A handheld mic is usually used for reporting. Certain mics, such as the ElectroVoice RE-50, are specially designed to reduce "handling noise" by reporters as they interview guests in the field.

Shotgun mic: A shotgun mic has a very directional polar pattern and can be used in noisy crowds, for example.

Lavalier mic: For more than one guest (for example, a panel of experts), it may be desirable to use lavalier microphones. These mics are usually clipped on the guests' clothing. However, in radio, it is more common to use mics on desk stands or fixed "boom arms."

Windscreen: This is usually a piece of foam that fits over the microphone to reduce the "rumbling" effects of wind on the microphone's diaphragm.

Zeppelin: This is a special windscreen that is shaped like a blimp or zeppelin. Usually used in windy conditions, they are far more effective than a foam windscreen for reducing the effects of blustery wind.

Announcer mics: Commonly used mics in a radio studio include:

- Shure SM-7
- Neumann U-87
- ElectroVoice RE-20

Vocal processing:

- De-esser: This helps to control sibilance from announcers, usually heard as a harsh "s" sound.
- Compressors and limiters: It's important for radio stations to maintain levels so that digital distortion is avoided. For example, compressors help keep the morning show hosts' banter and laughter under control.

Broadcast console: Unlike music consoles, a broadcast console has two "busses" (signal paths): PGM ("program": channels destined for air) and AUD ("audition": allows the engineer to audition material before it airs). It might also have a CUE function: a position for the fader at the bottom of the fader's throw. When the fader is pulled all the way down, audio from the source (such as a music bed or commercial break) is fed to the announcer or engineer's headphones. When the fader is pushed up, the audio begins. Much newer technology is described by **Ioana Barbu** in her profile.

Interruptible Fold Back (IFB): This allows the director or engineer to speak into the announcer's ear during a live broadcast. The audio that the announcer is listening to is "interrupted" by the director or engineer, who might say something like, "30 seconds to go"; or in the case of breaking news, "Julie is coming in with a new script," so the announcer is ready to receive written communication.

Cathy Hughes (USA)

"My mother bought me a transistor radio, and I had never heard Black radio before. It just hit me like a virus. I wanted to be on the radio!" (*Cloud-Fit Production Architecture*, 2019).

Cathy Hughes was inducted into the National Association of Broadcasters (NAB) Hall of Fame in April 2019. She is founder and chairperson of Urban One, Inc., the largest African American-owned and operated broadcast company in the United States, with 59 broadcast stations across the country (*Cathy Hughes Founder and Chairperson Urban One, Inc.*, 2018). When the company went public in 1999, she also became the first African American woman to head a publicly traded corporation (Hill, 2018). In addition, she was the first woman to own a radio station that was ranked number one in a major market (NAB, 2019).

In 2016, for its 45th anniversary, the Howard University School of Communications, where Hughes taught as a lecturer beginning in 1971, was renamed the "Cathy Hughes School of Communications."

Hughes recalls how she was inspired to start her business.

> "After I left Howard University, I worked for a group of investors that for nearly two decades had tried to get a signal which was dark, and they recruited me to get them on the air . . . they had been fighting legal battles trying to get on the air and they told me that they would like for me to prepare a package and shop it. And having a business background, I said to them 'Fine, I need an equity position,' and they very arrogantly and very unkindly said to me 'If you think that you are worthy of owning a radio station you should do your own.' And it was literally like a light bulb had gone off in my head. And so I thanked them and I resigned. And I went out and started structuring a package for me to go into business for myself. I didn't care that they were being sarcastic or being mean spirited. It was a revelation for me and I thank them to this day for it because that's why I have now been in business for myself almost 36 years."
>
> (*Cloud-Fit Production Architecture*, 2019)

You can find out more about Cathy Hughes at http://cathyhughes.com.

◼ Ann Charles (United Kingdom)

"Being a broadcast engineer isn't difficult if you know the basics," according to **Ann Charles**. "It's all signal flow – it's all just moving noise about!" (Charles, 2019).

She understands that for newcomers it might be difficult to know where to fit in, but she offers some encouraging words. "Everyone is a specialist," she continues. "You can become the world's expert at something really quickly because it's such a small industry. So, for example, if you like accessibility and radio systems, there aren't that many people doing it, so very quickly you can become the one who knows more about that than most people. And it's a really friendly and supportive community: people like to help and they like being asked questions and show off what they know. So you don't need to think, 'I don't know any of it', because your colleague doesn't know the stuff *you* know, either" (Ibid.).

If you are curious about working in broadcasting engineering, Charles says having previous experience in radio is an advantage – in fact, you might even be better equipped than someone who is purely trained in computers and information technology. "You know how to prioritize," explains Charles. " 'Stay safe! Stay on air!' That might mean moving someone to another studio, rather than trying to fix a problem right away. For a broadcast engineer, half the job is keeping everyone on air, the other half is planning projects" (Ibid.).

Charles has worked on some very large projects: approximately 80 million people listen to the work on the systems Charles and her colleagues rolled out. "The BBC was rebuilding Broadcasting House and I got a job as project manager doing audio edit and playout systems for the 'W1 programme' which was the largest project that the BBC had ever done. And then a few months into that, 'Project North' – which is the second largest project the BBC had ever done – was having some issues, and they said, 'Oh, your department is really good at delivery, so . . . congratulations! Everyone in your team is now going to work for the largest project the BBC has ever done *and* the second largest project the BBC has ever done – *at the same time!*' That was an intense five years!" (Ibid.).

The "W1 Programme" was a name given to a massive undertaking to centralize all BBC operations into one building. The Broadcasting House building was chosen to be the "home for the BBC's national and international journalism, BBC television services, BBC Network Radio, online teams and professional support services, bringing together thousands of staff under one roof" (APM Project Management Awards Winner's Case Study, 2013). The BBC's Project North built new studios in MediaCityUK, Salford, an "international hub for technology, innovation, and creativity" (MediaCityUK, 2018).

After leaving the BBC, Charles taught journalism in Wellington, New Zealand, and then went to South Sudan to train broadcast engineers and producers. Now she runs her own company; one of her clients is Broadcast Bionics, which is using new technologies to automatically transcribe broadcast content and social media interfaces using artificial intelligence and machine learning.

She is also Director of RadioTechCon, the UK radio and audio industry's technical and engineering event, which she runs with radio colleague Aradhna Tayal. "This year we also have the Radio Technology Masterclass which is an introduction to broadcasting engineering, designed for newcomers to come and have some hands-on training" (Charles, 2019).

You can find out more about Ann Charles at http://anncharles.tv and more about RadioTechCon at www.radiotechcon.com.

Ioana Barbu

"This is like the Starship Enterprise!" That is the typical reaction of new radio announcers when they first see the new technology in the radio studio, according to **Ioana Barbu**, former broadcast engineer with Global Radio and radio infrastructure coordinator at Bauer Media. "Others said, 'I could run a business on the side, because this is doing loads of stuff for me,'" she recalls (Barbu, 2019).

Today's radio consoles have amazing customizable and flexible features. "Some are touch screen; they also recall your console setup so you can 'hot seat' them. For example, some presenters prefer their microphone channel fader on the left, others prefer it on the right. So during a shift change, if there is music coming out one fader and the next announcer recalls a preset, the fader will move to a different place for the next announcer and will display a message that says 'pending' until you take the fader down. It is fail-safe, so you can't accidentally take yourself off air," she explains (Ibid.).

"Or if you are on-air and you have three songs before your next announcement, you can be interviewing a guest in the studio or on the phone using 'pre-fade record mode'. You can do the whole thing off-air while the on-air program continues through the same board, without having to swap studios" (Ibid.).

In the "old days" of analog radio broadcast, a console failure would be a catastrophe. At that time, engineers managed to keep things running, but Barbu's example gives a glimpse of what was involved. "I remember working in a recording studio where we had a 60-channel Neve desk and it needed 2 power supplies, and one of the power supplies died, but the way it was wired, that whole desk was down – which meant the whole studio was out of action. We had to have a Neve engineer come in and rewire the power supply so we could use *half* the desk . . . and that took days! Today with these radio desks, you walk in, plunk another module in, and you're done!" (Ibid.).

With new digital technology and audio over IP infrastructure (with protocols like AES67), it's also possible for any studio to emulate another, so if a board goes down, a presenter can walk to another studio, enter a few settings, and be back on the air in no time – *and* have their same console profile recalled.

Although Barbu is now a freelance music producer and mastering engineer (who also specializes in post-production), her experience with the broadcast signal processing chain, See Figure 3.1 gives her valuable insight into mastering – and the loudness wars.

"The diagram relates to a *broadcast processor*, which is applied to the whole output: Voice, music, and ads. But there is a *voice* processor in the studio for the announcers, so the voice, music and ads get merged in the brain of the desk in a stereo file. Then the broadcast processor gets applied to the output of this.

"But some broadcasters have the most ridiculous signal chain within the broadcast processor: eq-eq-compressor-multiand-eq-automated gain control (AGC) . . . you name it! But do we need all of what's in the diagram, guys?

"As mastering engineers, we've already made those tracks sound the best they can [on playback systems] because that's our job. So, I'm having this internal conflict! Except for the BBC, all the rest of the radio stations are commercial, and they have to sell adverts to make money. They believe one way to get audiences is to make it louder, because louder sounds better to our ears psychoacoustically – but the listeners can use the volume control. Obviously, factors like the content of what is played on the station matters more to a listener for tuning in, but loudness also makes a difference – and broadcasters make use of these massively complicated broadcast processors to give their stations a particular sound – like a station's 'signature' sound. It's a valid commercial reason to affect the audio and attract an audience – but audio quality-wise, from a technical point of view, it can degrade the audio quality, and there are no standards in radio for audio quality at present, at least in the UK" (Ibid.).

Caryl Owen (USA)

Caryl Owen is enjoying retirement at her home in Wisconsin. "I don't miss the faders," she says, which surprised her. She paints, gardens, and plays the bass. She agrees that all the best audio engineers are bass players. "We are background, we are the thing that holds it all together. I have met more frustrated bass players who are good engineers" (Owen).

When she was a child, she enjoyed spending time underneath the piano, listening to her mother play. "I still like the sounds underneath a grand piano!" She also liked listening to recorded music. "I had a little pink and grey Columbia record player and what seems to have been a complete set of Golden Books Records, but my real favorites were our recordings of *Scheherazade* and Richard Rogers' *Victory at Sea* (Gaston-Bird).

"I learned how to run a cheap PA mixer for the guys who had a garage band down the street in high school. I mixed for my bands when I wasn't singing, played around at college

FIGURE 3.1 Optimod broadcast processing signal flow

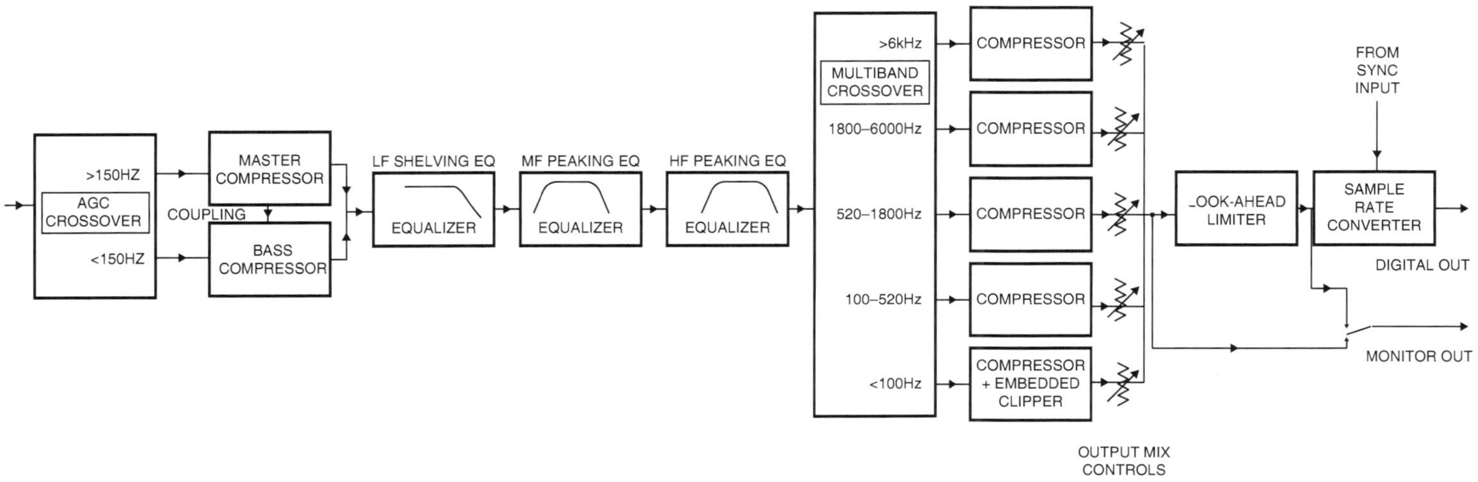

radio stations, shadowed a friend who was a monitor mixer for Kiss, and eventually took a class at a little local recording studio where I ended up running the office while teaching classes – sometimes I was only a week or so ahead of my students. That's a good way to learn fast!" (Gaston-Bird, 2017a).

Owen had worked in rock-and-roll studios where producers covered their patch bays so you couldn't see what they were doing. This seemed silly to her because "I had to take the patch bay apart and set it up the next day." She laughs. "Being in the studios in the early '80s was really tough. I would get, 'oh yeah, you have to haul the Mesa Boogie amplifier upstairs. By yourself.' And those things are heavy. Or, 'oh, you are going to move the B3 and the Leslie [speaker] up the stairs.' And we did because that's what we had to do" (Ibid.).

She met Carla Bandini at Wizard Sound Studio sound and later visited her at Associated Sound. "It was a jingle house and those guys could set up in 15 minutes, record three 30-second spots with a full band in one hour, *cut and mixed*. I shadowed her one time and I was amazed. Clients would book an hour and be done in an hour. Not like, 'we need to get the right drum sound on the first day' no, you had *three minutes*. It went from one session to the next. But it was like factory work" (Owen).

Owen began her career at National Public Radio in 1985. "The women at NPR were one-third of the tech staff. And they knew their shit. Everyone was generous with their knowledge. It was completely different from anything I'd experienced." In 1990 she moved to NPR's New York Bureau. During that time, she worked on StoryCorps with its founder, David Isay (College, 2015). "Of the things I've done in the past, I'm probably most proud of the many projects I worked on with David Isay. Even after all of this time, 'Ghetto Life 101' sounds amazing, especially when you consider that the original recordings were made to cassette by a couple of kids, and the mix was all analog, multi-machine choreography. We figured it took an hour to mix each minute of the final piece" (Owen).

Produced in 1993, "Ghetto Life 101" has achieved international acclaim and can be heard at https://storycorps.org/stories/ghetto-life-101 (Corps, 1993).

Caryl Owen
Source: Photo credit Jim Gill/WPT

In 2001 Caryl was recruited to become technical director at Wisconsin Public Radio, and in 2004 she took over as technical director for the show, *To the Best of our Knowledge* (distributed by PRI, Public Radio International [now PRX]). Altogether, Caryl has worked on four Peabody Award-winning projects, "three of them with David Isay and one with *To the Best of Our Knowledge* . . . It pays to work with the best in the business!" (College, 2015).

Having been one of the best in the business, she took seriously her role as a mentor. "Mentoring someone who has a deep desire to learn is one of the most satisfying things I've done. I'm so proud to have helped some very talented people get started" (Owen).

Suraya Mohamed (USA)

Suraya's mother, Muriel Mohamed, had a recording studio in the house. "She had a dBx machine, Otari 8 track, 12 guitars, patch bays. She was a composer, musician, singer, and guitarist" (Mohamed, 2019).

But Mohamed wanted to be a surgeon. That would change over time, and she would earn a bachelor of arts in viola performance and another bachelor's degree in recording arts and science from Peabody University. Near to Peabody's campus in Baltimore, Maryland, National Public Radio in Washington, D.C. was offering internships. Over time, she was hired as a broadcast/recording technician at NPR in 1990 (Ibid.).

A few years later, Mohamed became a mother around the time the industry began its transition to digital technology. "I was offered the opportunity to work in Studio 4A, NPR's new performance studio in 1993, and was invited to learn Sonic Solutions. Somehow, I knew that DAW would be my ticket to job flexibility. When my son was born in 1995, I began job sharing with Rima Snyder and went part-time until 1999. But when Rima moved, our job share ended, and by then I had three kids: a newborn, a one-year-old and a three-year-old. I opted not to do child care and did nine months on the overnight [shift] instead." She stopped engineering on New Year's Eve, 1999.

In 2000 she started working at the Smithsonian part time and worked from home on Pro Tools for part of the time. "I was a stay-at-home mom with my foot still in the market" (Ibid.).

She started freelancing on *Jazz Profiles* and *Billy Taylor's Jazz at the Kennedy*. "At that time, I was crossing the line between being a producer and an engineer, but I knew that would continue to be the trend. I would help people mix their pieces and make creative decisions. I was a 'creative engineer,'" she says. Then in 2004 she began a job share with NPR's Kerry Thompson editing the *World of Opera*.

Since 2014, Mohamed has been producer for *Jazz Night in America*, WBGO, NPR music, and *Jazz at Lincoln Center*. "I'm taking the medium to different places and making it accessible," she explains (Ibid.).

"I am proud that I have navigated in a man's world without losing my feminine identity . . . but also being tough and strong and walking that line. No one ever took advantage of me, nor did I experience sexual harassment, and I was held in high regard and respect even though I am a woman of color. But I had that determination that I wasn't going to let anyone stop me. That resolve showed in my attitude and demeanor. You had to assume otherwise people were going to step on you – but that could apply to any industry. I did it with grace and I was never a rebel about it. Not that I wanted people to like me, but it was important that I navigated those tricky relationships, so I didn't seem like a troublemaker or boat rocker. I continue to try to do that. You can do it! It's not always easy . . . but it's possible" (Ibid.).

Mohamed also produces NPR's holiday specials package, including *Tinsel Tales*, *Hanukkah Lights*, *Toast of The Nation*, *Pink Martini's Joy to The World: A Holiday Spectacular*, and most recently *Hamilton: A Story Of US*. She also works on the Tiny Desk as either a producer and engineer (NPR, c. 2019).

Lorna White (USA)

Since its inception 21 years ago, **Lorna White** has been the sound engineer for the popular radio quiz show, *Wait, Wait . . . Don't Tell Me!* The show is still going strong with no signs of slowing down (Gaston-Bird).

White began working at National Public Radio in 1984 and transitioned to working as an audio tech there in 1986. She enjoyed working with a large proportion of women on the technical staff. "Being with 40 techs was great – over a quarter of them were women. The competition for the good gigs was tough, and the stronger personalities usually won out. I didn't have much of a strong personality at the time, and I was satisfied with working in the studios and master control. I was the first female tech go through a pregnancy and have a child at NPR in Washington. That was an interesting learning experience for me and

Lorna White (rear of stage) mics up Paula Poundstone as Tom Hanks gets ready to host the NPR show, *Wait, Wait . . . Don't Tell Me!*
Source: Photo credit NPR/Andrew Gill

management! Two years later, the Chicago Bureau engineer position was open, and I applied so my family could move back to the Midwest. I was ready to be in charge of my own work atmosphere and do interesting things back in the Chicago area" (Ibid.).

Her earliest experience with audio was listening to cassettes, and she got to do sound effects at an early age. "In eighth grade we did a project with puppets made in art class. Two of my friends scripted a variation of the Tidy Bowl Man commercial, and I was excused from class to record the perfect toilet bowl flush." Her parents had a Hammond B3 with a Leslie speaker, and she took organ lessons starting at age five.

"When I was in high school, I was picked to become the high school accompanist for the choir. It was a small, rural high school and the girl who had previously played graduated. After a few months, my mom took me to a music store and together we chose a piano. That is one of my fondest memories. I was a piano major in college but hated practicing. I would rather record people who didn't mind practicing, so my piano professor took me to the Illinois State University radio station WGLT and I started there, filing jazz and classical records. That was my first introduction to NPR" (Ibid.).

You can listen to *Wait, Wait . . . Don't Tell Me!* at www.npr.org/programs/wait-wait-dont-tell-me.

Nadia Hassan-Garschagen (Germany)

"I've been into music my whole life," says **Nadia Hassan-Garschagen**, who works at Radio Berlin-Brandenburg in Germany. "When I was a little kid, I sang in the choir at school and I learned to play the guitar. I went to lots of concerts in concert halls and opera houses here in Berlin, and therefore I have been interested in music for a long time. On the other hand, I am interested in physics and mathematics as well." As far as her role models, Hassan-Garchagen points to Marie Curie, a Nobel Prize winner for physics (1903) and chemistry (1911). "When I finished school I was interested in music engineering, and so I looked for a program that could do both" (Hassan-Garschagen, 2019).

"In Germany, we have a Girl's Day where girls can go to companies that are more technical, and get information on how to study for those professions," she says. "We also have networking opportunities for women, not only within the company but within Germany; there are several radio and television stations that have an annual meeting, the Treffen der Medienfrauen ('conference for women in media'), for women technicians, editors and reporters" (Ibid.).

At RBB, Garschagen estimates one-quarter of the television and engineering staff are women, although far more management positions go to men. In the studios, the numbers of men and women are equal, whereas men tend to do remote gigs more often (Ibid.).

In 2017, Hassan-Garschagen helped set up and organize a remote reporting station for Kirchentag. "Kirchentag is a big event where all the Protestants come together, and there are a lot of meetings in the city, and RBB covers this event from our stations. So it's a big event. It's a big challenge to get all the equipment and people together, and organize everything on the fairground. We will build a little broadcasting station, and then the reporters and editors can come to this installation to make their recordings, file news pieces, or do live reports. It's a big task" (Ibid.).

Nadia Hassan-Garschagen in the entrance hall of RBB Studios, Berlin, Germany

Podcasts

The emergence of podcasts has provided a way for women to learn about recording technology without barriers (real or perceived) while having the freedom to choose subject matter independent of the need to appeal to a mass audience (see "Sound Women" on pages 85–86).

Author Robin Kinnie, in her article "The Growth of Women in Podcasting" writes, "Although podcasting has a low barrier of entry (most of the podcasters I spoke with record in their homes), there were still perceived obstacles shared. One of the obstacles includes the need to have expensive equipment, technical expertise and feelings of self-doubt. This was brought up by numerous respondents – lack of confidence. Many thought that they were not worthy of having a podcast due to lack of experience" (Kinnie, 2019).

"Werk It" is a women's podcast festival held at WNYC studios. The event is "for women+ in all stages of their careers currently working in or hoping to break into audio and digital media" (Studios, 2019). In 2018, Spotify began the Sound Up Bootcamp workshop, "a weeklong June intensive for aspiring female podcasters of color" (Spotify, 2018). There is also the "Google Podcasts Creator Program which 'seeks to increase the diversity of voices in the industry globally and lower barriers to podcasting.' Selected teams receive seed funding and participate in an intensive training program" (Blodgett, 2019) "Each of the programs include training, access to podcasting experts and the chance to get money to take your idea from concept to reality."

Bibliography

Barbu, Ioana. 2019. Interview with Ioana Barbu. *In:* Gaston-Bird, L. (ed.).

Blodgett, Sequoia. 2019. Google Is Once Again Looking for Podcasters and Creators of Color for 2019. *Black Enterprise* [Online]. Available: www.blackenterprise.com/google-podcasts-creator-program-2019/ [Accessed September 27, 2019].

Cathy Hughes Founder and Chairperson Urban One, Inc [Online]. 2018. Available: http://cathyhughes.com/about/ [Accessed September 27, 2019].

Charles, Ann. 2019. Interview with Ann Charles. *In:* Gaston-Bird, L. (ed.).

Cloud-Fit Production Architecture [Online]. 2019. BBC R&D. Available: https://www.bbc.co.uk/rd/projects/cloud-fit-production [Accessed September 27, 2019].

Gaston-Bird. 2017a. 31 Women in Audio: Caryl Owen. *31 Women in Audio* [Online]. Available: http://mixmessiahproductions.blogspot.com/2017/03/31-women-in-audio-caryl-owen.html [Accessed March 12, 2017].

Gaston-Bird. 2017b. 31 Women in Audio: Lorna White. *31 Women in Audio* [Online]. Available: http://mixmessiahproductions.blogspot.com/2017/03/31-women-in-audio-lorna-white.html [Accessed March 2, 2017].

Hassan-Garschagen, Nadia. 2019. Interview with Nadia Hassan-Garschagen. *In:* Gaston-Bird, L. (ed.).

Hill, Selena. 2018. Urban One Founder Cathy Hughes to be Inducted into NAB Broadcasting Hall of Fame. *Black Enterprise.*

Kinnie, Robin. 2019. The Growth of Women in Podcasting. *Podcast Business Journal.*

Mediacityuk. 2018. *Welcome to MediaCityUK* [Online]. MediaCityUK. Available: www.mediacityuk.co.uk [Accessed September 27, 2019].

Mohamed, Suraya. 2019. Interview with Suraya Mohamed. *In:* Gaston-Bird, L. (ed.).

National Association of Broadcasters. 2019. *Join NAB in Honoring This Year's Industry/Service, Radio, Television and Technology Award-Winners!* [Online]. Available: www.nabshow.com/happenings/awards-programs/nab-awards [Accessed September 27, 2019].

National Public Radio. c. 2019. *Suraya Mohamed: Producer, NPR Music* [Online]. Available: www.npr.org/people/505729293/suraya-mohamed [Accessed September 27, 2019].

Owen, Caryl. 2019. Interview with Caryl Owen. *In:* Gaston-Bird, L. (ed.).

"Programme of the Year, 2013: W1 Programme, BBC" APM Project Management Awards Winner's Case Study. 2013 [Online] Available: https://www.apm.org.uk/media/1197/programme-of-the-year-2013-bbc.pdf [Accessed September 27, 2019].

Ripion College. 2015. *Alumna Works with 'To the Best of Our Knowledge' on Wisconsin Public Radio* [Online]. Ripon, WI: Ripon College. Available: www.ripon.edu/2015/08/17/alumna-works-with-to-the-best-of-our-knowledge-on-wisconsin-public-radio/ [Accessed June 27, 2019].

Spotify. 2018. Amplifying Female Voices of Color Through the Power of Podcast. *For the Record* [Online]. Available: https://newsroom.spotify.com/2018-07-11/amplifying-female-voices-of-color-through-the-power-of-podcast/.

Story Corps. 1993. *Ghetto Life 101.* Available: https://storycorps.org/stories/ghetto-life-101/ [Accessed September 27, 2019].

WNYC Studios. 2019. *Werk It! A Women's Podcast Festival* [Online] Available: www.werkitfestival.com [Accessed September 27, 2019].

4 Sound for Television and Film

▮ Careers in Sound for Film and Television

Training for these roles is offered in a number of schools. Useful keywords when searching for college and university programs include "audio post production," "sound for film and television," and "sound design." Cinema Audio Society president **Karol Urban** has some advice for aspiring mixers: "Not everyone that I work with has gone to college. Maybe they will have gone to institutes, or Full Sail, or the New York Film Academy or that type of thing, which is fine. I find that my education in college is very precious: it gave me a structure in which to tackle new things that I need to learn; it taught me how to learn with other people on projects; and taught me collaboration which is very important. But the actual specifics of my job function are *on-the-job* learning functions.

"I find that the best way to get to the position of being a mixer is to simply be a mixer. Go find a student film to work on. You don't need somebody to give you permission to be a mixer. Be a mixer, until those people that you're working with will organically become better at what they do, and they will bring you along because you helped formulate what they create. Your filmmakers will take you with them. Your picture editors will continue to come back to you. And as that happens, you will be more experienced, you'll be more desirable to different companies. And that's how I found myself in the position of being a mixer" (Urban).

There are, very broadly speaking, three areas of specialization: location recording, post-production, and live broadcast. Below are some of the jobs you will find in each type.

Location Recording

▶ Production Sound Mixer: Depending on the size of the production, this person is responsible for miking the actors, possibly doing a mix for the cameras, syncing to timecode, and using metadata (text information added to a file about scene number, take, channel number, etc.) to pass on to the post-production team. On larger films, the production sound mixer supervises the sound crew, delegating these tasks. (See also: Tools of the Trade later in this chapter).

▶ Boom operator (first boom or utility sound technician): This person is responsible for getting sounds from the on-screen talent using a "fish pole" or "boom" with a shotgun microphone affixed to the end, and wears headphones to make sure sound is "on axis," which basically means that the mic is properly aimed at the talent and sounding good. They work with the camera and lighting crew to avoid shadows while getting as close to the talent as possible.

Post-Production

▶ Supervising Sound Editor: This person works with the director to understand the general direction of the film's mood and oversees the sound crew. They supervise the acquisition of sound and music with the sound team and may do mixing and editing as required.

▶ Foley Artist: This person is responsible for re-creating the sounds that are seen on screen. During the location recording, sounds such as footsteps are not picked up from the actors' mics (or might reveal problems in the set, like squeaky floors). Other detailed sounds like bottles opening, doors closing, clothes rustle, and the like are recorded so the scene has as much detail and realism as possible.

▶ Foley Editor: This person works to record the Foley artist and chooses the best, most realistic take for the scene. They also edit the sound, sometimes adding reverb to make the sound realistic and "fit" in the picture.

▶ Sound Effects Editor: This person is responsible for finding sounds from environmental ambience and "room tone" to special effects such as explosions. The sound effects editor also creates layers of effects to create convincing sounds. These could range from battles in outer space and alien life-forms, to car chases and shootouts, to the complete omission of sound for storytelling purposes.

▶ Dialog Editor: This person uses tools such as noise reduction, EQ, and compression to make dialog intelligible and pleasing, and to match dialog quality and characteristics from one scene to another.

▶ ADR Editor: For scenes that require Additional Dialog Replacement (some say the acronym stands for Automatic Dialog Replacement), the ADR editor will choose the right mic for the actor, set up a "looping session" where the actor can re-perform the scene multiple times for the director, and chooses the best ("hero") take. The ADR editor will also edit and process the dialog to blend in.

▶ Music supervisor: This role researches music that can be used in the film, sometimes in addition to original music. They make sure the cue sheets are done and music is properly licensed.

▶ Music mixer/editor: This person edits the music and music stems (e.g. percussion, brass, strings, and woodwinds). They decide how to balance and pan the instruments, perhaps to give a sense of envelopment.

▶ Composer: This person writes music according to the director's vision for the mood of the film, sometimes with an orchestra, or sometimes all "in the box" with sampled instruments, synths, drum machines, or experimental sounds.

Considerations of tempo are also important to help drive the film or add the right sense of drama, tension, or humor.

▶ <u>Re-recording Mixer / Dubbing mixer</u>: The re-recording mixer brings together dialog, music, and effects and balances the levels and adds signal processing to create the final "print master" of the film. Current Cinema Audio Society president **Karol Urban** compares the role to that of a chef who takes prepared ingredients and presents the final dish.

Live Broadcast

▶ For this category, the focus is on sound with picture (see also: Careers in Radio for sound without picture).

▶ <u>Broadcast engineer</u>: A broadcast engineer oversees transmissions of televised programming. This can include broadcast towers, studio-to-transmitter links (STLs), satellite feeds, and remote broadcasting equipment (for example, to cover special events like sports or political conventions).

▶ <u>Broadcast sound mixer</u>: A broadcast sound mixer runs the board for live programs. Their oversight of the feed usually ends where the broadcast engineer's job begins; in other words, the mixer makes sure the audio gets to the uplink, but what happens after that is the domain of the broadcast engineer.

▶ <u>Producer</u>: The producer might conceive of stories to cover for news and features. They might contact talent and guests, write scripts, and even record and edit the program.

A Master Control Room

▶ Master control engineer: At large networks, several channels of programming are happening at once; for example, at HBO there are various channels such as HBO Comedy, HBO Family, and HBO Latino, among others. In the master control room, engineers monitor the various channels on banks of television monitors to ensure that the quality is maintained and content is not interrupted on its way to the viewer. In case of an outage, the engineer must work under great pressure to restore services quickly to paying customers.

Tools of the Trade: Location Sound

In this section, **Lori Dovi** walks us through the field of location sound.

Sound cart: Some might call the location sound recordist's cart a work of art; Dovi says it's really a studio on wheels. "You are an owner/operator. You are investing vast amounts of money into creating a mobile recording studio."

The cart is designed with ergonomics and efficiency in mind: how fast you can get to what you need. "And you're going to create multiple carts," adds Dovi, "As you add and subtract equipment, the whole cart changes, and the equipment drives how we work."

"You're also creating the 'follow cart' for the rest of the film department with shelves and drawers; it's almost like building a house. How do you like to work? Well, how do you like to live? Where's the speed and efficiency of your layout?"

Laptop computer and recorder: Dovi uses a Cantar field recorder, "but there's also Zaxcom, Fostex and Sound Devices recorders. You also need a keyboard for typing in data to the recorder."

Microphones: There are a whole array of microphones based on what the scene is. "Boom mics, plant mics, wireless lavalier (lav) mics; they're your ingredients for basically creating a soundtrack; and then you have your wireless receiver setup: I use the Lectrosonic Venue system."

Communications or "comms": These are the various feeds that get sent to the director, video village, and any device used for playback. "Comms can go to the utility sound technician and boom operators; to the camera operators who listen to cues; also to the production assistants and script supervisor."

Video monitors: In the old days, the production sound mixer may have needed to work without the benefit of video monitors, "but now you're way far away from the soundstage and you are mixing live."

Boom pole: This is generally owned by the boom operator and utility sound technician, who also doubles as a second boom operator. They take their direction from the first boom operator, who is essentially running the set. The production sound mixer is head of the department, but they're located far away for sound isolation purposes. Although the production sound mixer is the boss, you're relying on the first boom to run the *set*.

"We also have the playback operator if there's music that has to be played back."

Gaffer tape: One can never have enough gaffer tape.

Women in Location Sound

Jan McLaughlin, CAS (USA)

In sound design, the goal is to be invisible: if your audience notices flaws in the sound, then the illusion is shattered. That philosophy makes it easy to understand why, when first starting out, an engineer might feel more comfortable "staying out of the way." But as **Jan McLaughlin** discovered, that approach is not necessarily helpful in the career of a location sound recordist. "It wasn't until after I got my first – and then the next year, the second – Emmy that I began to take myself seriously," she says. "I realized that, look – I've got input. And before that time, my strategy was to remain invisible. If nobody thinks about sound at any point in the process, then I'm successful. But it wasn't getting me bigger jobs and I decided that I don't have to be invisible anymore; that I have things to say and could go to the director and say, you know, I want to talk to you about how you want to handle this playback situation. And that was big for me; that I felt like I had a seat at the table after that" (McGlaughlin).

McLaughlin has captured two Emmys for her work *Nurse Jackie* in 2013 and 2014 and was nominated for another for the same series in 2012. She is now the owner of Sounds Good, LLC, and enjoys the freedom of running her own company. Although she had previously

Production Sound Mixer Jan McLaughlin, CAS
Source: Photo credit Diane Hounsell

employed the services of an agent, "I think I had learned what I needed to learn, which was that a question does not have to be answered right then and there. What the agents taught me is that I can buy time. 'Let me get back to you' . . . that bought me time to think and plan and strategize. So [now] I buy the time myself" (Ibid.).

It wasn't always that easy, of course. McLaughlin didn't get a degree in music, engineering, or recording; she started out as a paralegal doing real estate for around 10 years. "I decided I was going to change my career and part of my marriage was based on our love of doing poetry performances and I had been doing sound works, and we started doing film work with poetry: 'film poems.' . . . And in the meantime, I'd signed up for night school classes at the NYU School of Continuing Education in filmmaking because I wanted to make more poetry films and I knew that I didn't know the lexicon" (Ibid.).

Although she had gotten a job at a publishing house during the day, that company laid off two-thirds of its staff. McLaughlin focused on her filmmaking studies. "Everybody wanted to be a DP or a director. Nobody wanted to do sound, but I'm like, 'I'm a gearhead, I'd like it'. I'd had a band and I had my four-track recorder, so sound didn't intimidate me" (Ibid.).

Jan McLaughlin and her band, Sunstroke

The band McLaughlin refers to was Sunstroke. "Before the sound man, we used my gear and I did the stage mix. We did some live recordings from which only a couple sets survive – despite the fire that ate most of the tapes and the recorder." Even as a child, McLaughlin's grandfather had a wire recorder that he let her play with. Throughout her childhood and young adulthood, she continued to play music and record. "My next recorder was a 4-track Teac used for rehearsals of singing and playing guitar: the only way to get better. Later, I had various-sized bands with six members including drums, bass, saxophone, lead guitar and a sound man" (Gaston-Bird, 2017).

Fast-forward to McLaughlin's time at NYU. After finishing the program, she got her first job at Erskine Shapiro, a rental house in New York who handled rentals for the Olympics and the 4th of July celebrations and Macy's Tap-o-mania. "I was checking stuff in, troubleshooting, soldering, repairing, checking out stuff, and while I was there, I made a phone call." She was cold-calling several sound people looking for work and made contact with someone who would set her on the right path. "I got Joel Holland on the phone who invited me to the set of *Law and Order*. It was the beginning of season two of *Law and Order* (classic) and they gave me an invitation at the end of the day, but I didn't know it was a call sheet and I just kept coming back every day for four months. Nobody said not to, and nobody said anything until a couple weeks in, the shop steward – and this was a NABET (National Association of Broadcast Employees and Technicians, a union) show back then – said, 'well, you can you can start touching things. You can put some carpets out, touch sound blankets', and they let me inventory the truck – they'd let me clean the cables. . . . Then they had a two-week hiatus, and the mixer invited me back, and we took everything apart, cleaned it, put it back together and cleaned all the cases. I got to touch everything, and I was like 'Holy shit! This is great!'" (McGlaughlin). Bill Daly was the mixer, who along with Joel Holland became her mentors.

Recently McLaughlin was excited to work on season three of HBO's *Divorce*. This year, all the directors were women. "And they all were amazing," McLaughlin says. "In this group there were five [women] total: one directed two episodes, and every last one of them was efficient, decisive. . . . They'd done their homework. They knew what they wanted, and knew what they didn't need. And it was *incredible*. And uplifting" (Ibid.).

McLaughlin worked *When They See Us* (formerly *The Central Park Five*), a four-part limited series directed by Ava DuVernay and released in 2019 on Netflix. She is also working on the forthcoming motion picture, *Newark* (formerly *The Many Saints of Newark*).

Judi Lee-Headman (England)

"I wanted to be a solicitor and started studying for a law degree," says Judi Lee-Headman. "As part of my training I worked in an inner city law centre. It opened my eyes to the reality of the legal profession" (Lee-Headman, 2019).

Lee-Headman would not have guessed that years later, she would be a location sound recordist working in London. Back then, she says, her parents tried to protect her from that "other world," where they had to struggle, so she instead pursued a law degree. "I was very idealistic, so I went to work in a law center in the 1980s, helping people who couldn't afford a solicitor. There were cases dealing with social care: child care and benefits for clients, most of whom were poor. The turning point for me was a lady whose child was taken into care. She had needed help early on, but by the time she sought assistance it was too late.

"When I got disillusioned with law, I left the profession. However, I hadn't fully quali-fied at the stage I left to begin my career in the film industry. I started working as a researcher for a production company that made documentaries. Eventually, I was able to go out on location and watch the filming and it was then that I decided I wanted to do sound. I'll never forget the sound recordist taking the time to show me the sound equipment and having a genuine interest in my questions about the kit. I went back to the office and I told my boss I wanted to do sound and he laughed! That was the catalyst for me" (Ibid.).

Lee-Headman began looking for courses and applying for different schools. "I started to look for courses that would fit in with my childcare needs as by then I had a young child. I eventually got an interview for the National Film School, the most prestigious film school in the UK. The interview process was a daunting panel of 5 people and me! Most of the panel were pretty ok, apart from one guy who decided to play mind games by asking the most obtuse questions. Once I understood what was going on I retaliated by making my responses to his questions equally 'out-there': He asked me what I did in the year I had my child, so I told him I had been a 'domestic engineer.' He didn't know what that was!!

"Occasionally, other mixers used to talk jargon to see if I knew my stuff. At a seminar, I was approached by a fellow sound recordist. His opening line was, 'So you are the token woman!' I replied that as far as I knew a token was something that isn't real – and I am very real!" (Ibid.).

Production Sound Mixer Judi Lee-Headman

Today, she is an experienced production sound mixer with a list of credits that includes the NBCUniversal production, *The Capture*, and other drama credits such as *Homeland*, *Hidden*, *The Tunnel*, and *Brittania Season 2*. She assembles and manages sound teams, delivering the highest quality of location sound. She has also been a member of the panel for the BAFTA Craft Awards, judging the nominations for best sound.

Lori Dovi (USA)

Lori Dovi is a union production sound mixer of 25 years. "As a kid I was exposed to a lot of music in my house. Records were always playing after school from musicals to Nat King Cole, Ray Charles and Della Reese; my mom was very musically oriented. I studied piano, percussion, voice and guitar. When I went to San Francisco State University I thought I would get a degree in TV broadcasting. It made sense at the time" (Dovi).

"I knew in my heart that I loved music and I wanted to do live concert sound, but no women were doing it. I got talked out of it. I was told there were no women in it and no money in it either." Having been talked out of her dream profession – an unfortunately common occurrence among women – it would take some time for Dovi to find her niche.

"I dropped out of my program at SF State. I knew I did not want to do TV broadcasting and no one there was teaching production sound. I answered an ad to do sound design and board op at Theater Rhino (in San Francisco) for plays. Then an internship at KPFA in Berkeley became available. I got to mix the Eric Bauersfeld *Morning Show*. The lights started going off. I then found myself at Mobius Music (they did a lot of Windham Hill stuff). I started assisting and thought, 'This is it!' But it wasn't. Then I did some ADR at Zoetrope, and that wasn't 'it', either. Finally, I landed a job at a Big Zig video, a Betacam camera rental house where they rented out sound and video equipment. I was self-taught, though I had done sound design for theater, recording studios and radio production. I read every manual and book I could get my hands on and eventually landed a few friends who generously shared their production sound wisdom. So basically, it all clicked while I was working at Big Zig. I realized that I needed to be out in the world and on location – not in a small, dark, air-conditioned room doing post sound. It's basically a personality thing. And it's from documentary work that I found myself in the narrative space now doing features and episodic work many years later" (Ibid.).

Dovi found her way by working for free and low pay, paying her dues. She built her résumé and her experience, movie by movie. Today she is one of the more visible women production sound mixers. She recently wrapped production on David Fincher's *Mind Hunter* for Netflix; she has also worked on *Terminator: Salvation*, *Suicide Squad*, *Further Tales of the City*, *A Single Man*, *Nocturnal Animals*, and many more. In 2018, she founded the **International Women Working in Film** network. She is a member of the Cinema Audio Society (CAS) and a union member in IATSE Sound Local 695.

Tools of the Trade: Post-Production for Television and Film

Consoles: A variety of consoles are seen in larger post-production houses that work on feature films, such as the Avid S6 or the Neve DFC (digital film console). SSL manufactures the System T for live broadcast, and Calrec has a range of consoles from the Apollo (over 1000 channels) or the Type R.

Digital Audio Workstations: Pro Tools is still the DAW of choice in major studios, but some smaller houses (and facilities outside of the US) use Nuendo, Logic, and Adobe Audition.

Plug-ins: Plug-ins for broadcast include reverb and de-reverberation, noise and hum reduction, upmixing, downmixing, and loudness metering.

Audio over IP is used where there are a large number of channels and control rooms networked together. A number of protocols exist, such as Dante, AES67, and CobraNet.

Women in Post-Production

Ai-Ling Lee (Singapore)

In the movie musical *La La Land*, characters Mia (Emma Stone) and Sebastian (Ryan Gosling) sing a song together at the piano. But unlike most musicals, "sing a song" in this case doesn't mean "lip-sync a song." "That was all production," confirms **Ai-Ling Lee,** which means the actors were singing live. "And when Mia did her audition . . . the first few phrases were production sound, then gradually segued [to lip sync'ed] as she belted out the more dramatic moments in the song" (Lee).

In 2016, Lee was nominated for an Oscar for both Best Sound Mixing and Best Sound Editing (along with **Mildred Iatrow Morgan**), making her the first woman to earn nominations in two categories for the same film and the first Asian women to be nominated in the sound editing category. Morgan and Lee were the first all-woman team nominated for an Oscar for Sound Editing. Lee repeated her double-category nominations for *First Man* at the Oscars in 2018, and with Morgan, they repeated their all-woman-team Oscar nomination in Sound Editing for the same film.

Supervising Sound Editor Ai-Ling Lee

Source: ©2019 Mel Melcon/Los Angeles Times; used with permission

Bringing moments like Mia's audition in *La La Land* to an audience takes a lot of planning as well as some bravery on everyone's part, from the director to the sound crew to the actors. The production mixer was Steve Morrow, and composer Justin Hurwitz played the piano off-screen. The feed from the piano was sent to the actors' earpieces. "Personally, I see from the perspective of the viewer . . . it makes it feel so real and believable that they were really singing rather than recreating a song. So that's a whole skill in itself: to capture production singing."

Afterwards, Lee and her team cleaned up the production dialog and added Foley and backgrounds "so it segues naturally and more invisibly between those real moments [but we], still need to ground it so it doesn't sound suddenly sound like a music video" (Lee).

In addition to the singing for *La La Land*, there is the dancing, which required Foley. As supervising sound editor, Lee also oversaw those sessions. "[Characters] Mia and Seb were dancing and singing 'What a Lovely Night' along the hills in L.A. . . . because they are dancing, they had to do live playback of the music to them through the PA, so we couldn't use the dance steps. So, for those I had to recreate and experiment with the Foley walker, and also getting Mandy Moore the choreographer and her dancers involved, and [writer-director Damien Chazelle] involved, recording different shoes, different surfaces, different ways of dancing and make it deliverable. And in time to the music. The actors weren't professional dancers, and they were dancing upslope which made it challenging" (Ibid.).

Creative approaches to sound design are the most rewarding parts of the job, and Lee has quite a repertoire of amazing films under her belt from which to pull examples. In *First Man*, there is a tragic scene of the astronauts waiting in their capsule on the launchpad, unaware of their impending doom. Lee describes how she worked with Chazelle to develop the "arc" of the story. "It's kind of like the 'calm before the storm'. Damien wants the scene to seem very quiet and peaceful, and not knowing what's going to happen: they're just doing their just doing their regular tech check and training simulation. But for the spark that creates the fire, once that moment starts, [he wanted] to have a build-up, and then suddenly there's a leap in danger and tension and intensity, then it spikes to a maximum peak and then drops out. In one of our early conversations he just took a napkin and drew a graph, a visual representation of how much of an arc he wanted to have, because there's many ways of building out an arc. And during the building up, the flame should sound much more real, so it's more subtle, and once it leaps, there's a 'this-whole-flame-is-going-to-swallow-them-up' feeling. So we added animal vocals to sweeten some of the fire and flames and fireball, and ended with a tight explosion . . . but we cut the explosion midway so it's more abrupt. Then the picture cuts to the outside [of the capsule] and you get that muffled sound, and it goes back into silence, because life is gone" (Ibid.).

Similarly, Lee worked with director Jean-Marc Vallée for his film *Wild*. For this film, crickets and cicadas contribute to the story arc. "Reese Witherspoon's character was hiking the Pacific Crest Trail, figuring out her life. There's one scene she was hiking and she got confronted by two men, and anything bad could go wrong . . . we played with pitching the crickets and cicadas. [The insects] have a built-in frequency where it starts low and builds up higher and higher. [Vallée] doesn't want to be too heavy-handed with sound design, he doesn't want 'the Hollywood sound'. So in that scene I tried pitching up the cicadas and timing them at different moments of their conversation . . . just to amp up the danger and intensity without really telling you it's dangerous; without, for example, adding a crow," she says, laughing at the cliché. "Pitching up the cicada really high so it has this stinging, high tone that swells and ebbs and flows. We would place it a different moments in the film that can build the danger and make you feel like that this is not a safe place to be" (Ibid.).

Working with different directors has given Lee quite the toolbox to work with. On another film, *The Mountain Between Us* directed by Hany Abu-Assad, Lee uses wind to augment the story. "There will be scenes where the plane has crashed on a mountain that is covered with snow. So it's dead quiet. There's not much wind, but sometimes there is. So when they're hiding in the plane's broken fuselage and an argument between the two characters happens . . . there's no score under it, but every so often we build in a little wind buffeting through a cracked window, and we spot it very clearly at different moments in the argument, and it almost underscores the argument with sound. It might start off quiet, then gradually you hear a whispering wind, then building into low, buffeting wind, and the whole scene ends with one of the actors yelling out, and all of that wind cuts out when the argument hits a peak. [It's] like scoring with nature sounds, trying not to show our hand too much to the audience" (Ibid.).

"Or there's more 'in your face' sound design," she continues, "like on *Maze Runner*, giant maze walls that creak, moan, and crash. I would take pieces of styrofoam and record them and break them and crack them and play with different pitches so it sounds like big pieces of rock that would break" (Ibid.).

As a child, her father had a small home theater system with a laser disc player. "We watched *Terminator 2* dozens of times, or *Jurassic Park*. I was watching them and realized how I was brought into the environment and the world, and my feelings of surprise or fear were being driven by sound. You can even *not see* something, but *hear it*, and believe it. I was driven by my love for music and tried to be a part of filmmaking and this invisible tool that a filmmaker has" (Ibid.).

In Singapore during her early career, she worked at a sound post house doing radio and television commercials: recording, editing, and mixing everything. She also did music recording. But she really wanted to work in Los Angeles. "I wrote a lot of unsolicited letters to L.A. sound studio owners or film studio sound department heads to see if I could sit in and learn. They replied so I packed and went. I think [they hired me] because having some work in sound before and knowing I'm interested, and I made it clear I wanted to do sound effects." Her first job was at a place called Guaranteed Media. "We worked on a mix of movies like Wim Wenders' *Buena Vista Social Club*, Michael Almereyda's *Hamlet*, and *Waterboy*" (Ibid.).

Lee is seeing more women doing sound effects editing, departing from a little-known stereotype that more women are dialog editors than men. "Traditionally in Los Angeles, women are more in dialog and ADR, and some people just check into the assumption that's what women are more interested in; or thinking, subconsciously, if there's a gunfight sequence or explosions, maybe some women may not quite apprehend it as well or convey it as well.[1] But I have to say that I have been lucky enough to meet people who helped me out, especially early on. As long as you make it *clear* to people what you want to do – for me, sound effects, sound design, and mixing – and you are very passionate and do the best work you can, someone may take a chance and have you do that. And once you do that and keep doing it well, it's all by word of mouth. You start building a reputation. But I do have to say I have been seeing more and more young women wanting to do sound effects and mixing" (Ibid.).

April Tucker (USA)

April Tucker is a Los Angeles-based sound engineer specializing in post-production sound. She has been nominated for three MPSE Golden Reel awards for *The New Radical* (2017), *Motivation 3: The Next Generation* (2017), and *Squirrel Boy* (2006).

She is also a writer and has written dozens of articles for SoundGirls.org. She maintains a blog on her website, proaudiogirl.com, and has written for *Designing Sound*, the Cinema Audio Society's *CAS Quarterly* magazine, Reverb.com, and *Theater Art Life*. "My mom was a librarian for 15–20 years and my dad has written five textbooks. Even when I was in college, I was helping edit his textbooks. So I come from a family that is writing-oriented and that was my interest in blogging," she says (Tucker).

Recently she started the website SoundIsFun.com. "That came about because when I had my son in 2016, I was learning stuff about child development . . . completely random facts about sound. I didn't know that kids hear differently; that the way that a 6-month-old processes sound or the frequency response of the ear is different than that of an adult. No one ever taught me this stuff. And that's been a nice way to get my head out of engineering and mixing all the time." The site features advice for parents, book recommendations for toddlers, and activities involving sound (Ibid.).

In middle school, she loved math and music. As a violin player, she took private lessons and had somewhat of a competitive side. "You know, if you're a violinist in an orchestra with 16 violin players, that's incentive to practice and to be concertmaster. You want to be towards the front." However, during her college career as a music major at the University of

April Tucker

Northern Colorado Greeley, she became unhappy spending so much time in the practice room. "I remember going to the computer lab just looking for like what kind of careers might be in music other than performing. . . . It was down to three schools where I just liked the curriculum." Programs that combined electronics and music intrigued her, and she went for an interview at the Hartt School at the University of Hartford in Connecticut. "I remember going for an interview . . . and they asked me why I wanted to do it. I said, 'Honestly, I don't have any idea what you people do. But based on everything I've seen about what I would have to study, it looks like it'd be really cool.' And they gave me scholarship. Part of it was probably that they appreciated my curiosity, but I didn't have in my head that I was going to go be a producer" (Ibid.).

She went on to get a master's degree at McGill University in Canada, and while she was there, she worked for a summer at the Aspen Music Festival in Colorado, where she met someone who introduced her to Wes Dooley at AEA microphones in Los Angeles. She was eventually invited to work for them (Ibid.).

But then things began to change. One of her tasks at AEA was to help do inventory for a studio that was going out of business. Shortly before Tucker had moved to Los Angeles, the music industry was going through a cataclysmic upheaval, during which many major studios began to fold. "I'm coming in the industry super-gung-ho about getting into music, and I'm watching this major studio getting deconstructed, and these guys who work there saying, 'This industry is not in good shape now, we're screwed.' So I came into the industry at a time when music was really shifting. Even from when I first started school, they were talking about 5.1 DVD-Audio releases and mixing in 5.1 music, and then I come into the industry and we're seeing the recession happening. We're seeing people downsize. We're seeing music go into the home studios --not commercial studios where they're paying typical rates. I met engineers who said, 'my income has cut in half over the last five years'" (Ibid.).

Tucker had to become open-minded to other possibilities. "I want to do scoring, but then I heard there's five scoring stages in town, and I met a scoring engineer who said, 'my stage is going dark during the summer. I'm not working for three months out of the year'. So I just asked a lot of people for advice and said, 'If you were me what would you do?' And more than once, music guys were telling me 'Get into post-production'" (Ibid.).

She got her first job from the company who bought the studio for whom she had been doing inventory. "It was getting bought by a post-production studio run by two Canadians, and I just happened to meet them and talk to them . . . and they liked that I went to McGill. And they basically offered me a job because I knew which key went to which lock and how the alarm system worked" (Ibid.).

"I was the assistant scheduler so that meant that I was calling all the engineers. . . . So I got to know all the engineers, and I got to know who they picked and why – and I got to know the people that *wouldn't* get called back. I'm hearing all this stuff that you'd never hear if you were just an assistant [engineer]. I would also get lunches, take out the trash. . . . And I remember one time thinking, 'I've a master's degree. Why am I doing this?' And I'm so glad I stuck with it because it didn't take long: One day at work there was a computer that died, and they had a client coming in 20 minutes, and the session was only on this computer. All the techs and assistants and engineers were in the machine room trying to fix it. I was in the hallway thinking, 'I have an idea, but I don't know if I should say anything or not.' I decided it was about the team and not about me. I said, 'Has anyone tried booting the computer using this key command?' They tried it, and it worked! My boss came back later and said to me, 'You're in the machine room now.' I got a promotion just because of that. Over the next couple of years, they gave me opportunities to learn a lot of jobs from sound editing to ADR to Foley and mixing (which I've been doing ever since)" (Ibid.).

But in an environment where there is definitely a hierarchy, it's tricky to know when to offer help. "It's such a fine line between knowing when you should say something or not. And I think that's something women really have to deal with, too: how do you be a team player, and how do you act as support – how do you do that and not get walked on for doing it? . . . I remember working on a mix stage as a music editor with two other re-recording mixers. The mixers were working on a conform, and there was a *really* bad edit, but *it's not my place to say anything*. And I made a rule: if I'm a music editor, I'm *only listening for music*. If I hear a mistake in the dialog or I hear a mistake in the effects, that is for the sound supervisor to flag, because I had to be mindful of not overstepping my role. . . . Unless it was a case of a technical error or it wouldn't pass QC, then I might say, 'You might want to check that edit'" (Ibid.).

Today, Tucker continues her 15-year-long career as a post-production re-recording mixer and blogger. You can read more about her at proaudiogirl.com.

Karol Urban, CAS (USA)

"I knew I wanted to be working on sound for film and television when I was pretty young. I went to an art school, like in [the movie] *Fame* . . . you auditioned and you could study the arts as half of your days' curriculum all through high school. And immediately upon learning that you could actually make noise and not have to perform, I was like, 'Ooh, I want to do that!'" (Urban).

Karol Urban, current president of the Cinema Audio Society, would eventually find her way to Los Angeles, but only after a slight detour. "I wanted to go straight to the biggest market I could afford to go to. I was on my way to New York and I ran out of gas and ended up in D.C. No exaggeration: I was there [in D.C.] for 12 years. And for 10 of those years almost I was working at Discovery Channel in-house as their re-recording mixer" (Ibid.).

Between both D.C. and LA, she has done work on dozens of popular television series, such as *Grey's Anatomy*, *Station 19*, *New Girl*, Hulu's *Into the Dark*, *The Bachelor*, and *How it's Made*, as well as mixed a number of features such as *Band Aid* and Adam Rifkins' *Director's Cut*. In 2016, she was nominated along with her team for an MPSE Golden Reel award for Best Sound Editing for her work on *USS Indianapolis, Men of Courage*.

"I've always loved mixing. I've always loved working with musicians, and I've always loved telling stories. I'm really, really obsessed with stories and narratives. I'm the kid that never went outside to play. So I decided that that's what I wanted to do when I was around 13 years old. I was composing and engineering music in a recording studio at the time at Old Dominion University, which is where we had many of our classes for the Governors School for the Arts in Norfolk, Virginia. I went to college at James Madison University and I majored in Audio Production at the School of Media Arts and Design and I minored in music industry" (Urban).

Altogether with her current role as president, she has been with the Cinema Audio Society (CAS) for over 10 years. "Members are absolutely responsible for the fact that I'm working and doing well in Los Angeles," she says proudly. She's also proud of the fact that she has never experienced organizational gender bias in the CAS. "They've been wonderful.

I always felt they were asking 'Do you have knowledge, do you have something to add, do you have something to contribute? We're listening'" (Ibid.).

Urban has an elegant metaphor for what a re-recording mixer does. "I kind of equate it to being a chef in a kitchen. You've got yourself a *sous-chef* and/or a buyer that goes out and finds wonderful ingredients, and they cut them all into beautiful bruinoise, baton, julienned, even waffle cut or whatever is needed to make just the right style for the desired dish. And they are cut in these wonderful little shapes and sizes and arranged in a way where they begin to give you a sense of the goal of the client, and then it goes to the 'chef' or re-recording mixer. The chef takes all of those elements, he bakes them or sautés them, adds reverb, puts compression on it, cleans them up, noise reduces them, tweaks them so that they present nice and beautiful, and cooks it just right: puts it in space, pans it around the room, and places it nice and pretty, then records that as the final soundtrack" (Ibid.).

Just as the culinary industry has a very low number of women chefs, the film industry has a low number of women sound mixers, and both industries ascribe the problem to a lack of women in the field. "Unfortunately, I've never shared a console with another woman," says Urban. "I've had female sound supervisors and sound editors; I *know* other re-recording mixers, but we're so few that I've never had any ability to share a board with one. And actually – I believe this is also correct – the only commercially-backed major motion picture to be mixed by only women as re-recording mixers[2] was *World War Z* with Anna Behlmer and Lora Hirschberg. I don't think it's that people are standing in the way – although sometimes it's the case – I think generally it's just because there's not a ton of us. But it is changing. When I came here in 2011 in Los Angeles I think there was probably seven of us. Now I think there's at least 15. When you consider that was just eight years, that's a massive increase. We at least have to get a reservation to get brunch," she says, laughing (Ibid.).

Haniya Aslam (Pakistan)

Haniya Aslam is a "sound designer and composer" living in Islamabad. She studied at the Trebas Institute in Toronto. "I think I am the first audio engineer and studio owner," she says of her position in Pakistan. "Whenever I saw a woman doing something, it thrilled me. It makes you think that anything is possible. Follow your passion, and the universe has a way" (Aslam).

In a CBC (Canadian Broadcasting Corporation) interview, Aslam describes what got her interested in engineering. "I had a curiosity about – how does a song happen? I am an engineer's daughter. Things used to be lying around in the house, taken apart," she says. "I was always interested in how things worked, how you put them together. I wanted to reverse engineer a song" (Bhandari, 2019).

She has a basic studio setup: a medium room with a UAD Apollo interface and Mac tower computer. She uses ADAM S1X loudspeakers for her near fields as well as a surround controller for her 5.1 setup. She gets mixes from both Toronto and Pakistan, including lots of short films and news packages for abroad. "In Islamabad, very few people use Pro Tools!" she says. However, she uses Pro Tools for audio post-production and Logic for music production (Aslam).

Recently Aslam received acclaim for a song she wrote for Coke Studio (produced by Coca Cola Pakistan) called "Main Irada," an empowering women's anthem (Bhandari, 2019).

'Lisa' Xiang Li (China)

"This might sound a little bit arrogant, but I don't like people to think I'm lucky to become who I am," cautions **'Lisa' Xiang Li**. "I want to show everyone that I'm good enough, so I deserve all the good things happened to me. To achieve that, I always give myself pressure and push myself to work as hard as I can" (Li, 2019).

"I spent 4 years at Beijing Film Academy, which is a top-regarded place for studying film in China," she says. "That is where I began my study of film sound. After that, I didn't think I was ready for the real world, plus, I really wanted to see what the world most famous film world looks like."

So after getting her master's degree in film production (sound design emphasis) from Chapman University in Orange, California, Xiang had to obtain a visa in order to work in the United States. "It was a relatively tough period that made me grow up a lot," she says. "That was also the time I started doing a full-time internship at Monkeyland Audio, which turned out to be my favorite job and my visa sponsor. What's more, I was also doing a part-time morning shift job on weekends. During the whole year I needed to get up at 4:30 am on weekends. I even picked up a third job for one month which needed me to work 3 days per week at night, from 7 pm until 3 or 4 in the morning. The reason that I decided to do this job was because I felt I could learn more about post sound, and I could get to know more about how big projects work. The whole year, I got only 10 days' break, but I never regret even a second of the choices I made. Now, I get my hands on bigger and better projects compared to a year ago" (Ibid.).

Xiang's graduate program at Chapman was very hands-on. "I love the way the program encouraged relationships with other sound design students who were always willing to help

'Lisa' Xiang Li

you out. I tried everything about sound including production sound mixing, boom operating, post dialogue editing, Foley, ADR recording, and re-recording at school. While this is not a professional setting, it is the best chance to get familiar with all the process of sound. The film I designed at school in 2016, *It's Just a Gun*, won a silver Student Academy Award and I was nominated by the MPSE Verna Fields Student Awards in 2017" (Ibid.).

In 2018, she won the Cinema Audio Society student recognition award, "which is the biggest honor I've ever got in my life," she says proudly. This award would open many doors for her, and she also encountered some influential women in film sound. "I remember that night when I was leaving the award ceremony and there's a lady smoking outside the building. She called my name and spoke to me kindly. She said I would be the future, so just keep going and good luck. Later on, I realized she's Mary Ellis, production sound mixer of *Baby Driver*, who was nominated for an Oscar for Best Sound Mixing that year. What's more, I got a chance to shadow Sherry Klein, who's a great female re-recording mixer for so many famous shows such as *New Amsterdam*. Just by watching her working in front of the mixing console made me so excited that I cannot stop imagining maybe one day I can be as confident and smart and strong as she does. All of this has only made me want to work harder and be better" (Ibid.).

Soon after graduating, she returned to Monkeyland Audio to work full-time. "The people I met here, not only co-workers but clients, are very knowledgeable and more than willing to teach. Thanks to the many people who gave me those positive influences and experiences, and made the beginning of my [professional] life so bright and joyful." Already, Xiang is working steadily on some films that will be released on major streaming services. "*Pen15* on Hulu is my very first show as a sound effects editor and sound designer. *Trinkets* on Netflix is my first union project. I still remember as soon as I stepped out of school, I've been told numerous times that joining the union would be so helpful that I can open more possibilities for my career in the industry. And *Trinkets* became the first union show I was eligible to work on. I had a lot fun working on this project and I think everybody will get involved in the story, too" (Ibid.).

She will also have a theatrical release under her belt, called *Bad Trip*. "The whole experience working on this project was such an unforgettable memory that I learned a lot of editing skills, tried so many new plugins, and made new friends. [I'm] really looking forward to watching this movie in theaters" (Ibid.).

Xiang's story reveals how determination pays off at the beginning of a sound designer's career. "All my experiences became my own treasure and the guide of my future. It wasn't just lucky, I know I deserve it. And when the chance comes, I will catch it" (Ibid.).

Sajida Khan (India)

Music and sound engineering are both driving forces and a labor of love for **Sajida Khan**. After ten years in the industry, she was awarded with the coveted Rajiv Gandhi Excellence Award in 2015, which recognized her as being the "First Female Music Technician in India" (From Behind the Console).[3] She is among the top 200 famous Indian Women Personalities in Indian History and the top 14 First Indian Women in Cinema and Music. In 2018 she received the First Ladies Award by Indian President Ram Nath Kovind. Most recently, she received an honorary doctorate from the United Theological Research University for her

Sajida Khan, honored as India's first female music technician

"outstanding achievement and remarkable role in the field of Social Service and life achievement in various fields" (Kahn).

More awards and recognitions followed, even as Khan kept herself busy in recording sessions, live performances, and mixing in the studio. "I feel in India females won't get many opportunities," Sajida says, "and we have to do hard work." But Sajida never felt discouraged or down, and her parents were encouraging, as well as the people she worked with at the studio. "I was never depressed. My directors or my co-directors of the cinema never thought 'oh, she's a girl, she can't do it' – never. I never took a break, I completed the work that was assigned and stood like a man without assistance. I have taken a challenge. I have had to overcome all of this and by God's grace I have reached that level" (Kahn).

In 2018, Sajida headed to the nation's capital New Delhi to join a group of 112 women achievers from across the country, this time to receive the "First Ladies Award" from Indian President Ram Nath Kovind. In 2019, she earned an honorary doctorate from the United Theological Research University for her "outstanding achievement and remarkable role in the field of Social Service and Life Achievement in Various Fields."

Khan has been working for the last 12 years in Hyderabad, India, as an audio engineer for television and film, spending countless hours doing studio recording, mixing, and dubbing. "Unlike in Hollywood, things are very tedious in the Telugu film industry here. It takes 500–600 hours to dub one full movie. Then come the sound effects, music, then final mixing. I record the dubbing voices (ADR)," she explains (Ibid.).

"Women do not have many opportunities in this field in India," she says. She mentions that audio for film and television is a rare field, and some newcomers are put off by

Sajida Khan at work

the meticulous nature of the job. Precision is key, she emphasizes, stating that many young boys come to the studio with great interest to learn audio work, but don't stick around for too long and leave when they realize that it is no cakewalk. "They come for a few days and then go off. They see the mixer and the screen but they don't know they have to spend a lot of time after the recording to do all these things. This is one field which needs enormous patience before tasting success. Instant result is what the current generation looks for," she says (Ibid.).

In addition to her supportive parents, her studio colleagues also encouraged her. "I don't remember an occasion when I regretted my decision. I took my work as a challenge and spent long hours at my work place just like my male colleagues. All the hard work of the last many years has paid off," she says, contently (Ibid.).

Talking about her work in her first movie, *Slokam*, Khan recollects how excited she was working in the audio lab of Geetanjali studio with the music director and the whole production team. "The place was buzzing with dubbing, sound effects and music, all happening in tandem. I was as excited and confident of doing a good job. Using Behringer mixers and Pro Tools, and later Nuendo for sound effects, I loved every moment of the time spent there. It did not take very long for me to realise that the work is not as difficult as it seems," she says. "Pro Tools was easy to operate, not only for recording and mixing, but even scoring. But a lot of people don't compose in Pro Tools, they just use it for dubbing and mixing. In India, Logic is big, about 90% of studios use Logic. And everything is Mac only" (Ibid.).

Although she works primarily with dialog dubbing, she has also done background scores, sound effects editing, and pre-mixing, and plans to set up her own studio. "I hope I'll get sponsorship as there are bright chances for the recipients of the President's Award to get it. If not, I'll have to pool my own money. But I am keeping my fingers crossed," she says (Ibid.).

Catharine Wood (USA)

As a child, Catharine Wood enjoyed the arts and sports. When she was seven years old and living in New Canaan, Connecticut, her church took a trip to New York City to record a radio commercial at CBS studios. They sang a piece called "We've Got Love" – one line at a time – and created a harmony by overdubbing each line. She loved the experience: the smells of the studio and the sights of audio equipment, which she had never seen before. She was a shy child who barely spoke in school, but when she sang aloud and then heard a choir of herself during the overdubbing process, it was a very exciting and positive experience. It would be over 20 years later before she visited a professional recording studio again (Wood).

Growing up, she enjoyed listening to her parents' record collection and her own Beatles and Beach Boys cassette tapes in high school. In her early twenties, she developed a love of singing and playing guitar and piano. She took an aptitude test at the Johnson O'Connor Research Foundation in San Francisco and discovered her high aptitude for pitch, tone, and rhythm memory. After she completed a BA in art from Colorado College, she began songwriting, and in her early thirties, she began studying audio engineering at Full Sail's Hollywood location (Los Angeles Recording School [LARS]). It was then that her passion for audio exploded (Ibid.).

Catharine Wood, owner of Planetwood Studios, LLC

Wood's first studio job in Los Angeles was as an assistant engineer and dub room manager at an audio post-production mix house for commercials. She learned technological skills maintaining the high-end gear for that job and eventually built her own studio from the ground up, based on her own personal workflow. She designed the walls, wiring, outboard gear, signal flow, and networked digital audio workstations (DAWs). She considers the aesthetic and sonic design of the space to be professional and competitive. She had also worked at a prominent San Francisco ad agency and was an executive assistant to a producer on the Warner Brothers lot, so she brought a high level of corporate knowledge to her own operation (Ibid.).

Now she is a professional music producer, composer, multi-instrumentalist, and engineer who runs Planetwood Studios in Los Angeles. Among other high-profile work, she did audio engineering on the first Apple iPhone commercial. As the owner of a studio and an independent contractor, she handles most of the roles herself (Howe, 2018).

For several years, she was also the director of Southern California for the West Coast Songwriters organization and helped run their Hollywood Chapter events. This enabled her to learn the various levels of songwriter: from "newbie" to professional. As a result, she is able to communicate well in both art and business, create solutions, and get things done. She volunteers as a professional advisor to the Los Angeles Recording School's Recording Arts program and serves on the board of directors for the California Copyright Conference. She is a voting member of the Recording Academy and a member of their revered Producers and Engineers (P&E) Wing – as well as a voting member of both the Audio Engineering Society (AES) and the Society of Composers and Lyricists (SCL). These high-profile positions illustrate her philosophy that participation is a key and mandatory component of success. "In Los Angeles, it's very, very social," she explains. "It's all about who you know and what you know. I network as much as possible. The industry has never been more competitive than it is right now – but it's fun navigating and connecting with opportunities" (Wood).

In recent years, she's begun renting out her studio to select third-party producers and engineers in addition to her regular engineering, composition, and producer work. She doesn't think in terms of three- or five-year plans, explaining, "The industry keeps changing very rapidly. I have to be very flexible and constantly recalibrate." Every year, she reviews where her work has come from, how well it paid, how it went, and how she felt about it. Her goal is always to enjoy her work, focusing on commercial jobs as opposed to "art for art's sake" (Ibid.).

She points out that we don't say "men in audio" or "male engineer." "In my class, I was a number one student with 100 percent attendance and a 97 percent test average. I was the only woman in my class and also the only female engineer at my first big studio job, but it wasn't ever discussed or something I thought a lot about." She mentions that many men in her family are engineers, so she was always referred to as "an engineer in the family" (Ibid.).

"You never stop learning. You have to be so tough, you get knocked down a lot in this industry – but you have to pick yourself up again." At the end of the day, it boils down to respect, she says. "If someone doesn't already know my work, I gain respect by letting them know something they don't know already. Before I became more well known, people would come to my studio and say things like, 'is this your studio?' Or

Catharine Wood in the studio

worse, 'Is this your boyfriend's studio?' I have even had people tell me how to run my own sessions" (Ibid.).

It was an article by Rosie Howe that inspired Wood to become more accessible as a mentor. She'd already begun speaking on panels and doing interviews, but once she read the article, she realized she had a bigger responsibility. In the article, Howe highlights Wood as a knowledgeable and experienced woman working in the music business where so few women are visible. At the end of the article, Howe notes, "If you don't know where to start and who to follow, I have already mentioned a talented mentor above who can be that inspiration to live out your dreams." This was moving to Wood. "At first, I didn't think I had a responsibility, but now I realize I can't be so precious about what I know. My experience is worth sharing for the benefit of others" (Ibid.).

Anna Bertmark (Sweden)

Anna Bertmark is a sound designer and supervising sound editor. She studied music technology in London before discovering sound post-production, and has worked on films, TV drama, documentaries, commercials, and spatial audio projects (VR/360).

She started her career working for acclaimed sound designer Paul Davies where she learned dialogue and sound effects editing, while assisting on films like the award-winning *The Queen* and *The Proposition*. After freelancing for a couple of years, she set up Attic Sound Ltd, providing sound supervision and design.

She won a British Independent Film Award (BIFA) for Best Sound for Francis Lee's film *God's Own Country* and mentors up-and-coming sound designers (having been mentored herself by Dawn Airey, CEO of Getty Images). Bertmark is also a member of BIFA's Nomination Committee and has served as vice chair of the Association of Motion Picture Sound (AMPS). She works in Brighton and London, England.

Anna Bertmark AMPS, sound designer and supervising sound editor

Notes

1. In "Learning to listen: histories of women's soundwork in the British film industry," Melanie Bell writes, "A female dubbing editor reported being denied the opportunity to dub a battle scene because the subject matter was deemed by her male employer to be 'man's work'" (Bell, 2017). Learning to Listen: Histories of Women's Soundwork in the British Film Industry. *Screen*, 58, 437–457. originally in Sarah Benton. 1974. Women's Work: Dubbing Bird Song or Battle Scenes? *Film and Television Technician*, 22–23, March.
2. You may recall that Ai-Ling Lee and Mildred Morgan were an all-woman team who garnered an Oscar nomination for sound *editing*.
3. Meenakshi (or Meena) Narayanan is also credited with being the country's first and was active in the 1930s (see Chapter 1).

Bibliography

Aslam, Haniya. 2019. Interview with Haniya Aslam. *In:* Gaston-Bird, L. (ed.).

Bell, Melanie. 2017. Learning to Listen: Histories of Women's Soundwork in the British Film Industry. *Screen*, 58, 437–457.

Bhandari, Aparita. 2019. Haniya Aslam Never Expected to Be a Successful Musician. Now She's Inspiring Other Pakistani Women. *CBC Arts* [Online]. Available: www.cbc.ca/arts/haniya-aslam-never-expected-to-be-a-successful-musician-now-she-s-inspiring-other-pakistani-women-1.4979576 [Accessed June 24, 2019].

Dovi, Lori. 2019. Interview with Lori Dovi. *In:* Gaston-Bird, L. (ed.).

"From Behind the Console". *New Indian Express*, September 3, 2015 Thursday [Online]. Available: https://advance-lexis-com /api/document?collection=news&id=urn:contentItem:5GV3-5C91-JDKC-R4PC-00000-00&context=1516831 [Accessed October 3, 2019].

Gaston-Bird, Leslie. 2017. 31 Women in Audio: Jan JcLaughlin. *31 Women in Audio* [Online]. Available: http://mixmessiahproductions.blogspot.com/2017/03/mix-messiah-productions-presents-31.html [Accessed March 14, 2017].

Lee-Headman, Judi. 2019. Interview with Judi Lee-Headman. *In:* Gaston-Bird, L. (ed.).

Howe, Rosie. 2018. A Call to Action for 2018: Sound Girls. *Ms. In the Biz* [Online] [Accessed June 26, 2019].

Kahn, Sajida. 2019. Interview with Sajida Kahn. *In:* Gaston-Bird, L. (ed.).

Lee, Ai-Ling. 2019. Interview with Ai-Ling Lee. *In:* Gaston-Bird, L. (ed.).

Li, Lisa Xiang. 2019. Interview with 'Lisa' Xiang Li. *In:* Gaston-Bird, L. (ed.).

Mcglaughlin, Jan. 2019. Interview with Jan McGlaughlin. *In:* Gaston-Bird, L. (ed.).

Tucker, April. 2019. Interview with April Tucker. *In:* Gaston-Bird, L. (ed.).

Urban, Karol. 2019. Interview with Karol Urban. *In:* Gaston-Bird, L. (ed.).

Wood, Catharine. 2019. Interview with Catharine Wood. *In:* Gaston-Bird, L. (ed.).

5 Music Recording and Electronic Music

Defining Roles

So what's the difference between an artist, an engineer, and a producer? And what does a mastering engineer do?

Linda Briceño, the first woman to win a Latin Grammy for Producer of the Year offers a metaphor. "To me, the artist is like a professional athlete who is trying to run a marathon . . . the sound engineers are like doctors who are basically helping they artist sound the way they want to sound. The producer is like the trainer or coach, because they coach the athlete and give them the right mentorship to get them to the action" (Briceño).

Another metaphor for producer, engineer, and musician might be to equate the process with filmmaking: the producer is the director, the engineer is the camera operator, and the musician is the actor.

In regard to the mastering engineer, Anna Frick of Airshow Mastering explains, "A good mastering engineer understands the marketplace, consumer listening trends, delivery formats, and all the technical considerations associated with all of these. They are the bridge between what you have done in the studio and the way your fans hear your music" (*Get to Know Anna Frick*, 2017).

Electronic Music

Pamela Z

"I think of myself as an artist, and I think of the technology I use as my instruments or my tools," says **Pamela Z**. "I don't like the idea of just categorizing my work as 'tech work' or something like that. It's *art*, and these are the tools I use to do it" (Z).

In 2019, Z was named as a Rome Prize Fellow, which is a residency at the American Academy in Rome that will span several months of research and creative work. In 2016, Z was presented with the SEAMUS (The Society for Electric-Acoustic Music in the United States) Lifetime Achievement Award. She was also awarded a Guggenheim Fellowship in 2004, among numerous other prestigious honors. Regarding her work, in 2007 George E. Lewis wrote that it "raises important issues regarding transnationalism, Afrodiasporicism, and identity; acoustic ecologies; the articulation of race and ethnicity; and the place of women in technological media" (Lewis, 2007).

Looking at Z's YouTube channel, a casual observer will encounter a selection of electronically manipulated vocal pieces and "black box"-style performances with small sets featuring props and video screens. "The human voice is usually pretty central to that work," she

Pamela Z performing at Ars Electronica in Linz, Austria
Source: Photo rubra, courtesy of Ars Electronica

admits, "but some of it *doesn't* involve that. . . . I consider the combination of my voice and the electronics I use to be my instrument that I perform with. But I'm also a composer of works for other people" (Z).

Pamela Z has had many commissions in recent years and has composed a lot of chamber music, "although most of that time when I do that, there's some interesting component," she clarifies. "For example, if I compose a string quartet, it almost always includes a tape component, or processing on the instruments; or I perform one of the movements with them with my voice and electronics." But Z doesn't want to be defined as an "electronic musician." "My work is very interdisciplinary, and it combines a lot of elements. I don't want people to say, 'Oh yeah, she's the one who does that kind of work that involves a bunch of technology.' To me, even a concert pianist is using technology just as much as I am. It's just a different technology, and it's a really sophisticated piece of technology in a way" (Ibid.).

The instruments that Z uses include custom-made controllers. "The boards were created and designed by a dear friend and collaborator named Donald Swearingen who is a pretty brilliant kind of inventor, and also a composer and performer. Some of them he created for me, and some of them I've worked with him to develop, like the ones that I wear on my hands." In the photo above, Z models one of her hand sensors. "We worked together on figuring out how those would work. They involve accelerometer, gyro and magnetometer." The accelerometer detects the speed the device at which the device tilts, the gyro transmits information about how the device is turned on its X, Y, and Z axes, and the magnetometer measures the Earth's magnetic field. She also incorporates other devices that utilize ultrasound and infrared signals (Ibid.).

The information from these sensors generate MIDI messages, which can be read by software, where she can use the MIDI signals to manipulate sound. "The lion's share of the processing that I do in real time has to do with live sampling of my voice," she explains, "and layering and looping it through digital delay." For example, she might record a four-second sample. Using that sample, she could utilize granular synthesis or straightforward looping.

Pamela Z in her San Francisco Studio wearing a SensorPlay controller – 2014, during the filming of "All Sounds Considered" by Florence Müller & Goran Vejvoda

Not being familiar with granular synthesis, she takes a moment to elaborate for me. "So, I have this four second sample that I can loop," she explains. "If I use a granular synthesis plug-in to process my voice, it's grabbing a bit of sound, but instead of just taking a relatively recognizable long sample sound, it might be taking much shorter samples or really, really tiny fragments. They're not samples like individual, 48,000-per-second samples, but they are samples that are so tiny so that when you play them back, they sound kind of crunchy. With the granular synthesis processing, you can grab a sample and it chops it up into a bunch of grains, and then you can make decisions about how large or how small those grains are. You can make decisions about randomization, like when they get 'spit' back out, how randomly are they spit back in different orders . . . you can also add pitch shifting to them so that they come back out shifted at different pitches, and you can randomize it so that it's a lot of different pitch shifting and every little grain is different, or not."

When watching her performances, you might hear a "crunchy," "jittery," or "glitchy" sort of sound on her voice. "That's granulation," she says (Ibid.).

For visual effects, she uses Isadora (by Troikatronix). "Isadora is very like [Cycling 74's] Jitter in a way. When I first started doing interactive video in performance, I was using Jitter. But then when I started learning Isadora, I just realized that for me, I found it a lot more user friendly. So I switched to using Isadora quite some time ago." Like Jitter, Isadora uses objects and patch cords to build effects. "They call them 'actors,'" explains Z. "It's very easy to immediately make a patch playback video, like just dragging a movie player object onto the 'stage,' and then dragging in a 'projector' object, and then just taking a patch cord and connecting the two of them and – boom! you already have video. And then you have all these other objects that you can place in between those two processes that video or do different things with it." Jitter has since come out with some easier interfaces, but Isadora may have had them first, and at any rate it has been Z's mainstay for quite some time (Ibid.).

Growing up in Denver, Colorado, Z was already experimenting with "found sound." "My first performance was age 5 and we sang 'La Cucaracha' and used seed pods out of the trees in Denver. When they dry you would shake them and they had a shaker sound. That was performance with found objects!" Her father remained on the East Coast of the USA after her parents separated and her mother relocated to Colorado. "Dad sent us two guitars and I absconded with one of them, learned to play the guitar and became sort of a singer songwriter at a very early age (Ibid.).

Her father also sent some interesting new technology. "He sent us one of the very earliest cassette recorders when it was a new-fangled thing, and it was one of those Craig tape decks . . . they were sort of flat and maybe about seven inches wide and about nine or 10 inches long. They had a handle at one end and four buttons that went *ka-chunk!* It immediately went to my bedroom. I just started playing with the tape recorder and I loved it. And then he sent a second one because, there were four of us [kids]. I guess he thought that two of us would share one, but I took both of them into my room and I started teaching myself how to do [ping-pong] recording. You know, basically recording back and forth between these two tape decks. So I would have these extremely dirty recordings of like me doing four-part harmony" (Ibid.).

She also dabbled in being a radio disc jockey. "I would make these fake radio shows. I would do all the parts: I would be the DJ and the announcer, and I would make advertisements and I would write the songs and record the songs, and I would use the guitar and my voice to make all the layers of the songs. Then I would make up the pop songs that were sort of parodies of songs that were on the radio at that time. This would have been, I guess the late sixties, early seventies" (Ibid.).

By the time she was in music school at the University of Colorado in Boulder, she was earning a living playing music. In those days, every musician had to lug their own sound system around. "I don't know if you know this about Denver, but when I was there no venues had a sound system. Every musician had to own their own PA and haul it to the gig, set it up, play the gig, and then tear it down at the end of the night and take it home. So I had to buy a PA system and I had to learn how to use it. My first PA system that I bought had these big speaker columns that were four and a half to five feet high and this big, big old amp with a mixer that I would have to lug around to gigs. And I was playing like every week, making my living playing music. Sometimes if I was playing for the entire weekend, they would let me leave the PA set up for the weekend" (Ibid.).

Z also became a "real" DJ (recalling her childhood hobby) with a program on Tuesdays at KGNU in Boulder, Colorado. "I got to do board training with Fergus [Stone], and then I went to this local recording studio; they gave some sound engineering classes which I took just to learn some basics. I also bought all this equipment, because at the radio station we had access to these half-track, open reel tape decks because that was how we did everything in those days" (Ibid.).

All of these experiences would give the foundation Z needed to really break out of traditional music and find her calling; she just needed one last bit of inspiration that she would find in an unexpected place. "The music I was listening to was very experimental, and I had this really strong desire to change the music I was making and make it more experimental. I was trying so hard and I couldn't somehow manage to change my composing technique or my style of playing music. I couldn't budge it. I would just try and try, but it would come off like I was just trying to do quirky things. It was partially because I was just using the same tools I had always used. We get into these habits and it's really hard to break our mode of working when you just continue working with the same tools. And then one day I went to a concert and saw Jaco Pastorius" (Ibid.).

Pastorius was a prolific and famed bassist who played with the jazz group, "Weather Report." Z went to see them around 1982. "At one point the entire band left the stage and they left Jaco on stage by himself. And he began to do a solo with just his bass and a looper, and it was one of those guitar stomp box type loopers. And in those days, this was *not* a very common thing. I started hearing layers of his bass building up, and him playing over the top of it. And I was like, 'what on Earth is that? And what is happening?' I went to a music store the next day and I told the guy at the store, 'there was this *guy* and he had a *thing* and the sound was *coming back*' and he says, 'Oh, that's a digital delay'.

"I said, 'I'll take one'. And he said, 'You don't want what he had. He had a stomp box, they have extremely low sampling rate. Your voice is going to sound terrible. You should buy a rack mountable digital delay with a higher sampling rate'. So, I ended up buying an Ibanez digital delay. It was a DM 1000. It had 1000 milliseconds of delay because in those days memory was so expensive and this thing was hundreds of dollars for just one second of delay. And I took that thing home and – my poor neighbors! I never went to bed that night. I was just singing through this thing and making layers and loops and honestly, it was one of those unusual stories where things just change overnight. Literally my life changed overnight, and I found my voice as an artist that night and I never went back. Essentially from that point on I started collecting digital delays, investing in more of them and in a very short, few years memory got cheaper and cheaper (Ibid.).

"A few years later I had stacked up digital delays and some of them had as much as eight seconds of memory, which was, you know, a *whole lot* of memory. Eventually you could order a memory card that would bring up the length of the delay to 32 seconds. I started working with all these delays and long loops and, and out-of-phase loops and patterns and textures and timbres instead of just focusing on melody and harmony and rhythm like I had always done before. And my whole way of composing music changed. And the result of that was I just, I had to move to San Francisco because . . . Boulder (Colorado) people were like, 'Why? Why do you have to do this weird electronic thing? Why can't you just play Joni Mitchell songs?'" (Ibid.).

Z relocated to San Francisco where she now resides. "I found my tribe of experimental music, performance art, contemporary classical music improvisers, and avant-garde instrument builders. I literally changed what I was doing. It wasn't this gradual switch. It was overnight: I changed what I was doing with music and I am who I am because I made that radical turn" (Ibid.).

For more information about Pamela Z, visit pamelaz.com.

Suzanne Ciani

"Where we are now is to *uncover* the history and *make visible* the women who have been there," explains **Suzanne Ciani**, referring to the trending topic within the industry of women in audio (Ciani).

In 2018, Ciani attended a premiere of **Daphne Oram**'s piece "Still Point" performed by two orchestras with live electronic processing at the prestigious BBC Proms in London.

Suzanne Ciani featured in a promo flier
Source: 1977, photo credit Bob L.

"I cried, because I was never *allowed* to hear this," she says of the fact that Oram's work was composed in 1949 but took 70 years to receive its debut. "More and more women are getting involved in electronic music, but women were always there – and even *first* in many respects," she observes (Ibid.).

"This is a delicate interaction," she says about the artform. "When the range of the knob is 10 volts and you want a tenth of a volt, you have to be very sensitive. I think that women's hearing is more adapted to the broad range of electronics – we have very high frequency hearing. And that's what I loved! The range! How it sparkled in my ear! No acoustic sound could do that" (Ibid.).

In addition to the physiological aspect of electronic music making, there is a sociological one as well. "The other big turn on for women is: it's a personal thing. You don't have to interact sociologically with the powers that be and the business of music. It was very suitable to the opportunities that women had, which were none. I could make music by myself" (Ibid.)

Ciani is a classically trained musician who identified as a composer from the time she was a little girl. She came of age during the height of the early computer music area. "The first time I heard of [electronic music] I was in college in 1967. We went to MIT and there was a professor who was trying to coax sound out of his computer. Computers were room-sized at the time, and nothing was done in real time. But then I went to the West Coast, and all I can say is I was in the right place at the right time" (Ibid.).

Stanford University was near the campus of the University of California at Berkeley where she earned her master's degree in composition. ("It was very traditional, no technology involved – there wasn't any," she reminds me.) At Stanford she met two important figures in electronic music. "I met Max Matthews who had come to Stanford to teach 'Music V'. He was the father of computer music. And John Chowning was there," who was working with frequency modulation and would go on to invent and patent FM synthesis (Reid, 2015). She also met Don Buchla, who invented a voltage-controlled synthesizer (distinguishable from Robert Moog's synth partly because of the absence of a piano keyboard) (Pareles, 2016) (Ciani).

After graduating, she wanted to become an audio engineer. "And I was told, 'we don't hire women to do that.'" But one engineer, Bob DeSouza, welcomed her. "He said, 'You want to learn about engineering? I will teach you.' So I would go to his studio, and he taught me how to align tape recorders." It was then she decided she wanted to do electronic music. "My first album, *Voice of Packaged Souls*, was done at a radio station in Berkeley. They gave me the whole studio to use from midnight to 5 in the morning. It was a *musique concrète* approach with samples from the Stanford Computer and Music V, the Buchla 200, and field recordings" (Ciani).

The Buchla synth became central to her arsenal. "It's complex to perform it live. Is there ever a perfect performance? No. But it's not like I don't know what's going to happen when I plug this sequencer into this oscillator," she says, gesturing at the rig behind her. "And you *have* to know, because if you don't have an expectation, you can't control it. You've got to constantly be setting the parameters that you want. Buchla lowered the unpredictability of it by 'talking' to you. Early synths didn't have LEDs, so you didn't know what was happening. [But] with Buchla, everything talks. The intensity of a CV (Control Voltage). The position of a switch. You can see everything. Live performance is a subcategory of electronic music, and you have to design for it. And Buchla did. He even designed the road cases so they were light and functional" (Ibid.).

Ciani would go on to do high-profile work, such as sound effects for television. Perhaps her most famous sound effect was the "Coca-Cola Pop 'n Pour" she created with producer Billy Davis (Nemer, 2017). She received five Grammy nominations between 1988 and 1999 in the *Best New Age Album* and *Best New Age Performance* categories for her albums *Neverland* (1988), *Hotel Luna* (1991), *Dream Suite* (1995), *Pianissimo II* (1996), and *Turning* (1999). She is also credited with being the "first solo female composer hired to score a major Hollywood film" for her work on the 1981 film, *The Incredible Shrinking Woman*, starring Lily Tomlin (Ciani, c. 2017). In 2017, Ciani became the first woman to receive the Moog Innovation Award (Ibid.), an accolade she added to a long list, including the Alumnae Achievement Award from Wellesley College (her alma mater) and an induction into the Keyboard Hall of Fame, among several others (Ibid.).

More recently, Ciani continues to perform, collaborate, and experiment. In 2016 she performed in San Francisco – her first solo Buchla performance in 40 years – and has since toured performing in "quad sound": earlier in her career, she was into quadrophonic sound. "I didn't know that quadrophonic wasn't the norm. I am doing things 40 years later that seem new," she says. This live, quad recording can be found in the United States on vinyl – complete with a four-channel hardware decoder – and purchased from her website, www.sevwave.com.

"The essence of spatial [audio] is *movement*," she explains. "I like this software called Envelop. You can shrink and expand easily the number of speakers. But motion creates rhythm. So there's *spatial rhythm* and *musical rhythm*. I have control voltage placement using the 227 [module], the same one I used in the 60s and 70s. I integrate the spatial movement and the rhythm right here," she says during a Skype interview where her rig is visible in the background. Then she can assign the sound to multiple loudspeakers. "There's so many

Suzanne Ciani performing live at Terraforma, June 2017
Source: Photo credit Michela Di Savino

things you can do with distancing and coming closer, controlling the amount of reverb, controlling the sound from bottom to the top. The physical distance of the subwoofer, midrange, and tweeters help with a sense of directionality (bottom to top), as do the wave characteristics of the music: higher frequencies have shorter wavelengths and are easily localized by the psychoacoustic properties of Interaural Level Difference (ILD)" (Ibid.).

Ciani has some words of wisdom for aspiring musicians. "The most important thing is to follow the thread of possibility that unravels before you, and to be open." Stubbornness has its place, too, she advises. "Stubbornness is an attribute that will carry you forward no matter what. Music is a complex field, it's not like going into banking," where the parameters of what you do are much more well-defined. "My vision is to produce my record albums. But that took many detours. I started a furniture company but decided if I'm going to make money, I'm going to do music. I'm going to find a way to *make money in music*. So, think logically: what's your goal? It's like your child. Music has the possibility of being unique and personal because it's creative. You're not on an assembly line. If somebody else can do it, let them. *Find your unique voice*, what it is that *you* have to say. We all have a unique voice in music. And music wants to hear that" (Ibid.).

Cecilia Wu (China)

It just so happened a representative from Universal was in the audience at a band competition where **Cecilia Wu** was performing. "I got an intern job [with Universal] and an advance. . . . That's how I started my career as a songwriter and music producer." She worked in the music industry in China for 10 years, wrote several songs, did film scoring, and was executive assistant to the general manager of EMI Records (Wu).

At the Beijing Institute of Fashion Technology, she studied engineering and fashion design. "It's easier for me to do things by myself instead of paying other people," she laughs. She worked in the music industry in China for 10 years and did film scoring and produced albums for herself and for some friends. "I have very sensitive ears to good music, so I will be able to recognize those talents and they usually go pretty well; if I think a song will be a hit, it will. So people trusted me in that sense." In addition to EMI, she worked at Taihe Media Group, managing their publishing house and audio engineering team (Ibid.).

In 2009 she released an album based some work she was doing in Japan around meditation music and archival. Then in 2010, Wu came to the United States to present her work at an ASCAP (American Society of Composers, Authors, & Publishers) conference. "I randomly went to the other talks, and then found that my former advisor (who worked at Stanford and had worked with Peking University on a networked concert) was giving a lecture on how technology connects musicians from different countries and how they can do real time performance together through technology. Afterwards I went to visit CCRMA which is not far away and saw that it is a graduate school doing music, so I thought I would study orchestration there. I didn't even know what it was about!" (Ibid.).

Wu made the decision to quit her job at the publishing house to pursue studies at the Center for Computer Research in Music and Acoustics at Stanford University (CCRMA). She presented her meditation work to Chris Chafe, CCRMA's director. "He's a great musician and I worked with him when I was accepted. And one of the pieces was the *Mandala* piece" (Ibid.).

Eventually Chris advised Wu on her piece, *Mandala*, which was recently exhibited at the Denver Art Museum as part of a lecture called "Experiencing Embodied Sonic Meditation

Cecilia Wu, assistant professor, University of Colorado Denver

Through Body, Voice, and Multimedia Arts." As described on their website, "Rooted in the richness of Tibetan culture, embodied cognition, and 'deep listening,' Embodied Sonic Meditation practice invites new ways of using sensorimotor coupling through auditory feedback to create novel artistic expressions and to deepen our sonic awareness and engagement in the world" (*Experiencing Embodied Sonic Mediation Through Body, Voice, and Mulitmedia Arts*, 2019).

"After I finished my two years at Stanford, I started to think this is what I'm really passionate about. So I continued my PhD with Curtis Roads. He was the first person to bring granular synthesis into the digital world; He is kind of like the father of granular synthesis and wrote the book, 'Computer Music Tutorial' which is a huge book in our field. I was lucky to work with him.

"My background is not electrical engineering so it's hard for me to work in that very technical direction: you need to do math and linear algebra – I did a little bit, but it was 20 years ago! However, with my design background, I'm good at putting things together and making it work in a way that it people can appreciate it. That's how I got into the design of the technology, design of the interface, design of the interaction between sound, visuals and human perception as a way to give people an experience. Also, I have been practicing meditation for 2 decades since I was in college. So, I was trying to create a meditation experience through sound and visuals using technology. For my PhD I was focusing on human computer interaction design and working different with people who are good at math and signal processing, so we work as a team. And that is a very effective way to work with people, and I really enjoyed working with people" (Wu).

Producers

Kay Huang (Taiwan)

Kay Huang has been called the mother of Taiwanese pop music. She wrote the best-selling song, "Tempting Heart," and is the music director for "Super Mommy," a musical focused on family ties. Herself a mother, Huang connected personally with the story, which in part tackles the issue of maids having to leave their families to care for others. "The events have not happened to me personally, but my late grandmother was cared for by domestic helpers for about 20 years, right up to the time she left us." Her family still keeps in touch with her grandmother's helpers (Ang, 2018).

For those readers keeping track, several of the audio women featured in this book are "super moms," including Huang, who shared a bit about her relationship with her son for an article in the *Straits Times* (Singapore). "If he was playing computer games, I would be beside him, trying to understand what game he was playing and why. We have been through a lot. He now plays the guitar and when this musical was playing in Taiwan, I let him come on board as a musician. I think it was a good learning opportunity for him" (Ang, 2018).

During her childhood in Taipei city in Taiwan, she played classical piano and began writing songs when she was seven years old. She did not have any computer software; instead she played and composed on a Yamaha piano.

Around the time she was 13, she saw **Chan Chau Ha** (Chelsia Chan) playing piano on television as a featured composer combining pop and classical elements. At first she was

Kay Huang as a five-year-old child at the piano

jealous, but then she was inspired. "[I] was dreaming about to be the first one for doing this, so I thought, 'Why is this woman is ahead of me for doing these kinds of things?'" Over time, Huang realized that seeing this rare performance by another woman was empowering in a way, and she began to feel more confident in her compositions and in her career (Huang).

Huang graduated from the National Taiwan University of Arts in 1985. Afterwards, she joined Rock Records in 1986, a Taiwanese record label (founded in 1980). One of the artists from Rock Records had gone to Japan and brought back the technology of sequencing and taught her. Huang thus became the first known woman learning MIDI sequencing in Taiwan's pop music industry during the mid-1980s. For example, they used a Roland MC 300 sequencer, MC 500, and then its successor the MC 50. With regard to the technology, "everything started getting better and better," she says through a translator, "but most of the time I still use a piano for writing everything" (Huang).

By 1986, Huang had released her first album, *Blue Boy*, and would continue to write successful hits, such as "I Am Ugly But Tender" (1988, sung by Zhao Chuan) and "Tempting Heart" (1999, sung by Shino Lin) (Ang, 2018).

Over time, Huang preferred not to be connected to a record label, which required quotas and deadlines. She began to expand into writing her own songs and writing for film. Huang uses Pro Tools now and has begun some long-distance collaborations, using technology such as Source Connect. Most recently, she collaborated with a Japanese artist on a VR installation (Huang).

The role of the recording engineer cannot be understated, she says. "A really good recording engineer is super important. They're really the one helping make it sound even better, helping to maintain the quality, and even preserving the audio recordings. They're important not only for their work and the processing, but for the record industry as well" (Ibid.).

Kay Huang in 1987, composing with the Korg M1, Two Roland MC 500 sequencers, and Yamaha DX7

Elaine Martone (USA)

From her childhood days playing oboe to her professional career as a music producer, **Elaine Martone** has always been passionate about music. "I still love the way music makes me feel, and the way it brings people together. Music is access to the eternal. And my path is to have fun while making it!" (Martone, 2019).

She didn't set out to be in the audio business, however. Born in Rochester and raised in Long Island, New York, she moved to Cleveland because she wanted to study with the late oboist, John Mack, after she graduated with a B.M. from Ithaca College (Ibid.).

Martone was successful at getting into the Canton Symphony. However, she discovered she had some performance anxiety. In Cleveland, she studied with Pamela Pecha, whose then-husband Robert Woods (who co-owned Telarc Records) invited her to work at Telarc. Martone started as a production manager there, and her first task was to organize test pressings in a large room. "The room was covered in stacks of boxes, so I figured out an organization system" (Ibid.).

Next, Martone was taught how to edit. Telarc's first digital editing system was a JVC editing system, followed by the Sony DAE 1100. Telarc was one of the first to record with a sample rate of 50 kHz with the Soundstream Digital System, invented by Thomas Stockham. Later, digital systems for compact discs would use 44.1 kHz as the standard, "Red Book" rate, but this higher sample rate initially showed what was possible (Ibid.).

A female colleague once asked Martone if she had ever been held back because she was a woman. "I always use the skills I have of communication and meeting people where they are. I never felt dismissed, and Telarc was a great place to spread our wings. We were given freedom and flexible schedules; we created a positive atmosphere and had happy employees" (Ibid.).

Martone would become Executive Vice President of Production for Telarc, where she worked for 29 years. During her career, Martone has had a number of Grammy wins and nominations. As a producer, she has five Grammy wins and has been nominated for another seven.

Grammy Wins:

(2009) Best Surround Album: "Transmigration"
(2006) Producer of The Year, Classical
(2006) Latin Grammy, Best Classical Album: "Gershwin: Rhapsody in Blue"
(2004) Best Jazz Instrumental Album, Individual or Group: "Illuminations"
(2004) Best Choral Performance: Berlioz, "Requiem"

Music producer Elaine Martone

Grammy Nominations:

(2014) Best Surround Album: Mahler, "Symphony No. 2 'Resurrection'"
(2014) Producer of The Year, Classical
(2007) Best Surround Album: Vaughan Williams, "Symphony No. 5; Fantasia on A Theme
   by Thomas Tallis; Serenade To Music"
(2004) Best Classical Album: "Higdon: City Scape; Concerto for Orchestra"
(1991) Best Musical Show Album: "The Music Man"

In 2019, Martone is planning on producing more recordings for orchestras and chamber musicians (Martone, 2019).

"I believe in the power of teamwork. I love the collaborative process of putting together teams of people who have a vision for something bigger than themselves. One plus one is greater than two." She also loves mentoring. "We stand on the shoulders of the women who have come before us. In return, I will do my best to help mentor other women. After all these years of being in the music business, I want to pass on my hard-won knowledge to others. I feel it's the very least I can do" (Ibid.).

Linda Briceño (Venezuela)

As **Linda Briceño** accepted the Latin Grammy for "Producer of the Year" for her single *11* and the album *Segundo Piso* by MV Caldera, she cradled the award and tearfully addressed the audience by saying, "I'm the first woman to win this award. . . . I hope this is the start of women winning in this category." She also paid homage to two women who were nominated in the category before her: Maria Rita (2006) and Laura Pausini (2001) (Venezolana Linda Briceño ganó Latin Grammy en la categoría Mejor Productor del Año, 2016) (Aguila, 2018). During the awards ceremony, she held up her award with both hands: She had walked the path that had been paved before her; now she was going to lay some bricks.

Trained as a classical musician, Linda Briceño is also known as Ella Bric: "ella" is Spanish for "she," and "bric" is translates roughly as "strong" in this context. She began learning engineering and producing at an early age on her own, using her Ella Bric as her alter ego. "It started as an act of rebellion against the system because I wanted to do stuff on my own; I didn't want to be directed by someone who had a different vision than mine. So I wanted to protect my ideas and somehow project *what* I wanted, in the *way* I wanted" (Briceno). As a youth, she participated in El Sistema de Orquetas (also known as "El Sistema") and is an accomplished trumpet player. She also explored music engineering as a teenager. "I basically started doing engineering by creating my stuff in Garage Band 20 years ago. I had no idea how to use those programs, but I also had Fruity Loops and those really small programs that at least gave me an idea of how to record myself" (Ibid.).

In time, she worked hard and became a renowned jazz trumpeter, vocalist, and recording artist. She received two nominations for Latin Grammys in the "Best Pop Traditional Album" and "Best New Artist" categories for her album, *Tiempo*. "However," she adds, "I was not being awarded by my work as a producer," although she has credits as a producer on those two albums along with Pablo Gil. Still, the nominations were valuable, and she stressed their importance in an interview published by CNN Español: "In

Linda Briceño, first woman to win a Latin Grammy for "Producer of the Year"
Source: Photo credit Carlos Rosario

a time where there is a globalization of genres, such as the 'urban' genre, there are very small genres that have been overshadowed. And this globalization has also perhaps cut off the opportunities for people to get to know the work of many independent artists." She went on to say that she does not have a recording studio and has self-financed all of her projects, thus highlighting the gap between independent producers and big labels (Huston-Crespo, 2018).

Her newfound success also led her to discover other women in similar situations. "I find out that there were so many females around the world – not only South America – dealing with the same issue as me: not finding a right space to create and communicate with others, which is the purpose of music: sharing ideas" (Briceno).

Now that she has achieved success and notoriety, she wants to help others. "I feel like the most important thing that I've been able to offer other generations is by receiving them at home and talking to them about the importance of: first, being aware of how important your ideas and your creations are and how to protect them; second, how to be able to do the things you own and take responsibility for what you want to. What do you want to portray on that record? Why do you want to say? How do you want to say it? Then okay, let's help you find that sound for that specific purpose. The third part is just inspiring them to not having excuses, like, 'nobody believes in me so I'm not going to do anything'. To take responsibility for their own music and start taking the risk of getting into the world of music production" (Briceño). In this way, she is nurturing independent talent in an intimate way that big labels are not equipped to do.

As part of her effort to continue to inspire, she produced a short film, *Solitude*. The emotionally stirring piece tells the story of a young girl who loves playing trumpet and leaves her conflict-ridden country to pursue her music. She pounds the pavement to make a name for herself but is rejected by the popular jazz clubs in New York. One day she sees a television news report of horrific violence and suffering in her home country. In the darkness of "solitude," a light shines upon her: a kindly man comforts her and welcomes her to play at his club, where she finds a community of support and fame. The true spirit of the film is the nurturing environment, which allows for a magical moment at the end which you can experience for yourself: You are encouraged to view the video which is linked in the "Filmography" section at the end of this book.

Briceño tells the story of why she was inspired to make the film. "I remember watching cartoons when I was really young, and one of the things that really drives me about them is the great music that was in the background. There wasn't any dialog, the music was doing the work. Basically, they're giving the audience the opportunity to finish the sentence, and that's what I'm trying to do with these short films" (Ibid.).

She talks about her mentoring of others in the same vein: "I think I do have an obligation with the next generation. Maybe I'm not going to be able to take all the pain away of what I went through. They may have their own challenges, but I want to make sure like at least I leave part of the road a little more clear for them" (Ibid.).

She would also like to bring light to the role that Latinos play in the music industry, and is pushing for her accomplishments to be recognized outside of the Latino community as a win for all female producers by highlighting that a Latina won the Producer of the Year category in 2018. "And I even spoke to the Grammys, because although that I won a Latin Grammy, this is *still* a Grammy, and it is important that people respect the work of *everyone*, no matter if they are signed with a big label or are an independent artist. I feel like that award – besides myself and my ego – equals a big win for all of us! I feel like not only do I represent the Latinos, but I represent all the females in the academy. I want to be in a room full of diverse people from all over the world and learn from them," she continues, "and for them to get to know my story" (Ibid.).

Erica Brenner (USA)

"It is not necessarily about notoriety, it is about how the recording *moved* people. We were up against heavy hitters. We got emails from people who were blown away. And I'm not gonna say it doesn't feel good!" (Brenner).

"I'm the last in a long line of colleagues to win one," says classical music producer **Erica Brenner** of her Grammy win in 2019 for Cleveland-based baroque orchestra Apollo's Fire's album, *Songs of Orpheus*. "I feel privileged to walk into that 'club'. We are all the same person before the win, and the way the public sees you afterwards is remarkable in how much weight it carries. Part of me thinks it really matters, and I'm proud, and I'm happy. But I wish there was a way to explain what I have *always* brought to the table. So many people do so much good work whether or not they get an award. But sometimes, something special happens. Karim Sulayman's vocals and expressiveness were key ingredients and we captured his emotion and translated it. Apollo's Fire plays with such energy and weight and effervescence. They know the style on period instruments and have a well-deserved reputation for what

(L-R): Jeannette Sorrell, harpsichord and conductor; Camilla Tassi, production and diction assistance; Ian Dobie, recording engineer; **Erica Brenner, producer**; Karim Sulayman, tenor, project concept; Brian Kay, lute; René Schiffer, cello; William Sims, lute

they bring to their performances. All the stars aligned: the sound, the expressiveness. It has been a nice trajectory" (Ibid.).

"Actually, my mom was a huge influence – a pianist and a vocalist," she recalls. "We had classical music on 24 hours a day. I took great pride knowing classical music when I was young. There's a piece by P.D.Q. Bach called the 'The Unbegun Symphony' (a parody of Haydn's 'Farewell Symphony') which cobbled together a lot of classical works. Only one of my other friends and I could name all the pieces that were in there – a source of great pride! So I was basically immersed in classical music growing up.

"Later, I received my Bachelor of Music degree as a flutist, and then I came out to Cleveland and got a job in the Canton Symphony Orchestra. While there, I met **Elaine Martone**, who played second oboe in Canton and was also the director of production at Telarc Records. She asked me if I would ever be interested in editing for their record label. I hadn't thought of the audio industry in any way, shape, or form until that moment. I trained on the job; I didn't get any real schooling in audio at all. She wanted a classical musician to edit their CDs because she knew a classical musician would understand phrasing and we had similar ideas about music. In fact, I was given a lot of freedom by the producers to make edit decisions. Elaine was a mentor every step of the way and had a huge influence in my success in the recording industry (Ibid.).

"I was trained almost completely at Telarc. While getting my master's at Yale, I was a recording work-study student and did a small amount of razor-blade editing, but otherwise, I learned to edit at Telarc on a Sony 3000 digital editor. As I got proficient at editing, I was asked to start producing. In addition to producing, I became the Director of Audio Production, overseeing the scheduling of all audio post-production toward getting the final master to the manufacturer. In the 'wheel of production' which involved graphics and visual concepts, licensing, the CD booklet notes, and audio, I took care of making sure the audio 'spoke' of that 'wheel' was delivered on deadline. In addition to owners Bob Woods and Jack Renner (the original producer/engineer team at Telarc), and Elaine, who became Executive VP of Production and oversaw the entire production department, there were 3 recording and mastering engineers, 2 producers, 3 editors and an assistant engineer, in addition to occasional freelance producers, engineers and editors. It was a very well-oiled machine (Ibid.).

"And then the record industry – as we all know! – took a turn, and Telarc was sold to Concord Music Group. After Telarc was sold, they decided that the production department was no longer needed. So everyone got booted. Now Telarc still releases maybe 3–4 CDs a year, but in our heyday it was 3–4 or more per month! (Ibid.).

"A few of the guys formed their own company; Elaine Martone and Bob Woods formed their own company; and I formed my own company. And what ended up happening is that we still work together. I am not an audio engineer, I'm a producer, so when I'm contacted to do a recording, I call the guys I had worked with all those years, even though we are not in the same company any more. As producer, I coordinate studio rates, plan schedules and logistics like piano tuning, etc. and manage the session from beginning to end working everybody through the recording process itself, following the score and making sure everything is covered. But I always rely on an engineer to bring the audio expertise. It's a very collaborative process between engineer, musicians, and myself. I usually prefer to edit my own projects. A lot of producers don't but I love it. That's where I started. I have the most control over the project, and I can work so closely with musicians to get what they want.

"I feel lucky to work with the engineers that I have – they are artists in their own right, getting the perfect sound for any given project. I rely on them completely. In the end, it's really all about the music, and about what the artist wants to say. During sessions, I try to tell artists that they might hear me over the talkback telling them that things are out of tune, or not quite right, but just know that I am always listening, questioning, 'would the artist like what just happened'? If the answer is 'no,' then I have them do it over. It's not about me, but what is going to represent their artistry most completely" (Ibid.).

Brenner now runs her own production company, Erica Brenner Productions. I asked her what she wants readers to know about her. "The most important thing to know about Erica Brenner is **Elaine Martone**," she responds (Ibid.).

You can learn more about Erica Brenner at ericabrennerproductions.com.

Virginia Read (Australia)

Virginia Read is the first female sound engineer to win an Australian Recording Industry Association (ARIA) music award. "I produce and engineer classical music CDs," she explains. "I started out doing a music degree. My major was composition. I got into electronic music and liked the technical stuff. I ended up enjoying that kind of thing."

Engineer and producer Virginia Read

She got a job at the National Film and Sound Archive doing preservation work on old recordings from the 1930s to 1970s. "I enjoyed it but after a while I got more interested in making recordings rather than preserving them. I missed being involved with music. My supervisor at the Film and Sound archive was a woman from Poland named Wanda Lazar; she had had an amazing career in the Polish film industry but eventually emigrated to Australia in the 80s. I was really lucky to meet her, she was a great mentor. She encouraged me to go to McGill." Lazar had attended the same university as Wieslaw Woszczyk, who directed the program at McGill University (Ibid.).

Read focused on recording classical music at McGill and then worked as an engineer in New York for a while. "I was just an engineer, I wasn't a producer. But when I came back to Australia, I found there was a lack of producers who were specifically trained in, or who could really [be a producer] at a serious level. There was a record label at ABC and I started working for them and we could then raise our expectations as to what we could do for recordings and the kinds of projects we could take on. I became a producer as well as an engineer" (Ibid.).

Read now works for the record label run by the Australian Broadcasting Company (ABC). "It is a broadcaster but has a record label as well. It's a lot like the CBC in Canada. The profits we make are fed back into the ABC. In Australia, we are pretty well the last functioning record label doing classical music; it's only ABC Classics that regularly publishes the music of Australian composers and performers" (Ibid.).

I asked Read what it's like being a classical music producer – what happens on a typical day? "I come to work, I set the microphones, organize for piano tunings; I don't plan the repertoire, I am given a project with certain musicians. I then follow the score and tell them where the mistakes are. My goal is to get a great performance. I am focused more on performance as much as technical issues. I found that if you go for the good performance, the musicians self-correct their technical playing issues" (Ibid.).

"I also do all the mastering and 'deliverables', which are all the files you need to create. I don't get involved in negotiating with musicians or choosing repertoire or collaborators,

Recording session with the saxophonist Amy Dickson (left), didgeridoo player William Barton (center left), producer Virginia Read (center right), and Australian composer Ross Edwards (right) at Eugene Goossens Hall at the ABC in Sydney

but once those have been made, I do every step in the chain through to the master. My job is probably 1/4 recording and 3/4 post production, and endless edits!" (Ibid.).

When asked about what her deliverable formats are, she laughs. "We're still making CDs! We want to do that for as long as we can. These days it's a DDP file to make the CD. Then I supply wave files of each track for Apple at the resolution I recorded at, which is mostly 48KHz, 24 bit. Then I also supply wave files of each track at 44.1KHz 16 bit to all the other streaming services." When I ask why she works at 48 kHz instead of a higher sampling rate, her answer is simple: "The workload is such that I have to be fast . . . dealing with drive space and backing up and so on. With my time constraints and the large amount of material I have I need things to be simple. And I reckon' my stuff sounds just as good!" (Ibid.).

"I'd like to do proper high res (high resolution) stuff. I switched to Pyramix in 2003 with the expectation of doing multichannel sound and DSD. I had it all ready to go but high res didn't take off in Australia. And the direction of the industry made it less imperative. For me it's the high bit rate that makes the difference. Once you listen to it, it sounds good. You can't tell. I remember in the 90s the promise and the technology for high res and multichannel was terrific, but now it's gone in the opposite direction. This isn't what we were hoping for" (Ibid.).

When recording, she uses Grace microphone preamplifiers. "I favor minimal mic techniques, I use fewer and fewer mics. Being careful with the mic placement gets a lovely sound quality that has depth and nice accurate color. So I will work at getting a good mic set up and the musicians' placement. I keep my cable runs short. And I mix in the box. I add reverb, sometimes a little bit of compression. And that's all I will do. Sometimes I use a bit of EQ judiciously, and only if there's a problem. You can mix something so carefully these days with automation so that you don't need to do much with mastering" (Ibid.).

"We release a lot of output and so I have to meet a lot of requests, and we can't waste musicians' time with the technology. So they don't feel the impact of the tech: I'm the producer and engineer and I try to keep things running as smooth as possible. I'm more about pulling a good performance. That's just as important as your high sample rate! If their playing is tired or tentative you have a 'low res' performance; well, if you can get the opposite of that – energetic and full of life – then you get a high res performance!" (Ibid.).

Selected Discography for Virginia Read:

In Circles
Amy Dickson (saxophone), William Barton (didgeridoo), Daniel de Borah (piano),
    Adelaide Symphony Orchestra, Nicholas Carter
April 2019

Romantic Bach: From Intimate to Epic
Jayson Gillham (piano)
January 2019

Shining Knight
Stuart Skelton (tenor)
West Australian Symphony Orchestra, Asher Fisch
September 2018

Brahms: Tones of Romantic Extravagance
Daniel Yeadon (cello), Neal Peres Da Costa (piano), Nicole Forsyth (viola), Robin Wilson
    (violin), Rachel Beasley (violin)
Ironwood
March 2017

Handel: Keyboard Suites & Babell: Suite
Erin Helyard (harpsichord)
September 2017

Heard This and Thought of You
James Crabb (classical accordion) and Genevieve Lacey (recorders)
January 2016

The Kiss
Nicole Car (soprano)
Australian Opera and Ballet Orchestra, Andrea Molino
March 2016

Beethoven: Piano Trios
Seraphim Trio
April 2016

The Haydn Album
Skye McIntosh (violin), Daniel Yeadon (cello), Erin Helyard (harpsichord)
Australian Haydn Ensemble
July 2016

Gershwin: Take Two
Simon Tedeschi (piano), James Morrison (trumpet), Sarah McKenzie (voice)
July 2014

JS Bach: Brandenburg Concertos & Sinfonias from the Cantatas
Orchestra of the Antipodes, Neal Peres Da Costa, Anthony Walker (Nos. 1 & 2),
    Anna McDonald (Nos. 3 & 4), Erin Helyard (Nos. 5 & 6)
March 2012

Tapas: Tastes of the Baroque
Australian Brandenburg Orchestra, Paul Dyer
September 2010

Handel: Concerti Grossi Op. 6 Nos. 1–12 HWV319–330
Australian Brandenburg Orchestra, Paul Dyer
November 2009

Bach – Violin Concertos
Richard Tognetti (violin), with Helena Rathbone (violin), Diana Doherty (oboe)
Australian Chamber Orchestra
January 2007

Ye Banks and Braes – Folksongs
Cantillation, Sinfonia Australis, Antony Walker, Paul Stanhope,
    Teddy Tahu-Rhodes (voice)
January 2007

Rameau: Dardanus
Paul Agnew, Paul Whelan, Kathryn McCusker, Stephen Bennet, Damian Whiteley,
    Penelope Mills
Orchestra of the Antipodes, Antony Walker
April 2007

Bach, J S: Sonatas for Violin & Harpsichord Nos. 1–6, BWV1014–1019
Richard Tognetti (violin), Neal Peres Da Costa (harpsichord), Daniel Yeadon (cello)
September 2007

Rodrigo: Works for Guitar
Slava Grigoryan (guitar), Leonard Grigoryan (guitar)
The Queensland Orchestra, Brett Kelly
April 2006

New Light New Hope
Gondwana Voices
Lyn Williams, Mark O'Leary, Sally Whitwell (piano)
March 2003

Handel: Messiah
Sara Macliver, Alexandra Sherman, Paul McMahon, Teddy Tahu Rhodes, Christopher Field
Orchestra of the Antipodes, Cantillation, Antony Walker
November 2002

■ Mastering Engineers

Emily Lazar (USA)

Emily Lazar made history as first woman mastering engineer to win a Grammy in the Best Engineered Album (Non-Classical) category for Beck's *Colors* album. Previously she was nominated for *Wasting Light* by Foo Fighters (Album of the Year, 2012), *Chandelier* by Sia (Record of the Year, 2015), and *Recreational Love* by The Bird and Bee (Best Engineered Album – Non-Classical, 2016). She also mastered **Fanny's** 2018 release, *Fanny Walked the Earth*.

Emily Lazar accepts the Grammy for Best Engineered Album (Non-Classical)

I asked her about her childhood experiences, which led to her love for audio. "I was fortunate to grow up in a home filled with music," she says. "My mom was an amazing guitar teacher and taught students out of the house, so she was always singing and playing guitar and our home was filled with music. Meanwhile, my father was exceptional at playing the stereo. He revered obscure vinyl recordings, had an abiding love of tube equipment and speakers, and was always an early adopter of the newest digital audiophile formats available at the time, so we always had a pretty amazing audio system in our home. I just didn't know any other kids that listened to music on DVD-A, Laserdisc, SACD, DSD, or Blu-ray players in their living room, and he made a point of trying them all! Looking back, I took it for granted and thought everyone had music constantly around them and listened to music in this pretty unique way. Both of my parents listened to a lot of rock, esoteric folk and blue-grass, and of course R&B. Whether they intended to or not, in retrospect, they served as guides who trained me to experience and listen to music intently.

"Specifically, my mother inspired the musician, performer, and songwriter in me, while my father fostered my interest in the way things sound. For example, my father's insistence on high-fidelity audio was a massive influence on my career. He loved playing albums for me with gorgeous and complicated production. I can remember his face while he listened and how he taught me to notice and isolate the particular qualities of sounds in music. We had so many discussions evaluating and dissecting recordings. Notably, it was in those deep conversations that I became aware that my ears and brain processed sound in a unique way. While I've never gone to the trouble of being formally diagnosed, I think I'm lucky enough to enjoy what's known as *synesthesia*. In the simplest terms, the ability to 'see' sounds as color. I appreciate 'seeing sound' may sound incredibly bizarre for folks who can't relate, but it is the best way I can describe a lot of what I do when I'm working in the studio. I'm grateful my parents helped me hone in on it because that gift provided clarity on my calling to work in music.

"In the end, I guess I'm an amalgamation of my parents' influence. That combination of creativity and love of making music along with the fascination for sound and all things audio created the foundation for what I do today" (Lazar, 2019).

When asked about her earliest experience with recording, Lazar had this to say: "I started as a songwriter and musician, and I was always intensely frustrated by the fact that what I was hearing in my head didn't sonically match what was being recorded. That's really what brought me to the other side of the glass. Once I began to understand the language of recording and the whole world of physics behind sound, I was hooked.

"I was also very fortunate to have some amazing early experiences with engineers who opened doors. They helped me realize this was something I could do. Before that, it never occurred to me that mastering was a possible career path. As I was in the thick of it, it dawned on me that I could turn my frustration and anger into something incredibly positive and constructive.

"That first experience led to a much deeper love affair with understanding sound and how we can manipulate sound with technology, including earning a master's degree in Music Technology from NYU at a time when people didn't really do that.

"But whether it's a first experience or my latest work, I think one of the most exciting things about recording is that there's always more to learn. Chasing new 'firsts' and discoveries is a challenge I love about our work" (Ibid.).

Emily Lazar is founder and chief mastering engineer at The Lodge in New York City, an audio mastering and specialized recording facility that has operated in New York since

1997. I asked her to talk about her leadership role there compared to her past roles in audio. "I started my career in a traditional studio environment," she replied, "doing mastering the way it had been done for decades. It occurred to me that this approach was too clinical and technical. For artists, it felt like going to the dentist. Artists would come in for something they knew was necessary, they would go through this slightly uncomfortable process without really knowing what was happening, and they would leave with an album that was cleaned up.

"But this focus on the technical overlooked the nature of music. Great music is emotional at its core. It's a creative art form where collaboration, feedback, and engagement result in a better final product. So when artists were observers, and not really understanding what was happening or meaningfully engaged, we would end up with a product that wasn't consistent with the emotional story artists were trying to tell. Sometimes being in tune emotionally with the artist or the story is important because it's counterintuitive to what the normal technical 'rules' tell you that you should do. Being too technical keeps you from the emotional result.

"As someone who began my career as an artist, I empathize with the artist's perspective. That's why the traditional engineers approach didn't work for me. I like the weird comradery that comes from feeling as if I'm an extension of the band, rather than an outside service provider.

"When I created The Lodge in 1997, we sought to create a more relaxed and collaborative where artists could ask questions, have a dialogue with the engineer, and even continue the creative process. We think it creates an atmosphere that's not only more comfortable for artists, but a place where people can learn, exchange ideas more freely, and ultimately create music that's most reflective of the artists' creative vision. That's differentiation has been a big part of our success" (Ibid.).

With her Grammy win, Lazar has been asked to speak on the topic of women in audio. I asked her what questions she dislikes being asked in relation to the topic. "This is a field where only 3% of the engineers are women," she points out. "Meanwhile, when you think about professions like computer programming or investment banking where women are also famously scarce, they comprise around 20% of their respective fields. That said, I certainly understand the fixation some people have: when you're as small of a minority as we are, gender should not be ignored. We're unicorns.

"But when I walk into a studio ready to work, I don't walk in thinking, 'I'm a female mastering engineer.' I approach my work as an experienced mastering engineer. Don't get me wrong – female representation is important if not vital, and we 100% need to have our seats at the table. I love that young female engineers are now starting to see other women succeed and are in turn inspired to be the next wave of music-makers. When it comes down to accolades and discussion topics I'm more interested in being included on a list of great mastering engineers than great female mastering engineers. I didn't go into this field to be the first female this or that. I do it because I love making great music sound as good as it possibly can. Gender is not the lens through which I view my work" (Ibid.).

Lazar prefers to talk about music and technology rather than what it's like to be a woman in the field. "I love talking about ideas for making music sound better. Talk to me about new technologies, old technologies, the latest research about sound in virtual reality environments – anything but what it feels like to be a woman 'in a man's world.' There are countless exciting new innovations in audio and artists, engineers and producers are

Emily Lazar

consistently looking for the best ways to capture and improve their sounds. My artistic choices and techniques – I think that's the interesting stuff. To a large extent, I think the most successful engineers are perfectionists who do not waver from their mission to make music sound incredible. I am lucky enough to do what I love for my career, so, ask me about why I love doing what I do, or ask about my passion for enhancing the sonic value of a piece of a music and you'll probably have a hard time getting me to be quiet" (Ibid.).

I asked her about any role models she has. "As cliché as this might sound," she responds, "my role model is, most definitely, my mother. Aside from being my best friend, she is a consistent source of inspiration, and the person who instilled in me my love for music" (Ibid.).

In late 2019, Lazar has been continuing her high-profile work and looking forward to new projects as well. "I'm always working on a number of projects at any given time, and all are exciting and special in their own way. Recently, I've had the distinct honor of mastering the 5.1 surround sound mixes for the 50th Anniversary release of the iconic Beatles' album, *Abbey Road*. To be able to play a role in how this new mix of the album will be heard by Beatles fans around the globe was truly unbelievable" (Ibid.).

Lazar has worked on over 4,000 albums with such artists as The Beatles, Paul McCartney, Coldplay, David Bowie, Foo Fighters, Beck, Dolly Parton, Lou Reed, Panic! At The Disco, Sia, Little Big Town, Linkin Park, Third Eye Blind, Wu-Tang Clan, John Mayer, The Velvet Underground, Depeche Mode, Destiny's Child, Anti-Flag, The Prodigy, The Killers, Morrissey, Vampire Weekend, Sinéad O'Connor, Alanis Morissette, Tiësto, Chainsmokers, Haim, Björk, Moby, Garbage, Sonic Youth, Sinéad O'Connor, RZA, Tegan and Sara, The Raveonettes, The Shins, Thievery Corporation, Death Cab for Cutie, Tiësto, The All American Rejects, Vanessa Carlton, and many more. A current trustee of the Recording

Academy, Lazar is also a member of the board of governors for their New York chapter. She currently serves on both the Producer & Engineers Wing's National Steering Committee and National Advisory Council and is also the co-chair of the New York Chapter's Producer & Engineer's Wing. She has participated on numerous industry conferences panels as well as delivered keynote addresses at many events including the Audio Engineering Society (AES/NY) Conference and the NY Mayor's Office Music Month Conference. Lazar is a graduate of Skidmore College with a BA in English/creative writing and also earned a master's degree in music technology from New York University (Ibid.).

You can read more about The Lodge at www.thelodge.com.

Jett Galindo (Philippines)

"Back in the Philippines, I remember doing mix tapes from MTV using my dad's pro-audio tape deck. I've always had a love for audio from growing up being surrounded by music. My family runs a music management business. When I was a kid, I would always hear my dad teaching a band to play cover songs from the rehearsal studio in our home. My dad also encouraged me to tinker around with computers at a young age, so I grew up being passionate about both tech and music. It all clicked when I realized I could do both by pursuing a career in the field of audio. I've never looked back since then" (Gaston-Bird, 2017b).

Jett Galindo is a mastering engineer at The Bakery in Los Angeles. She graduated from the Berklee College of Music, summa cum laude, and was a recipient of the **Robin Coxe-Yeldham** audio scholar award. Among her mentors **are Susan Rogers** (Prince's engineer), iZotope's Education Director Jonathan Wyner, Jerry Barnes (Nile Rodgers and Chic's

Jett Galindo on the mic

bassist and producer), and AES Lifetime Achievement Awardee and Grammy-winning mastering engineer Doug Sax. Among her credits are the *La La Land OST*, *The Carpenters with the Royal Philharmonic Orchestra*, Colbie Caillat, and Haley Reinhart, among others.

"I started getting serious with recording when I went on a 2-month European tour with my college glee club. Portable recording devices weren't accessible in the Philippines back then, so I had to carry around a heavy backpack with my laptop and a Tascam US-122, and pray that the venue we were performing at had a mic stand I could borrow for my condenser mic. It was a lot of effort but definitely worth it! This was around the same time I got started on my internship in a recording studio in the Philippines (Gaston-Bird, 2017b).

"Prior to The Bakery, my role primarily focused on the audio production/engineering aspect. I focused on being the recording engineer of Jerry Barnes at Avatar Studios in New York (now known as the Power Station at Berklee NYC), and then Doug Sax' right hand woman at the famed Mastering Lab in Ojai, CA until his death in 2015. Once I joined fellow Mastering Lab engineer Eric (Boulanger) in his venture of opening up The Bakery later the same year, Eric and I both had to wear many hats besides being mastering engineers. When not mastering or cutting vinyl, Eric works on studio electronics & maintenance while I work on website development. These are among several responsibilities we share to this day. Especially when you're in the audio/music industry, you have to make an effort to stay well-rounded and abreast of trends on music, technology, etc. to stay competitive" (Ibid.).

Galindo specializes in vinyl mastering and has written on the subject for iZotope in a blog called "Mastering for Vinyl: Tips for Digital Mastering Engineers."

Mastering engineer Jett Galindo inspects a vinyl master

In 2019, The Bakery has been working on some exciting projects. "We're very proud of our work with Weezer: I worked with Eric on the mastering of both the *Black* and *Teal* albums simultaneously. The *Black* album was Weezer's latest release featuring original music. The *Teal* album, however, was a surprise album drop and featured a diverse selection of cover songs including 'Mr. Blue Sky' from Electric Light Orchestra, 'No Scrubs' by TLC. The *Teal* album was released after Weezer's cover of 'Africa' (by Toto) went viral (Ibid.).

"Another exciting development is the growing amount of work we do on vinyl mastering for the video game music genre. Some recent releases include a Kingdom Hearts EDM LP entitled *Kingdom Heartbeats* by DJ RoboRob, and a *Legend of Zelda* tribute LP entitled *Children of Termina* by Rozen. Both albums charted on Billboard, a feat which was recently unprecedented for the video game music genre" (Ibid.).

Piper Payne (USA)

"I was convinced my house was haunted," remembers **Piper Payne**, "so I would bug my house with little recorders to catch the ghost. I had one of those mini cassette recorders. Do you remember the toys that were little microphones with a little spring in it and sounded like a spring reverb? I was fascinated by that" (Gaston-Bird, 2017c).

Piper Payne is a mastering engineer at Infrasonic Mastering and owner of Neato! Mastering in San Francisco. "When I was in 1st grade, I really wanted to play drums, but my dad

Mastering engineer Piper Payne

wouldn't let me learn drums until I learned how to play the piano. So I took piano lessons on the computer, and at the same time I was learning how to build my own computer, and program it using Red Hat Linux when I was 6 or 7. I've been into computers, robotics and things like that since forever" (Ibid.).

Payne earned a BFA in audio at the University of Michigan and a graduate studies certificate in music production and recording from the University of Stavanger in Norway. She was also a senior audio associate at the Banff Centre, where **Theresa Leonard** invited her to work (Leonard).

After working for Coast Mastering, Payne opened Neato Mastering in San Francisco, California, in 2017 and soon afterwards announced that she was joining the team at Infrasonic Mastering, merging Neato with the Nashville-based company, owned by Pete Lyman. Her clients include Third Eye Blind, Madame Gandhi, Geographer, Elettrodomestico (Jane Wiedlin/Go-Go's), ANIIML, Shamir, Betsy, Between You & Me, and Fritz Montana, as well as Bay Area favorites Kat Robichaud, Sioux City Kid, The She's, Emily Afton, Lia Rose, Abbot Kinney, Travis Hayes, Kendra McKinley, Van Goat, and Diana Gameros (Infrasonic, 2019). She is also co-founder and co-chair of the AES Diversity and Inclusion Committee.

In an article about Payne in *Sound on Sound* magazine, she encourages "bedroom producers" to make use of the most important tool for judging mixes: a community of peers. "If you build a studio of your own, you don't get to check out other studios and you don't get to check your mixes. The best-sounding records come from people who ask for feedback and perspective from other people. And that's about not having an ego: I ask opinions from people I trust and from my peers" (Gaston, 2018).

"I love talking about sound, about presentation of records, and about the way records make people feel. I love physical or somatic reactions (when your body does something without your brain knowing) to music. I love talking about other awesome women in audio." Among her role models, Payne lists Hillary Clinton, Susan B Anthony, and her mother (Gaston-Bird).

Anna Frick (USA)

Anna Frick is a mastering engineer at Airshow Mastering, and in 2019 she was elected to the Recording Academy's Board of Governors for the San Francisco chapter. She recently mastered the 2018 Grammy-nominated album *North of Despair* by the bluegrass group, Wood & Wire (Curtin, 2018).

As a child, she loved to record her parents on a Fisher Price tape recorder. "I wanted to show my dad how dumb he sounded yelling at football refs . . . it didn't help," she jokes. She also enjoyed making "mix tapes" and recording conversations with her friends. "I remember one time attending a big concert. We happened to have seats a few rows behind the soundboard. I remember looking at all the knobs and buttons and faders and thinking that whoever knew what all those buttons did must have the coolest job. Then I looked over the sound guy was playing a Gameboy. And I thought, yeah I want that job" (Gaston-Bird).

In high school, she visited a recording studio for the first time. As do many other young producers, she recounts producing her first band. "Some friends of mine were in a band at the time and had just released an album. I got to talking to them about where they recorded it, called the studio and got my friend in to record. I produced the album, though I had

Mastering engineer Anna Frick

absolutely no clue what I was doing. Taking it from conception to fruition was quite the education – everything from getting the best performance from the musicians, learning the process of how an album comes together in the studio to completing the artwork and burning CDs (we didn't manufacture it). I was flying by the seat of my pants, but my desire to learn was insatiable. At the end of the project the engineer said to me 'Hey, you've got a knack for this. You could make a career out of it.' That set the wheels in motion" (Ibid.).

She looks up to her mother who encouraged her to pursue a career in math and science, saying to her, "we need more women in those fields, so it's up to you to change that." She also has some role models in the field. "I admire any woman who is able to rise above the status quo, carve her own path and still stay true to herself throughout the process. In the audio world, **Leslie Ann Jones, Piper Payne, EveAnna Manley** and about a zillion kick-ass female musicians. I probably get inspired by someone new every day" (Ibid.).

Her daily inspiration comes from her work as well; in 2017 she had just completed an album for The Last Revel, an acoustic trio from Minnesota. "Their energy, songs and voices are just awesome. And they recorded all live while renting a cabin in the woods. And of course, I always have to mention the Rise & Fall of Paramount Records. That project was just so intense and awesome and took about a year to complete both volumes. I don't think it got nearly the recognition it deserved. It literally tells the history of the blues" (Ibid.).

You can find out more about Anna at airshowmastering.com, and on Airshow's Facebook page, you can see video of Anna speaking about mastering and listen to some of her work.

Darcy Proper (USA)

Is there any hope for surround music since DVD-Audio and Super Audio CD went by the wayside a decade ago? **Darcy Proper** thinks so. "It's a matter of us putting out good, interesting content – not just repurposed for immersive, but the results are far more inspiring when the project is conceived with the immersive canvas in surround. Once the music industry puts out interesting content and the format is convenient, it's a matter of time" (Proper).

In 2008, Proper was the first woman to win a Grammy in the category "Best Surround Sound Album" for Donald Fagen's *Morph the Cat*. Since then, she continues to master surround albums for Blu-ray Disc and holds out hope for surround music formats, especially given the development of VR technology. "There must be 3D sound to go with these (VR) elements. And because of that, people's homes will be equipped for immersive playback, whether it's standard surround or 3D playback. I think where we have been running behind is behind having an easy system to embrace. A lot of people are equipping themselves with modest, immersive listening environments, and that will get easier as the headphones are developed to be immersive audio friendly. A lot is being done in binaural with the idea of being able to express those relationships" (Ibid.).

Proper played saxophone and clarinet in high school, but she didn't want to be a professional musician and had yet to learn about audio as a career. "We used to do rock and roll shows in my high school's music program to raise money for uniforms and field trips. One day a guy came with a 16 channel Soundcraft board and a PA system. And I thought, 'I want to know what that is!'" (Ibid.).

Eventually she found the program at NYU, which was in its second year (now known as NYU Steinhardt.) She received a bachelor of music with a specialization in music technology in 1990. She considered pursuing a master's degree, but ended up working as an assistant technician at Soundworks (in a basement under the old Studio 54 building). "My first full-time job was as an assistant technician at dance remix studio, and while working there a friend from school was working at Sony Classical productions: the studios for classical music for Sony and every other non-classical genre were separate (for a few years). They were looking for people to do QC (Quality Control). When masters had to be generated, they were done on 1630 U-matic tapes which needed 6 copies, and each had to be listened to for digital errors and glitches. They had a handful of people working for 4–5 hours doing QC work, and that was the job that led to me becoming a mastering engineer: hours and hours of listening with no distractions" (Ibid.).

She eventually got a full-time position as an assistant recording engineer and then later did classical music editing. Around 1992 she got her first mastering job. "My first mastering work was re-mastering works from the Sony catalog, and then the Broadway catalog and even classic jazz like Frank Sinatra, Tony Bennet, Doris Day, and Rosemary Cloooney. About 1995, classical merged with non-classical, and became Sony Music Studios at 54th and 10th in New York – which I think is now condos or something" (Ibid.).

Around 1999 she began working on her first surround sound projects with Elliot Scheiner, a early participant in the field of surround. She worked there until 2005 when she was offered a position in Belgium at Galaxy Studios with Wilfried van Baelen (inventor of the Auro 3D format.) "They had a great surround set up there, and were looking for someone to 'drive their race car.'" Eventually, with film work taking over at Galaxy, Proper and her husband Ronald moved to the Netherlands to start Proper Print Sound (Ibid.).

"In 2018, we did a Grammy-nominated project; Ronald was the engineer I did the mastering. It was the Engine Earz Experiment album called *Symbol*." (London-based musician Prash Mistry is the group's founder.) "It's electronic-based music with environment and the freedom to pan when it suits the music, whereas with acoustic projects you're aiming for realism (recreating the space and enhancing it)" (Ibid.).

Proper has also mastered pieces for Stephan Both at MSM in Munich, which specializes in Blu-Ray audio-only discs for the Pure Audio Blu-Ray label, including surround sound with height channels (Ibid.).

Fun Facts: Vinyl Mastering

"Streaming is great for discovery," says mastering engineer **Jett Galindo**, "but really the best way to own music is to buy a record. Even if you have a premium membership to Spotify and have the ability to download music you'd be surprised – especially for obscure, niche artists – give it 5 years: sometimes the music disappears from your download folder. You don't own the music" (Galindo).

Vinyl is making a comeback as a way to own a physical product. In fact, the profession of "mastering" really came into being when vinyl was introduced: In the early days, a good mastering engineer knew how to operate the lathe, knew about RIAA equalization curves, and knew how to maximize the dynamic range (soft and loud sections) on a record. They might need also to know how to maximize the amount of time on a record, which had a fixed physical size (Ibid.).

"There are only two companies in the world that produce lacquer masters," she explains. "Apollo [Masters] in the USA, and MDC in Japan. Doug Sachs says, 'lacquer is a glorified name for nail polish'. It's the same material: nitrocellulose lacquer. Apollo produces two kinds: Apollo and Transco. Transco has a more natural top end (Ibid.).

"So altogether there are three types of lacquer," she clarifies, "MDC, Transco, and Apollo. The consensus is that MDC is superior – and it's true. Even in terms of quality control. It's a clean, smooth surface with no unevenness. You want lacquer masters to be as smooth as possible" (Ibid.).

The record will be 12 inches, but it starts as 14 inches in diameter. She explains why: "The material is an aluminum disc, but it gets coated with nitrocellulose lacquer. So it starts as liquid and then it dries. So because it starts off as liquid there is surface tension on the edges which causes unevenness, which is why you expand the size of the disc so that by the time you cut it down to the 12" size, it is as flat as possible. And it's also good for handling, because you're not only worrying about unevenness at the edges, you are worried about fingerprints touching the edges of the lacquer master. So 14" is perfect size for cutting the 12" master" (Ibid.).

"The lacquer master plays on any standard turntable. That's why when a client wants to hear how it sounds – before we start the whole vinyl manufacturing process – we can cut their music on a 12" lacquer disc just so they can take it home and listen to it. But when we cut the actual vinyl master, *no one is allowed to play it*, because any stylus that gets dropped on the vinyl master will leave an imprint. It's going to be part of the final LP that comes out. So, you don't want any of that imperfection. Assuming the lacquer master is approved, it gets put into a box for shipping for stage 2 of the vinyl master manufacturing process which is plating.

"There are a lot of plating companies all over the world. Many pressing plants have their own plating setup. The machine where the pressing is done is called a **matrix**. There's a lot of pressing plants that have their own matrix and plating setups, but there are also companies *only* do plating and they tend to do a really good job at it. In the USA there are 2 plating companies: Mastercraft in New Jersey and NiPro in Santa Ana, California. NiPro distributes MDC lacquers, so it's just the perfect situation for us in LA. But Ni Pro and Mastercraft do a great of plating.

"In the 2nd stage, once we cut the lacquer master and it gets shipped to plating, the lacquer gets cleaned and gets dipped onto a bath of liquid nickel. The whole bath is charged, and because the lacquer master is made of aluminum disc, the liquid nickel sticks like a magnet in all of those crevices and all of the grooves. Then when it hardens, you end up with a negative of the lacquer master. And that is called the **metal master**.

"From there the process of making the metal stamper is done. It's just a repeat process of the plating: The stamper is like a negative of the lacquer master and it's molded to work depending on the kind of stamper that's being used by the pressing plant. Then once the stamper gets made then it moves to stage three which is the **press**. The stamper for Side A goes on the top facing down, and the Side B is lying down facing up. The label for each side is then put on and sandwiched in between the two stampers, and in the very center is a vinyl chip. It looks like cookie dough, and is hard at first, but then press starts, it uses tons of pressure and heat and instantly melts the chip that molds to side A and Side B stamper, and you end up with the vinyl record. Then they trim it to the final 12" record album" (Ibid.).

If you'd like to know more, Galindo wrote an article for iZotope, called "Mastering for Vinyl: Tips for Digital Mastering Engineers," which gets into more technical detail about the best processes for preparing a master for vinyl cutting.

Jett Galindo displays a vinyl master

Fun Facts: Loudness Wars

The Loudness Wars have been around in a way since music was first recorded. The perception that "louder is better" is a well-documented psychoacoustic effect, and mastering engineers are often asked to make the recording as loud as possible without distorting or sometimes even *with* a tiny bit of distortion being acceptable. (Vickers, 2011). The first time the general public rejected the damage done by Loudness Wars was with the video game *Guitar Hero*, in which Metallica fans noticed that the game version of the audio sounded better than the single released on compact disc (ibid).

FIGURE 5.1   Average RMS levels of hottest pop music CDs 1980–2000 (Vickers, 2011)

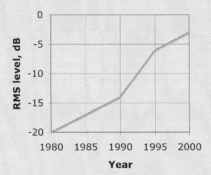

FIGURE 5.2   Percentage of adolescent females with HFHL by degree and year (1985–2008). Note: HFHL = high frequency hearing loss; total N with HFHL = 1,071 (Berg and Serpanos, 2011)

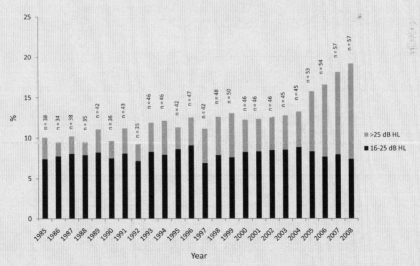

Source: Reprinted from the Journal of Adolescent Health, Volume 48, Berg & Serpanos, High Frequency Hearing Sensitivity in Adolescent Females of a Lower Socioeconomic Status Over a Period of 24 Years (1985–2008) 6 pages, 2011, with permission from Elsevier

Not only is reduced dynamic range fatiguing to listen to, there is evidence that hearing damage is on the rise because of this trend. A study of at-risk Hispanic and African American teenagers shows that "High frequency hearing loss (HFHL) doubled over the 24-year period from 10.1% in 1985 to 19.2% in 2008" (Berg and Serpanos, 2011). Comparing these two figures likely demonstrates a correlation between loudness levels and hearing loss (Vickers, 2011). See figures 5.1 and 5.2.

Recording Engineers

Lenise Bent (USA)

"I often hear, 'there are so few women in audio'. Well, there are more than you think," says **Lenise Bent**. "The women in the audio industry deserve more recognition and acknowledgement" (Bent, 2019).

Bent was the first woman to win a platinum album for her engineering on Blondie's *Autoamerican* in 1980, which featured the hits "Rapture" and "The Tide is High" (a cover of a John Holt song originally performed by the Paragons in 1966). Other profile projects she worked on include Steely Dan's *Aja* (assistant engineer) and Supertramp's *Breakfast in America*.

Bent has always been at home in the entertainment industry. A Los Angeles native, she was just eight years old when she was signed to the Screen Children's Guild and worked

Recording engineer Lenise Bent's first session

as an extra in film and television along with her brother Richard. She also played the flute performing in the Compton Festival Orchestra (SoundGirls.org).

The first studio at which she worked was the Village Recorder, which later became Village Studios.

"What was so fun as well as challenging about *Automerican* was that each song was uniquely different from the other. 'Walk Like Me' was real punk. 'Rapture', starts with tubular bells, then funk guitar and saxophones. 'Go Through it' channels the theme song to 'Rawhide' and 'Here's Looking at You' is a 1920's era pop song. 'Faces' is dark, moody jazz featuring Ray Brown, Jr. on acoustic bass and Tom Scott on tenor sax. To capture each song's unique sound I would use different mics and different vocal chains so when I mixed the record every song had a different tonal quality to it. The signature sound on the record was Debbie [Harry's] voice. I love the whole album but my favorite song is 'T-Birds', a raucous, multi-layered, 'wall of sound' song about a roller derby queen and the ancient Mayans. Flo and Eddy from the Turtles and Frank Zappa sang background vocals. Check it out . . . and play it really loud! On 'The Tide is High' we had three of the best percussionists in the world: Alex Acuña, Emil Richards, and Ollie Brown. I told them, 'just bring everything! and play whatever you want to play', and I tracked them three times, the equivalent of nine percussionists! I had a blast!" (Bent, 2019).

Bent likes to use a variety of devices when she mixes. "Most importantly, whatever format I'm working in, I use my ears. I know what I want to achieve so I choose from what's in my personal audio toolbox. I also like to create different colors of sound based on the ambiences/reverbs I add to them. I learned this by working on the Steely Dan projects and at the Village in general. I like plate reverb, EMT 140's specifically. The Village had maybe 8 of them and a couple were stereo (the EMT 140 is mono but the 240's are stereo). All major studios had these, and usually more than one."

Recording engineer Lenise Bent at Avatar

Recently Bent was doing a project with the rock/alt band Primal Kings. "Actual EMT's are pretty pricey so, for our analog studio, we decided to build our own. One of the guys in the band had a friend who was a welder and he got the schematics online and built not one, but two plate reverbs for me for around 110 dollars each. I told Paul Wolfe of API fame at the NAMM show about them and he said he had some genuine EMT electronics called Martek kits, and he gave them to me! A Martek kit has a transducer that sticks on the plate to pick up the reverberation from the signal sent to the plate. Paul generously guided us through the creation of these plate reverbs. We called the first one 'Frankenstein', because when the guys tested it and it worked they yelled, 'IT'S ALIVE!' The second one was to have a brighter sound and shorter reverb. It's called 'Bride'" (Bent, 2019).

"It's important that everyone realizes that audio engineering is genderless. It's about being a good engineer and being passionate about it; you can't be 'ho-hum'. It's not about the money (well, not anymore!). We are recording artists, too; the equipment and software are the instruments with which we play, the palette with which we paint. Audio recording requires passion and artistry along with the technical expertise. Don't just throw a pre-set of a plug-in to your session: know *why* you're choosing it" (Bent, 2019).

Jill Zimmermann (Germany)

While still in high school, Jill Zimmermann's career path shifted quite suddenly from math and physics to music engineering. "One day I was riding my bike home from school and I was listening to music. Out of nowhere I just closed my eyes, took my hands off the handlebars and started head banging" (Zimmermann).

Recording engineer Jill Zimmermann

What happened next would change her trajectory – literally and figuratively. "I hit a curb, fell over against a tree, and I remember my first thought was, 'what is interrupting me listening to music right now?' And then I opened my eyes and I was like, 'Oh . . . damn. I could have driven against a car and just . . . died'. And for me it didn't even matter in this moment because I just wanted to listen to the end of the song. When I noticed this, I thought, 'okay I have to do something with music in life'" (Ibid.).

Zimmermann was born in Bochum, Germany, and grew up in nearby Krefeld. As she was looking for careers that interested her, she went to the job center to inquire about jobs in her field of interest. She convinced her parents to take her to an open house at the School of Audio Engineering (SAE) in Cologne (Ibid.).

She received a diploma of audio engineering and bachelor of art in audio producing from SAE. She went on to the University of Applied Sciences in Düsseldorf and received a bachelor of engineering degree in media technology. But then she decided to move to Canada and began a job at Jukasa Studios (Ibid.).

"Germany is competitive," she says. "Even German artists, once they reach a threshold, they leave. I didn't want to work on demos for the rest of my life. And I do like the Canadian music industry. I didn't even know how many artists I liked who were Canadian: Billy Talent, Cancer Bats, Neil Young – I mean there are so many artists coming from Canada that I like the sound of" (Ibid.).

She was also drawn to the competitive nature of the industry. "Competition means people try to do better and you can learn more. And for me it was very important that after school I didn't want to just open my own studio and just try something out because I think that even after you are done with your studies, you have so much more to learn. That's when the fun starts (Ibid.).

"Even after five years of being here, you get a new producer and a new band, and all of a sudden, everything's different: they're going for a different sound or somebody has a certain way to work where you just realize, 'oh yeah, that sounds *really good*, I should adopt that'. I just find this whole industry is something where you never stop learning" (Ibid.).

Zimmermann has worked on two albums that have won the prestigious Canadian Juno award: she engineered July Talk's *Touch* (2017) and was assistant engineer on Harrison Kennedy's *This Is From Here* (2016). It is quite an accomplishment, since not once in the Juno's 49-year history has a woman been nominated for Recording Engineer of the Year (Dunn). "I work on every project that goes through the studio, so I have more involvement by default really. I remember one week I worked 96 hours, and it was . . . *good*, actually! Mainly because it's just all so new. In the beginning everything is so exciting, and every band you work with is so new: with different styles and different people, because in our studio we don't specialize in genre – that's one thing I really like, because this way you get better in everything" (Ibid.).

Analog technology is at the center of her work, and she has grown fond of her favorite outboard gear. "I absolutely love using Neve preamps. The 1073's I work with especially. I do love the 1176 'Blue Stripe' Compressor we have, I put that on the vocals and all those frequencies I usually have to take out are gone, and it sounds automatically better; even though it's 'just a compressor', it kind of EQs the sound, too. But when it comes to mixing or mastering, the Manley Massive Passive is just . . ." she swoons.

"We had it out for six weeks because we needed it repaired but that's when I noticed . . . 'oh my God, I *need* it.' I tried using plug-ins, I tried using other EQs . . . they just didn't give this three-dimensional life to it. I even tried the plug-ins that emulate that certain unit and it doesn't sound anything like it. I love real outboard gear. You can't tell if you haven't used it,

Jill Zimmermann makes adjustments to the Manley Massive Passive EQ

but once you've grown with it, you need it. It's substantial, because it just adds so much to it, and you also have to do less work." She admits, however, that digital has its time and place. "Some artists take a long time to complete their projects, so just having the recall is easier so that the sound is repeatable" (Ibid.).

As for the future, she doesn't have a five-year plan, per se. "I do have a 'to do' list, but I don't want things to be structured. I don't like having boundaries. I don't want to miss opportunities because they don't fit in 'the plan'. I do want to get 'best engineered' on the Junos, or a Grammy at some point in my life." In the meantime, she still does not listen to music while driving (Ibid.).

Erin Tonkon (USA)

As a child, **Erin Tonkon** enjoyed playing piano and guitar. However, she was far more interested in getting cool guitar tones than practicing. At age 16, she started engineering at her high school (Fusion Academy in San Diego, which had a recording studio) and had an epiphany – she realized recording is what she wanted to do.

She began interning at radio stations and in time got an internship with Sony in New York. After returning home and discovering there weren't many opportunities, she applied to and was accepted by the Clive Davis Institute NYU Tisch, which was a dream come true for her. Her teachers included Bob Power, ?uestlove, Nick Sansano, and Jim Anderson – legendary producers and engineers. It was Sansano who encouraged her to find a mentor, and Tonkon suggested Tony Visconti. Sansano was able to connect her with him, and in 2013 Visconti hired her.

The first day on the job with Visconti was working with David Bowie. "David was larger than life," says Tonkon, "but what it taught me is that people are still – at their core – just artists who want to be able to express themselves in an authentic way and get across what they're trying to say, and speak to people, and have meaning. And he wasn't an exception to that." She also learned the value of diverse opinions. "I thought my opinion had no value or perspective because I was younger. And Bowie was *constantly* asking for my opinion and my input – constantly! – because I was the youngest person in the room. And that's how he stayed relevant and had such a long career. Now I'm watching the kids. What do they think is cool? Just because it's pop and it's not 'high art' doesn't mean it's not good. That was a huge lesson – to have some value for my own perspective, and that if you have a different background than everyone else in the room then your opinion is super valuable" (Tonkon, 2019).

With respect to the technology, Tonkon has learned to keep it simple. "It's important to listen to hi-fi and lo-fi," she says. "Velvet Underground holds up because it's a good arrangement. The arrangement is what's most important. If you can throw up two microphones – a stereo pair – in a room and it sounds good, then it's gonna be good! Unless you screw it up, and you can definitely screw it up by over-engineering it. For sure. [For example:] too many microphones. Over the years I've really cut back on what I use. Less microphones, less plug-ins . . . learn stuff really well and try to cut back, because you introduce phase issues, and if you have too many options you get into trouble, too" (Tonkon, 2019).

Her personal favorites include her work with Lady Lamb, who had been trying to find a producer for a year. Tonkon's insistence in allowing Lady Lamb to express her own unique voice allowed the collaboration to flourish. In fact, Tonkon is adamant in her belief that she should connect with and serve the artist without an ego. She is also excited about working with Pixx, who was signed to 4AD Records – a label for whom Tonkon has immense respect.

Her philosophy is to involve the artist in the recording process. After the recording, she doesn't push the artist out of the studio so she can "do her magic." Artists want to be heard,

Recording engineer Erin Tonkon

listened to, and validated. "A lot of artists I work with don't know what compression is, or don't know what plate reverb sounds like, or delay, or the midrange, because no one has ever explained it to them. And I like to show them so that they feel empowered. There shouldn't be any secrets" (Tonkon, 2019).

Wendy Beines (Argentina)

Wendy Beines is from the south of Argentina. Her career in audio began a little earlier than that of many other engineers, starting with quite a whirlwind of training beginning at SonarCTS, an audio engineering school in Neuquen, Patagonia, Argentina. She was only 16 when she spotted a brochure for the school. "I went there one day and said, 'Hi, I'm interested in this – what is it?' And they said, 'well this is recording in a studio, editing audio, working with Pro Tools,' and I'm like 'great, I'm 16, can I take the course?' I was still in high school. But they let me enroll." After the course was over, she needed to finish high school. As soon as she graduated, she moved to Buenos Aires to further her studies at an audio school called Escuela Tecson. Her early training gave her an advantage; she was allowed to bypass the first year of courses, and she took the second year as a two-month, intensive course. In the third year, she took electives. "At the end of that year – it was all in a year – I did the exam for the 1st year and 2nd year and got the degree like if I had done the 2 years plus all the

Recording engineer Wendy Beines
Source: Photo credit Nando Carbonell

specializations. During that year I found a job in a studio. My father mentioned I was studying sound, and there was this guy who said, 'I have a studio, send her over' and . . . it was all really fast. I was 17 and I was already in a studio, working" (Beines).

She worked at Spector Studios for five years as a recording engineer and studio manager and produced concerts as well. "I even did a degree in music business," she says, at the Universidad de Palermo in Buenos Aires. "I'm passionate about the whole process of audio. There always has to be sort of like . . . I don't want to call it a spiritual kind of thing, but that thing that makes you feel good, and for me it's seeing the bands going into the studio – maybe it's their first time recording an album – all that happiness and being part of that, helping them make that dream come true. Every time a band comes to record an album it's a dream for them, and it's something that they've been working on a lot: seeing all that and being part of it, I believe is the greatest thing. Beyond the fact that I love the technical part, it's the feeling that it gives you . . . it is that experience: what you make people feel at the studios. Afterwards, they release the album and present it at a concert, and you go there and you see people singing the songs that you know already from working with them in studio. That experience. I love that part. When I worked in movies, I loved going to the premiere and watching people's reactions. I'm not watching the movie, I've watched it 1,000 times already. I'm watching how people react" (Ibid.).

"And got to a point where I was doing pretty well in Argentina. Then I came to Europe on holidays and loved London. It felt like home, I wanted to try something new and was

Wendy Beines at Spector Studios
Source: Photo credit Tian Firpo

willing to start from zero. I thought my experience would be helpful, but in London, it wasn't" (Ibid.).

She worked in retail for two months and then got a job at Not Now, a record label in London. "I feel coming from a third world country where maybe our resources aren't as good as we have here, and you have one the best universities in the world for this in England – the University of Surrey –it's really hard to compete with all that. At this point I haven't been working in audio for a couple of years (three so far) and they feel I'm not updated, so it will take me longer to start again" (Ibid.).

Nonetheless, she is determined to work in audio again and wants to inspire women with her story. "I don't know if I did it, I don't know if it's going to happen during my life, but the contribution I'd like to make – or hope I did – is inspiring more women to be part of this so they can see they can do it as well. Too many years ago when women couldn't be lawyers or doctors, one [woman] said, 'I can do that' and did. I hope it inspires more women to get involved. I don't know if it *has happened* or *will happen* but, I would like that to be my contribution" (Ibid.).

■ Other Women in Music Recording

Jane Clark (USA) was the engineer for the 1970's R&B group, the Commodores. (Her sister, Chris Clark, received an Academy Award nomination in 1972 for *Lady Sings the Blues*) (Cruise, 2016).

Janet Jackson: American pop star Janet Jackson was the first woman to be nominated for a Grammy for Producer of the Year, Non-Classical along with Jimmy Jam and Terry Lewis for *Rhythm Nation 1814* in 1989.

WondaGurl: This Canadian record producer achieved fame at a young age of 16, working with Jay-Z and other hip-hop celebrities.

Trina Shoemaker was the first woman to win a Grammy for Best Engineered Album (non-classical) for Sheryl Crow's *The Globe Sessions* in 1999.

Missy Elliot: This American producer, rapper, singer, and dancer has produced dozens of hits for Aaliyah, MC Lyte, Whitney Houston, and Mary J. Blige and wrote and produced her own hits.

Sylvia Massy: This American engineer and producer is famous for her work with Tool, The Melvins, Johnny Cash, and hundreds of others. She is the author of *Recording Unhinged: Creative and Unconventional Music Recording Techniques*. Massey also holds dozens of gold and platinum album awards.

Marta Salogni: An Italian producer and engineer famous for her work with Björk and Glass Animals, among several others. In 2018, she won the Music Producers Guild (MPG) "Breakthrough Engineer of the Year" award.

Mariah Carey: This American singer Mariah Carey received a Grammy nomination for "Producer of the Year (Non-Classical)" in 1991 at age 21.

Imogen Heap: Imogen Heap was the second woman to win a Grammy for Best Engineered Album, Non-Classical for her album, *Ellipse*, in 2010.

Mandy Parnell: This British mastering engineer won the MPG awards "Mastering Engineer of the Year" in 2012, 2015, and 2017. Famous clients include Sigur Ros, Björk, and Brian Eno.

Sylvia Robinson: This American record producer and label executive is known as the "mother of hip-hop" for starting the Sugar Hill Records label (a subsidiary of All Platinum Records, which Robinson started in 1966 with her husband, Joe Robinson), putting together the Sugar Hill Gang and producing their hit, "Rapper's Delight."

Bibliography

Aguila, Justino. 2018. The Latin Grammys' Reggaeton Problem Is a Lot Like Grammys' Hip-Hop Problem. *Los Angeles Times* [Online]. Available: www.latimes.com/entertainment/music/la-et-ms-latin-grammys-news-20181115-story.html [Accessed June 24, 2019].

Ang, Benson. 2018. Singer-Songwriter Kay Huang Is Music Director of Mandarin Musical, Super Mommy. *The Straits Times* [Online]. Available: www.straitstimes.com/lifestyle/entertainment/this-mommy-loves-making-music.

"Artist: Elaine L. Martone" [Online]. Recording Academy. Available: www.grammy.com/grammys/artists/elaine-l-martone [Accessed July 10, 2019].

Beines, Wendy. 2019. Interview with Wendy Beines. *In:* Gaston-Bird, L. (ed.).

Bent, Lenise. 2019. 2019. Interview with Lenise Bent. *In:* Gaston-Bird, L. (ed.).

Berg, Abbey and Serpanos, Yula. 2011. High Frequency Hearing Sensitivity in Adolescent Females of a Lower Socioeconomic Status Over a Period of 24 Years (1985–2008). *Journal of Adolescent Health*, 48, 203–208.

Brenner, Erica. 2019. Interview with Erica Brenner. *In:* Gaston-Bird, L. (ed.).

Briceno, Linda. 2019. Interview with Linda Briceno. *In:* Gaston-Bird, L. (ed.).

Ciani, Suzanne. 2019. Interview with Suzanne Ciani. *In:* Gaston-Bird, L. (ed.).

Ciani. c. 2017. *Suzanne Ciani: Bio* [Online]. Available: www.sevwave.com/bio [Accessed July 10, 2019].

A Word with the Commodores. Soul Train Cruise. 2016 [Online]. StarVista LIVE, LLC Available: soultraincruise.com/news/a-word-with-the-commodores [Accessed July 10, 2019].

Curtin, Kevin. 2018. Wood & Wire Nominated for Grammy. *The Austin Chronicle*, December 8.

Experiencing Embodied Sonic Mediation Through Body, Voice, and Mulitmedia Arts [Online]. 2019. Denver Art Museum. Available: https://denverartmuseum.org/calendar/curators-circle-dr-jiayue-cecilia-wu [Accessed July 10, 2019].

Galindo, Jett. 2019. Interview with Jett Galindo. *In:* Gaston-Bird, L. (ed.).

Gaston-Bird, Leslie. 2017a. 31 Women in Audio: Anna Frick. *31 Women in Audio* [Online]. Available: http://mixmessiahproductions.blogspot.com/2017/03/31-women-in-audio-anna-frick.html [Accessed March 5, 2017].

Gaston-Bird, Leslie. 2017b. 31 Women in Audio: Jett Galindo. *31 Women in Audio* [Online]. Available: http://mixmessiahproductions.blogspot.com/2017/03/31-women-in-audio-jett-galindo.html [Accessed March 6, 2017].

Gaston-Bird, Leslie. 2017c. 31 Women in Audio: Piper Payne. *31 Women in Audio* [Online]. Available: http://mixmessiahproductions.blogspot.com/2017/03/31-women-in-audio-piper-payne.html [Accessed March 8, 2017].

Gaston-Bird, Leslie. 2018. Piper Payne Mastering Engineer *In: Sound on Sound.* Cambridge: Sound on Sound, Ltd.

Get to Know Anna Frick [Online]. 2017. Airshow Mastering. Available: www.facebook.com/airshowmastering/videos/1283329601720218/ [Accessed July 5, 2019].

Huang, Kay. 2019. Interview with Kay Huang. *In:* Gaston-Bird, L. (ed.).

Huston-Crespo, Marysabel. 2018. Linda Briceño, la venezolana que hizo historia en los Latin Grammys 2018 al llevarse el premio "Productora del año". *CNN Español* [Online]. Available: https://cnnespanol.cnn.com/2018/11/17/linda-briceno-la-venezolana-que-hizo-historia-en-los-latin-grammys-2018-al-llevarse-el-premio-productora-del-ano/.

Infrasonic. 2019. *Piper Payne Joins Mastering Team at Infrasonic Mastering* [Online]. Available: www.infrasonicsound.com/news/cool-stuff-coming-soon [Accessed July 10, 2019].

Lazar, Emily. 2019. Interview with Emily Lazar. *In:* Gaston-Bird, L. (ed.).

Leonard, Theresa. 2019. Interview with Theresa Leonard. *In:* Gaston-Bird, L. (ed.).

Lewis, G.E. 2007. The Virtual Discourses of Pamela Z. *Journal of the Society for American Music*, 1, 57–77.

Martone, Elaine. 2019. Interview with Elaine Martone. *In:* Gaston-Bird, L. (ed.).

Nemer, Hannah. 2017. *'Pop 'n Pour': This Electronic Music Pioneer Created the Sound of Coke's Beloved Bubbles* [Online]. The Coca Cola Company. Available: www.coca-colacompany.com/stories/meet-suzanne-ciani-the-legendary-creator-of-cokes-pop-n-pour [Accessed July 10, 2019].

Pareles, Jon. 2016. Don Buchla, Inventor, Composer and Electronic Music Maverick, Dies at 79. *The New York Times*, September 17.

Proper, Darcy. 2019. Interview with Darcy Proper. *In:* Gaston-Bird, L. (ed.).

Reid, Gordon. 2015. John Chowning: Pioneer of Electronic Music & Digital Synthesis. *In: Sound on Sound*. Cambridge, UK [Online]. Available: https://www.soundonsound.com/people/john-chowning [Accessed September 27, 2019].

Soundgirls.org. *Giving Back to the Audio Community – Lenise Bent* [Online]. Available: https://soundgirls. org/giving-back-to-the-audio-community-lenise-bent [Accessed May 4, 2019].

Tonkon, Erin. 2019. Interview with Erin Tonkon. *In:* Gaston-Bird, L. (ed.).

Venezolana Linda Briceño ganó Latin Grammy en la categoría Mejor Productor del Año. 2016. *Tenemos Noticias* [Online]. Available: https://tenemosnoticias.com/noticia/latin-grammy-gan-ao-467344/1046061 [Accessed June 24, 2019].

Vickers, Earl. 2011. The Loudness War: Do Louder, Hypercompressed Recordings Sell Better? *Journal of the Audio Engineering Society*, 59, 5.

Wu, Cecilia. 2019. Interview with Cecilia Wu. *In:* Gaston-Bird, L. (ed.).

Z, Pamela. 2019. Interview with Pamela Z. *In:* Gaston-Bird, L. (ed.).

Zimmermann, Jill. 2019. Interview with Jill Zimmermann. *In:* Gaston-Bird, L. (ed.).

6

Hardware and Software Design

Careers in Hardware and Software

When it comes to finding a job in audio engineering, **Dawn Birr** has this to say: "Who we are looking for are people who have a background in electrical engineering, psychoacoustics, or acoustics: people that spend their time measuring audio and understanding how it translates practically in the world. But the number one thing we hire for is attitude – and then we'll train you. That's a big deal to us: if we find the right person who's willing to learn, we're going to make sure that they learn. It never hurts to have the credentials, and education is an important tool, but it's some of the soft skills, too, that we're definitely looking for" (Birr, 2019).

A survey of job openings at aes.org bring to light the following desirable skills:

- ▶ AutoCAD
- ▶ DSP
- ▶ Digital Filter Design
- ▶ Nonlinear Audio Processing
- ▶ CODEC and other Sound Quality Evaluation
- ▶ Quality Control
- ▶ Sales/demos

A degree in electrical engineering, computer science, physics, or acoustics would be a good place to start. There may be "maker" or "builder" clubs in your community if you want to explore the subject further. Organizations such as SoundGirls.org, Women's Audio Mission, and the Yorkshire Sound Women Network host workshops in gear building, soldering, and cable repair, among other activities.

Fun Facts: Make Your Own Audio Plug-in with MATLAB

Each year, the Audio Engineering Society hosts a student design competition at each of its international conventions. In 2019, they added another competition specifically for designing plug-ins that can be used with a Digital Audio Workstation (DAW).

The tutorial at https://uk.mathworks.com/help/audio/gs/design-an-audio-plugin.html guides the user through the steps of creating a simple code, which then gradually becomes more complex.

For example, the first step is to "Define a Basic Plugin Class," whose function is only to pass audio through. The code given is:

```
classdef myEchoPlugin < audioPlugin
  methods
    function out = process(~, in)
      out = in;
    end
  end
end
```

(MathWorks, 2019)

You can then add a property to your code. The following property, "gain," increases the level of the input.

```
classdef myEchoPlugin < audioPlugin
  properties %<---
    Gain = 1.5; %<---
  end %<---
  methods
    function out = process(plugin, in) %<---
      out = in*plugin.Gain; %<---
    end
  end
end
```

(MathWorks, 2019)

You will of course want the DAW user to be able to adjust the gain using an "interface" such as a slider or knob and so on. In order to use the tutorial, you will need access to MATLAB® and the Audio Toolbox® environment. If you are a university student, check to see if your school has a license. There are also tutorials to complete before you jump straight in; this entry in the book is only meant to excite your curiosity.

Check the AES student competition site at http://www.aes.org/students/ for more information about this year's competition.

Tools of the Trade: Hardware and Software Design

Ears: As with any audio profession, the most important tools are your ears. **Orla Murphy** emphasizes the need for critical listening skills, saying, "Your computer graph might say one thing, but your ears say another."

Other tools and skills you might use in the profession include:

- MATLAB®: There are courses, workshops, and tutorials in MATLAB, which can be used to design audio software or plug-ins, or to measure signals. The software is usually part of a university's license, so students can readily take advantage of the software. (For individual use, the software is rather expensive.)
- C+ programming: This is especially useful as a starting place to understand the proprietary software used in some companies.
- Finite Element Method (FEM) software: This is a way to predict how products react to vibration, heat, fluid flow, and other physical effects. Finite Element Analysis (FEA) shows whether a product will break, wear out, or work the way it was designed.
- LabVIEW®: This software is used to acquire data from a device and analyze the results using a plethora of mathematical functions including calculus and Fast Fourier Transform (FFT).
- COMSOL®: The optional "Acoustics Module" is designed specifically for those who work with devices that produce, measure, and utilize acoustic waves. It can also simulate acoustic pressure, wave propagation in air, water, and other fluids. Application areas include speakers; microphones; hearing aids; sonar devices; building and room acoustics; as well as noise control in mufflers, sound barriers, diffusers, and absorbers.
- Oscilloscope: This is used for measuring frequency, amplitude, and phase.
- Soldering iron: This is used for connecting physical electronic components. It works at a very high temperature to create a permanent connection between wires and electrical components.
- Gaffer tape: One can never have enough gaffer tape.

Laurie Spiegel (USA)

"It gives me a feeling of awe that our species has been capable of launching such a mission, of reaching the edge of the solar system," writes **Laurie Spiegel** about the Voyager mission. "It makes me think that we might be able to face some of the challenges we ourselves have created, what we are doing to the environment and to other species, as well as to members of our own human race. We are, after all, all one race: human. These feel like dark times, but the Voyager mission inspires us that we can do better" (Spiegel).

In 1977, two gold-plated copper phonographs with several audio recordings were launched into space on the Voyager spacecraft. One of the recordings was Spiegel's piece, "Harmonices Mundi," which she retitled in 2012 as "Kepler's Harmony of the Worlds." Johannes Kepler was a seventeenth-century astronomer and mathematician who proposed laws of planetary motion and attempted to explain them in terms of music. Of her realization, Spiegel writes,

> "I have chosen the variant title used here to embody that – to Kepler – this work might not have represented a simple sonification of astronomical data. I speculate that Kepler looked for deeper meanings and implications, as would have been natural to one who at times was employed as an astrologer and whose mother was nearly burned as a witch. I think that Kepler just might have envisioned a truly universal

harmony among diverse living beings as well. In Chapter 10. 'Epilogue . . . by way of Conjecture,' Kepler makes a case for the existence of life on other worlds besides the Earth. [Harmonies of the World, by Johannes Kepler, tr. Charles Glenn Wallis [1939], pp. 1084–5.]

(Spiegel, 2012)

Spiegel created the work at **Bell Labs**, where she had been working since 1973. She used a DDP-224 computer running Generating Realtime Operations On Voltage-controlled Equipment (GROOVE), developed by Max Mathews, that filled an entire room. She had to

FIGURE 6.1   Geometrical harmonies in the perfect solids from Harmonices Mundi (1619)

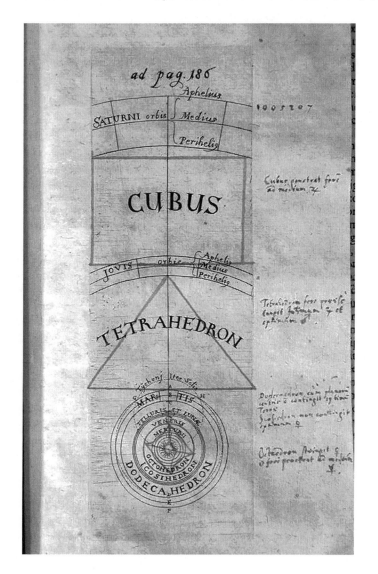

Laurie Spiegel at Bell Telephone Labs
Source: Photo credit Emmanuel Ghent

walk back and forth between rooms down long hallways to control the analog part of the system and then interact with the computers' input devices, whose control voltages were calculated by the software she wrote and sent down long trunk lines (Spiegel, 2012).

She soon got the idea to "turn the GROOVE into a VAMPIRE," standing for "Video and Music Program for Interactive Realtime Exploration/Experimentation." Using her program, she was able to use parameters to draw and move colored shapes around a screen.

> "I reasoned that just as GROOVE's computer control of analog modules had made interaction with relatively complex logic systems a realtime process, permitting realtime interactive computer control of musical materials for the first time, realtime interactive computer graphics should be possible as well by similar means. Instead of recording the image on film frame by frame, I should be able to code myself a visual musical instrument that would let me play and compose image pieces by recording the control data as time functions and playing back the time functions as visual compositions."
>
> (Spiegel, 2012)

As computer technology became smaller and relatively more affordable, Spiegel, along with her colleagues Charlie Kellner and Ellen Latham, presented a paper in 1980 about Kellner's alphaSyntauri system titled "A Keyboard Based Digital Playing and Recording System with a Microcomputer Interface." The peripherals and software for the Apple II computer were intended to be an affordable, accessible way for creators to "generate and control timbres and effects" (Kellner, 1980). Spiegel developed the Pitch Sweep effects and other features for their AlphaPlus operating system (Syntauri Corp alphaSyntauri Digital Synthesizer, 1981), (encyclotronic, c. 2017).

In 1980, she released her debut album, *The Expanding Universe*. In the album's liner notes, she takes notice of the number of women in electronic music. "I don't think it's a coincidence that there seems to be a relatively high percentage of women and other composers who the media might discriminate against working in electronic media" (Reynolds, 2012). Her comment deserves more than a passing mention, and more about the climate of feminism and music technology in the 1970s and 1980s is discussed in Chapter 1.

Spiegel, in 1981, predicted that music creators could eventually make their work available digitally to fans "through public archives where masters are available for duplication (for a small royalty to the maker)" and wondered about "what kind of credit, royalty, or bookkeeping system will ensure that creators get something out of the use of their work," given the world's new Information Age, where people could make cassette copies of records or videotapes of broadcasts (Spiegel, 1981). In this way, she was among the very first to portend the advent of digital music distribution via streaming[1] (Reynolds, 2012).

Spiegel continued to work with computer software, still focused on real-time interaction to create music. She created the "Music Mouse – An Intelligent Instrument," one of the first publicly available computer-based music composition programs. The software was an early example of "algorithmic composition and intelligent automation" (*Female pioneers in audio engineering*, 2017). Music Mouse was introduced in 1986 for Macintosh, and soon after for Amiga and Atari computers as well. In the manual's introduction, Spiegel writes,

> "This is a very exciting time for music. With the advent of computers, many of music's past restrictions can begin to fall away, so that it becomes possible for more people to make more satisfying music, more enjoyably and easily, regardless of physical coordination or theoretical study, of keyboard skills or fluency with notation. This doesn't imply a dilution of musical quality. On the contrary, it frees us to go further, and raises the base-level at which music making begins."
>
> (Spiegel, 1986)

Music Mouse allowed the user to manipulate many variables in real time to control orchestration, harmony, contrapuntal motion, articulation, and other musical parameters. "Just move around however you feel like it," the manual advises. "Listen. Some places will sound like points of arrival, while you'll find yourself wanting to keep moving on when you pass through other places" (Spiegel, 1986).

Recently I asked Spiegel if she thought about reprising Music Mouse for iOS and/or Android, which she assures me she has. However, the approach would have to be a bit different: "For the multitouch human interfaces such as iOS or Android devices provide, one would want to do a different kind of control interface to the parameter set, to that model of navigable musical space. Music Mouse was designed in an era when a single x-y location was 'pointed to' using the computer's mouse. While you moved the mouse with one hand, the other hand manipulated a very full qwerty keyboard of parameters that controlled how the movement of the mouse would be translated into musical sound. iOS provides up to 11 simultaneous x-y inputs. I have thought about how that might interface to a musical parameter set at the same level as Music Mouse, for harmony and counterpoint, but not to the point of implementing anything yet" (Spiegel).

I also wondered if she has worked with Max/MSP. "I keep thinking I should revisit and try to get into it," she replies. "So many people I know use Max/MSP/Jitter. When I first tried it, I didn't like it, but I was coming from having designed my own modular patch language, which was quite different, but which never ended up being released. That was going to have been the control language for the successor to the McLeyvier, and after that project fell apart, I wanted what it was supposed to have been, and Max/MSP just wasn't that. But I might give Max another try at some point. It's named after Max Mathews, one of my incredible mentors, and derives from his 'unit generator' concept, which is an extremely productive category of design for music creation technology" (Ibid.) (Spiegel).

Spiegel's *Universe* album was reprised in 2012, around the time Voyager I entered interstellar space, and as of 2019, Voyager 1 has been traveling for 41 years. Spiegel muses upon its fate. "That little computer, Apple II level tech, is still running after all of these years. But I also feel sad, that at that distance from the sun it's too far for the solar panels to keep up the charge needed for continued communications. The Voyager craft will soon be all alone in the far cold reaches of space, incommunicado. Will any intelligent beings ever find that gold record? I doubt it" (Ibid.).

Laurie Spiegel
Source: Photo credit Irina Prosser

Carla Scaletti (USA)

"Everything is waves. Matter is waves. Sound is waves. And sound can affect us in a way that most other things can't," says **Carla Scaletti**, referring to a revelation she had after seeing the work of Hans Jenny's book on cymatics. The subject matter of that book is closely related to the works by three researchers mentioned in Chapter One of this text: Ernest Chladni and his vibrating plate, Margaret Watts Hughes and her eidophone, and Mary Waller's further study on Chladni figures. Cymatics also deals with the observation of material upon vibrating surfaces. Like these scientists, Scaletti was intrigued by these forms (Scaletti).

But her search for patterns was broader and began in high school when she began obsessing upon her interest. "It was literally one of these moments where I couldn't sleep and I was staying up all night just thinking, 'what am I going to contribute [to society]? What am I going to do?' And I just had the sense that, okay, there are patterns out there in the universe. And maybe the best way I could contribute would be to turn those patterns into sound and try and understand and communicate those patterns. So that was my kind of teenaged, very dramatic feeling about what I wanted to do (Ibid.).

Carla Scaletti
Source: Photo credit Kurt J. Hebel

"Even as little kids we learn how to read graphs, take data, make a bar graph or a line graph. But we don't learn about sound: *listening* to those numbers is very natural. You can just listen to the stream of numbers as an audio signal . . . you could take that stream of numbers and modulate another signal. Modulation is the message: you're hearing how that sound is changing, and basically that's also the same thing as music, right? There's not just one long tone, but you're hearing *changes*, and that's where the meaning is" (Ibid.).

Scaletti specializes in data sonification: transforming numbers into sounds. As she put it in her lecture, "Sonification is a Joke," she clarifies that sonification is *not* a joke; rather it is the "third leg of human auditory communication, equal in importance to the other ways we have to convey meaning through sound: speech and music" (Scaletti, 2017).

Consider the stethoscope: doctors use it to diagnose heart and lung problems in the human body. An untrained listener might only hear a "whoosh, whoosh," but doctors "use their ears to look inside your body in a non-invasive way," Scaletti explains. "There's an interesting paper where they interviewed medical students who were first learning to use the stethoscope [Harris]. It doesn't come naturally. It takes some training, and what they found was that a lot of people said, 'well, I just hear all this sound and I can't [detect] anything'. But the people who got it faster were people who were musicians. The theory is that it was because they had some training in listening. *Analytical* listening; not just putting in your earbuds just to go running or something" (Scaletti).

Scaletti also designed the Kyma (pronounced "KEE-ma") sound design language. "The idea of Kyma is to be a language of sound: for specifying sound, for generating sound, for modifying sound and for manipulating sound, and also for creating an environment where you can perform live, electronic sound. From the beginning [I] always wanted to blur the boundaries between the idea of sampling and synthesis and processing. Within a computer, sound is a stream of numbers. You don't have to have these boundaries between things. Music is sound. It's audio; it's not notes on paper. A lot of people, when they say 'music' they mean a piece of paper with dots on it but music to me is *sound*. So, it's not based on *notes* as the basic element that you build things out of. It's based on *sounds* or *sound objects* as the things you build everything from" (Ibid.).

Programs like Kyma, Max, and Supercollider might seem intimidating at first, and the temptation is to say they have steep learning curves. "I think it's actually not true," Scaletti

FIGURE 6.2   A screenshot of a Kyma patch

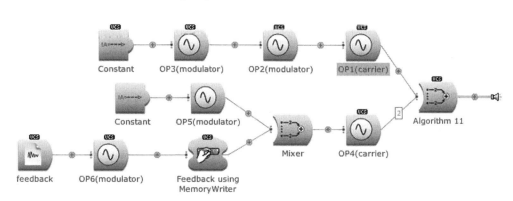

Carla Scaletti
Source: Photo credit Bryan Dematteis of Foveal Media

insists. "I think Kyma is super easy because all it is is sounds. There are these elements you click on; press the spacebar and you hear a sound. And if you want to see how the sound was made you double click on it. But because it's open – and this is true of Max and Supercollider, too – and because they don't limit your options, and they don't hide things, and they make it modular, you make your own re-combinations; this opens up this whole world of, 'Oh! I can learn about acoustics; I can learn about psychoacoustics and signal processing! I can learn synthesis!' and suddenly it's like a lifetime project. But that's OK because it's fun!" she laughs. "I don't mind that it's a lifetime project. It's a good thing" (Ibid.).

For those readers who are looking for direction in their lives and careers, Scaletti has the following advice: "Follow your curiosity. And trust it. Even if it seems to be leading you in an unusual path or something different from everyone else. You should trust your intellectual curiosity and your creative curiosity because it's not going to take you to the wrong place" (Ibid.).

You can find out more about Carla Scaletti at carlascaletti.com.

▉ EveAnna Dauray Manley (USA)

EveAnna Dauray Manley is the president of Manley Laboratories, world-renowned manufacturer of microphones and microphone preamplifiers, compressors, and equalizers.

"My first experience recording was with a Dictaphone. I would put the mic up to the speaker and record Casey Kasem's Top 40 show," she recalls. Manley's stepfather, Albert J. Dauray, owned Ampeg, the guitar amp manufacturing company. [www.manley.com/history]. "He sold it to Magnavox in 1972. He was a business man with Unimusic – it's all in the *History of Ampeg* book. I remember hearing the stories of the late 60s when the Rolling Stones took the Ampeg amps out [on tour]. He met my mom in the 70s, after his first wife had died. He sold Ampeg . . . and he was scrambling for all those years after that. So we grew up in an unstable environment. That was a major influence on my life: I didn't want to live

EveAnna Manley displays her 2016 TEC award for the Manley FORCE
Source: Courtesy of Manley Labs

with that kind of financial". Instead she credits her first boss, Terry Jones, with fostering a more positive business acumen. "I worked at a picture framing shop in high school. He was a small business owner. I learned by observing him and applied that to how to run my own business. He didn't directly teach me, but I learned from him and his wife Patsy how to enjoy life. Work hard and enjoy life" (Gaston-Bird, 2017).

After attending Columbia University and majoring in music, she decided to head to the West Coast in 1989 (Manley.com). At the time, Vacuum Tube Logic of America was owned by David Manley and his son, Luke. "I was looking for a job in the music industry, and I interviewed at Vacuum Tube Logic. I remember looking at the vacuum tube amplifiers and wondering, 'What the hell is that? A receiver?' I didn't have electronics training at school, so I didn't get to play with circuits, and in physics class I only learned simple circuits (Ibid.).

"I made coffee, tea and PCBs (printed circuit boards). I learned how to solder. . . . I trained other girls to stuff circuit boards. That company was running by the seat of its pants! This was all before computers: they had no purchasing system in those days. So we would self-kit our parts. They'd say, 'build 30 of these' and then you had to count your own parts. If you looked in the bin and there were no resistors you had to ask someone to buy them. So one day I took a ruler and a marker and created a purchasing system with part number, description, PO (Purchase Order) number, date, etc. And then I developed an inventory system" (Ibid.).

David and EveAnna founded Manley Laboratories, Inc. in 1993 (Manley.com). "[We] got married in the early 90s . . . and I ended up buying the company after he left in 1996. The transition took place within several years and I learned things, created systems, implemented systems, and got into the job by doing it. I saw something that needed to be done and took care of it. The hardest part about owning a small company is you can't afford to hire experts, so you *have* to learn how to do this shit yourself. It is creative, [although] I might solve problems in a different way than what we typically consider to be 'creative'. But you are always in a catch-22: you're too small to hire so you have to learn to be a versatile person" (Ibid.).

EveAnna Manley was the de facto operating CEO of Manley after David's departure in 1996 and eventually assigned his total shares in the company to her in the course of their divorce agreement. In July 1999 his resignation was announced in *Pro Sound News Europe*. During that three-year period and under EveAnna's leadership, Manley Laboratories achieved a 75 percent growth in sales. They released the Manley VOXBOX˙ in 1998 and the Manley Massive Passive stereo equalizer in 1999 (Stack, 1999). In 2016, the VOXBOX was inducted into the "TECnology Hall of Fame," and the Massive Passive was nominated for a 1999 TEC Award in the Signal Processing Technology/Hardware category.

"I was always a leader," she says. "Whether it was playing first chair or serving as band president." But she has also learned the art of collaboration. "The audio business is not always competitive. I am always reaching out to Josh Thomas of Rupert Neve, Erica McDaniel of Universal Audio, John Jennings of Royer, Paul Wolff . . . these guys are total best friends even if we are in the same field. They are my peers. And they reach out to me, too. It's a wonderful industry for that. We often interface and help each other" (Gaston-Bird, 2017).

In 2016, the Manley FORCE˙, a four-channel, high-voltage vacuum tube microphone preamplifier won the TEC award for Technical Achievement (Labs, 2016).

At the NAMM show in 2017, Manley Labs debuted the Reference Silver microphone, their first in 27 years, and successor to their Reference Gold and Reference Cardiod mics. "It uses the David Josephson C3 capsule (developed for the C37A tube mic). It's a capsule with a rich middle tonality recalling the original Sony mics from the 1950s. A bunch of LA engineers say the Sonys are their favorites. We are building a new mic with rich middle tonality, which is great for woodwinds and brass. Combining that with the latest switching power supply custom designed for vacuum tubes is a *tour de force*. And it has this interesting metal finish on the body. It was hard to do, but it's great. It looks like reptile scales. My crew was mad at me but I insisted!" (Gaston-Bird, 2017).

She has a bit of advice for those just starting out in the profession. "The most important trait of all is not gender but honesty," she advises. "I would caution anyone, not to be afraid to say 'I don't know'. It's okay to be humble and truthful and ask for clarification. Then you have an opportunity to learn something new! Most people are happy to share and contribute to your personal growth" (Gaston-Bird, 2017).

The Manley Force® four-channel microphone preamplifier

The Reference Silver microphone by Manley Labs

You can learn more about Manley Labs, EveAnna, and their line of products at www. manley.com.

Marina Bosi (Italy)

Born and raised in Italy, **Marina Bosi** attended both the Conservatory of Music and University of Florence, where she received degrees in music and in physics. Her dissertation, which she carried out at the Institut de Recherche et de Coordination Acoustique/Musique (IRCAM), was "Design and Simulation of the Control Unit of a Computer for the Numerical Treatment of Musical Signals" (Bosi, 1987).

In Italy, Dr. Bosi worked at Tempo Reale under the direction of the Italian composer Luciano Berio. Tempo Reale, founded in Florence by Berio in 1987, is an important European center for research, production, and educational activities in the field of new musical technologies and electronic music. Shortly thereafter, Berio was commissioned to write a piece for the inauguration of the performing art center "Il Lingotto" in Turin (Torino), Italy, which was originally owned by FIAT, the car manufacturer. "FIAT had a strong influence in Italy, but in Torino even more so, and one space they owned was this huge building . . . where, believe it or not, on the roof they were test driving newly designed cars. So, you can imagine the dimension of this place" (Bosi).

"It was transformed into a media center with a movie theater, etc., and Luciano Berio was commissioned to write a piece for the newly restructured space. Luciano was envisioning a piece where the sound was moving around through space and asked for my help from the technical point of view. At the time, I felt I needed to get more familiar with the work that was being done in the USA, specifically at Stanford's CCRMA (Center for Computer Research in Music and Acoustics) where John Chowning wrote the multichannel electronic piece 'Turenas'. So, I found myself packing to go to Stanford for 6 months (Ibid.).

Marina Bosi

"When I first arrived at CCRMA I worked on a 'test' project implementing, with the help of David Zicarelli for the interface, a real-time system that allowed a sound source to move in space following the motion of the mouse. I called this system Quadrifoglio (Bosi, 1988) and composer/violinist Mari Kimura employed Quadrifoglio during the annual Summer CCRMA concert to give the real-time illusion of moving her violin sound around Stanford's Frost Amphitheater."

About 30 years later, Dr. Bosi is still at Stanford. "I never went back [to Italy]," she laughs. "One of the things that attracted me once I arrived here was the vitality of Silicon Valley. After my initial 6 months expired, I found a job at Digidesign which was one of the very few companies at the time dealing with high quality audio and computers. As one of their first employees, I was designing and implementing DSP modules for Sound Designer, Digidesign's very first product and the precursor to ProTools" (Ibid.).

At the time, the CD format was becoming the de facto standard for high-quality audio. Limited transmission speeds and storage sizes raised the challenge of transmitting (or storing) a library of high-quality audio. To transmit a mere four minutes of CD-quality audio, it would have taken almost 10 hours at the then-typical modem speed of 9.6 kb/s. (Compare that to an "average" broadband connection today of about 20 Mb/s!) "When they asked me at Digidesign to work on data compression, my reaction was, 'oh, why me?' I liked sound, I didn't want to alter it in a disruptive way, but the need for compression was certainly there, and we were starting to see new technology that enabled things unimaginable only a few years earlier.

"I started working on perceptual audio coding at the end of the 1980s when the field was still in its infancy. I quickly got so interested in it since I realized that I could exploit the advances in DSP to represent audio signals in ever more compact and efficient ways, and by applying psychoacoustics models to identify irrelevant components, we could make transmission/storage of high-quality audio a reality. But I have to admit, few of us at the time would have dared imagine the revolutionary impact that audio coding would have on the general consumption of digital media. Then I was hired in the research department at Dolby" (Ibid.).

Dr. Bosi was part of the research team that designed and developed coding schemes at Dolby Labs, including AC-2 and AC-3, which is now known as Dolby Digital. While at Dolby, she also led the development of MPEG-2 AAC, which was later adopted by Apple iTunes. In fact, she was recognized by ISO/IEC with the Award as Project Editor in the development of International Standard ISO/IEC 13818–7: 1997.

After Dolby, Dr. Bosi joined DTS as Vice President of Technology, Standards, and Strategies and then became Chief Technology Officer at MPEG LA. Dr. Bosi was also actively involved in the development of the standards for DVD-Audio and the ATSC (Advanced Television Systems Committee) (Ibid.).

Meanwhile at Stanford, she created the first known North American university course on perceptual audio coding and MP3 technologies. Based on this graduate class, Dr. Bosi wrote the acclaimed textbook *Introduction to Digital Audio Coding and Standards* (Kluwer/Springer December 2002) translated into Chinese and Korean.

She also was a cofounder of the Digital Media Project (DMP) along with Dr. Leonardo Chiariglione, chair of MPEG, a nonprofit technical group focusing on Digital Rights Management (DRM). "Once you shift the model from physical things like vinyl or CD to streaming and different technology where you don't have a 'physical' medium – you just have the data – [there are issues] related to rights management in terms of authoring, distributing and usage" (Ibid.).

Dr. Bosi is a senior member of IEEE and a fellow and past president of the Audio Engineering Society (AES). She has been active in the AES since she arrived in the USA almost 30 years ago and has worked through its ranks from member/chair of the San Francisco local chapter to Western Region VP and AES convention chair as well as co-chairing AES' first international conference on high-quality audio coding. In speaking about women in audio, Dr. Bosi agrees that there are more women in the field than people perceive, but who are often treated as if they are "invisible." "Women for some reason are not as publicized . . . women do the same things as men, but their accomplishments often just don't seem to catch the attention as they should.

"We should educate young women in becoming more assertive and, at the same time, we should educate men to better respect and listen to other people. It seems that there is kind of a 'fork in the road' where boys are sweet and gentle until a certain age, and then many feel they must follow a 'male model', even if it is constraining to them. So, I see the two sides as well. It's really a matter of nurturing the boys and letting them express their feelings as we do with girls. Honestly, we kind of shortchange boys as well as girls. It's not 'us versus them'; we have to be together at some point" (Ibid.).

Marina Bosi's panel on Perceptual Audio Coding from the 2008 AES 125th Convention in San Francisco (Pictured L-R: Louis Fielder, Marina Bosi, Bernd Edler, Gerhard Stoll, John Princen, J.D. Johnston, Karlheinz Brandenburg, Ken Sugiyama

Erisa Sato (Japan)

Erisa Sato is chief product planner of TASCAM® audio equipment, leading projects of a wide range of consumer and professional products in Tokyo, Japan.

She started with sound recording at the Tokyo University of the Arts, which was one of very few courses for audio engineering at the time. "I wanted to be a composer, but you have to know engineering things to be a composer." She took sound recording classes and loved the scientific approach to music (Sato).

Before starting her studies at the university, she was composing and playing classical piano. "Twelve years ago, during my first year at the university, I noticed that playing piano is not my thing to live with. I got a chance to work at the Science & Technology Research Laboratories of NHK (a public broadcaster) for a 22.2 channel public live-viewing project. Since then, my part-time job was all related to audio experimentation and audio recording, which helped me to see the actual happenings in the industry as well as the study and classes at the university" (Ibid.).

She worked as a recording assistant, audio engineer for archiving digital tape, and an assistant PA engineer from a small café to a huge venue. "All of those experiences made me think about the essence of 'audio engineering' and what role I could take. I'm very happy that I could find my passion so easily and seamlessly. This industry has tons of extraordinary interesting technology and research so I never get bored" (Ibid.).

Erisa Sato, chief product planner of TASCAM audio equipment

Among her influences, she cites Mrs. Hiroko Fukaya (classmate of famed composer Riyuchi Sakmoto), who was her biggest mentor. She also credits Toru Kamekawa, who taught her about sound recording (Ibid.).

She went on to pursue a master's degree, and her thesis was "The Effect of Reverberation on Music Performance," which describes how performers adjust their playing based on the information they are getting through reverberant signals in the space in which they play. The paper was also published as an e-Brief at the 2011 AES Convention in New York (Ibid.).

After graduating, she joined TASCAM as a product planner. "I've always wanted to be a product planner, because there are classes and special events where very famous sound engineers come – even George Massenburg came to our university for special class for recording which was quite exciting – and from those events I found there are so many engineers today who are having trouble with their gear: price-wise, spec-wise, everything. I thought, 'that is simply not the ideal relationship of creativity and tools', but they still have problems all the time. So I just wanted to coordinate the environment for those engineers" (Ibid.).

She is proud of her work on the TASCAM DA-3000 stereo master recorder, which can record PCM 192 kHz and DSD. "It's still one of the biggest projects I was ever involved with. My role was a planner, but I was also on the team to decide every detail: how to place the physical diagrams and structure of the menus." She was also doing sound evaluation to define the color and position of the sound. "I asked Professor Kamekawa and Professor Marui to help me out, so we did a joint experiment with Kamekawa and Marui on the difference between PCM and DSD. The paper, 'Subjective Evaluation of High Resolution Recordings in PCM and DSD Audio Formats' [www.aes.org/e-lib/browse.cfm?elib=17166], is based on a study with 45 people. Most of the people say that DSD has more spatial expressions and the coloration of the sound is brighter than PCM, not just based on musical objects but sound effects stimulus like jingling keys" (Ibid.).

Her dream is to become a consultant or integrator. "I think there are many ways I can help engineers besides just making the gear" (Ibid.).

◼ Elisabeth Löchen (France)

In 1994, when the security guard of the Bally's Hotel in Las Vegas did his sweep of the ladies' toilets outside the green room where Steven Spielberg was scheduled to make an appearance, **Elisabeth Löchen** hid in a stall with her feet lifted up so she wouldn't be seen. She was broke, her children were over 5,000 miles away, and it was her last chance to meet Spielberg and tell him her story.

Things did not start out this way for Dr. Löchen, who holds a PhD in psychopathology. In the past, she regarded films as "something you do to pass time at the weekend." But when some friends approached her about investing in a floundering recording studio, she decided to join them, but only under the condition that she be allowed to run it. In less than a year, the studio became profitable (Chanel, 2013). She recalls that she was the only woman running a studio in France in 1986 (Löchen).

Eventually her business associate, Pascal Chedeville, came up with a system for digital audio playback in cinemas. Unsure of how to implement the idea, he partnered with Löchen to create the Löchen/Chedeville (LC) Concept System, the world's first dual-medium, digital sound system for film and filed for a French patent in 1989 (Pendleton, 1994).

Löchen worked night and day, trying to find investors, films, and partners; installing the equipment in cinemas; and testing it. The format used MUSICAM compression to fit the audio onto magneto-optical discs, which were synchronized with a SMPTE timecode track printed on the film. [Karagosian/Löchen] LC Concept sound was used on over 30 films, including *Cyrano de Bergerac*, *Basic Instinct*, *Bis and Ends der Welt*, *Backbeat*, *The Accompanist*, *Heaven and Earth*, *Cliffhanger*, and others.

A 1991 news release in the *Hollywood Reporter* with the headline "French firm enters digital audio derby" reports that Löchen and Chedeville visited Skywalker Sound post-production studio in Santa Monica to run demonstrations of the format. At the time, Cinema Digital Sound (developed by Eastman Kodak) was the only competing format, printing its digital audio information on the film (replacing the classic analog optical soundtrack) as opposed to a separate disc, as did the audio in the LC Concept format. "Because the sound isn't on the film, it isn't subject to the wear and tear and abuse that a release print typically goes through," told a spokesman for LC Concepts to the *Reporter* (Parisi). (Contrary to the

Elisabeth Löchen

Source: ©2019 Randy Leffingwell/Los Angeles Times; used with permission

FIGURE 6.3   List of films released with LC Concept sound

L. C. CONCEPT

FILMS IN LC DIGITAL SOUND

1991	CYRANO DE BERGERAC	JEAN-PAUL RAPPENEAU
	UNTIL THE END OF THE WORLD (VO)	WIM WENDERS
	JESUIT JOE	OLIVIER AUSTEN
1992	TOUS LES MATINS DU MONDE	ALAIN CORNEAU
	THE LOVER (VO)	JEAN-JACQUES ANNAUD
	BLANC D'EBENE	CHEIK DOUKOURE
	LA BELLE HISTOIRE	CLAUDE LELOUCH
	BASIC INSTINCT (VO)	PAUL VERHOEVEN
	IP5	JEAN-JACQUES BEINEIX
	L627	BERTRAND TAVERNIER
	BITTER MOON (VO)	ROMAN POLANSKI
	L'ACCOMPAGNATRICE	CLAUDE MILLER
	LA FABLE DES CONTINENTS	INA
1993	ARIZONA DREAM (VO)	EMIR KUSTURIKA
	L'ARCHE ET LES DELUGES	FRANCOIS BEL
	FALLING DOWN (VF)	JOEL SCHUMACHER
	MATINEE (VF)	JOEL DANTE
1994	CLIFFHANGER (VO)	RENNY HARLIN
	FREE WILLY (VF)	SIMON WINCER
	HEAVEN AND EARTH (VF)	OLIVER STONE

NEXT RELEASES :

1995	SILENT TONGUE (VO)	SAM SHEPARD
	BOILING POINT (VO)	JAMES B. HARRIS
	BACK BEAT (THE BEATLES)	IAN SOTLEY
	L'AFFAIRE DU SIECLE	JEAN-JACQUES BEINEIX

belief that Dolby was a contender with its SR-D format, it was not until 1992 that SR-D was officially released with the film *Batman Returns*.)

The next year in France, Elisabeth received a visit from three lawyers from Universal. "They wanted to run *Jurassic Park* on their digital sound system (DTS) using the LC patent." Rather than accept a very low, one-time payout for the patent, Elisabeth insisted on a partnership but was turned down (Chanel, 2013).

By 1993 a bitter legal battle was underway between LC Concepts and DTS. The European issue of *Variety* magazine reported that police were stationed in the projection booth when *Jurassic Park* premiered at the Deauville's Festival du Cinema Americain in Deauville, France, ready to confiscate any DTS equipment. This was a measure made possible by the Grand Tribunal of Lisieux, who had issued a restraining order. In the article, Löchen is quoted as saying, "I regret having to do this because Spielberg is someone I personally admire. *Jurassic Park*, I find a very good film and it makes me heartsick to have to attack it this way. I sent [Mr. Spielberg] a personal letter explaining everything – not one from lawyers or companies – but from a little French woman, who has been working on something for eight years. [The lawyers] answered me saying Mr. Spielberg has no time to read such a letter and that I'll only hear from lawyers in the future" (Mikelbank).

This brings us to 1994 and the bathroom at Bally's. Löchen had, by this time, spent all of her money on lawyers, determined to prevail. Having been invited to do her annual lecture in the United States, she left behind her two children, aged 10 and 13, in Paris. When she arrived in Las Vegas, Nevada, she realized no one was aware of the litigation happening in Europe and that she had to bring her fight to the States (Löchen).

"So I called my children and told them, 'I can't come back right away,'" she remembers. "They ended up spending 4 months by themselves. My mother almost died one week after my departure and that's why she wasn't able to babysit them. And [my kids] were supporting me, they said 'Mom, Mom don't worry, it's going to be okay. So I spent 3–4 months to try to get lawyers to fight for me and I had absolutely not a dime to pay them, so I had to find a law firm to work for me on contingency." She also had lawyers in Washington who were *not* on contingency, and things were getting very expensive. "It was a nightmare because I had no more money, and it was like, 'oh my God? Where do I go?' I was sleeping on friends' couches and my kids were alone and I had to find a solution. I had to meet him," she says, referring to Steven Spielberg, who owned an interest in DTS (Ibid.).

Then she learned that Spielberg would be at the National Association of Theater Operators (NATO) Western Regional Conference, "Showest." She contacted the conference organizers, and as a hardware manufacturer, she was able to secure a ticket. "But I couldn't buy the dinner ticket because it was too expensive for me." By chance, she met a Frenchman working for NATO and pleaded for help meeting Spielberg. Her friend arranged a 7:00 p.m. meeting. "He said, 'you have to be 5 meters away from the door of the green room, and I can enter with you'. So, I put on a long dress and put a big sweatshirt over it, and took a book with me, and at 2 pm went to the corridor outside the green room." Luckily she spotted a ladies' room five meters away (Ibid.).

"I went in the restroom and locked the door of my stall and waited till 7 pm. Even the guard came and said, 'everybody out!' I didn't say anything I just put my feet up and prayed, 'I hope nobody saw me'. So I hid in the stall. [My friend and I] didn't sync our watches like in *Mission Impossible*. He told me he'd be there 'around 7'. I swear to God I was waiting 5 hours." Then at 7:00 p.m., she put on her evening dress and walked out of the bathroom, and just at that moment, her friend was passing by with two other producers. "I slipped in between my friend and the two producers without saying anything, and we were able to go in front of the big bodyguards – they thought I was their little girlfriend or something – and here I am in the green room" (Ibid.).

Among the celebrities in the green room were Robin Williams, Michelle Pfeiffer, and several other big Hollywood names. One of the NATO organizers who had previously told Löchen she was forbidden to meet Spielberg was there, too. She tried to hide herself, nibbling on hors d'oeuvres, when she began to choke on her food. Frantically, she sputtered, trying to shoo people away who had come to her aid. "And suddenly I saw Spielberg," she says. "And I waited for him to finish speaking with the person he was talking to, and I went to him and I said, 'Hi, Mr. Spielberg, I am Elisabeth from LC Concept', and he looked at me and said, 'Oh, I'm Steven Spielberg'. 'Yeah, I know who you are,'" she says, laughing (Ibid.).

"We started to talk and he said, 'You know, we're going to have many machines, many units in Europe this summer', and I said, 'Well, you know this summer [my] patent will be granted for Europe so you won't be able to use your units in Europe', and he was looking at me . . ." Löchen leans forward, demonstrating Spielberg's posture of concern (Ibid.).

A single thought was going through her head. "I was like, this is my last opportunity. I just told him everything. I had to tell him. I wasn't mean, I was not aggressive, I was just telling him what happened. I didn't ask for anything, I didn't say anything wrong, I didn't say he was wrong. I was just like, this is it. This is the situation. And at one point he had to go on the stage, and he said, 'Sorry, I have to leave' and he was looking for a business card in his nonexistent tux pockets. And I gave him my business card and I said, 'I guess you are easier to find than I am'. He looked at the business card and he said, 'you're a doctor? A doctor of what?' And I said, 'Make an appointment and I'll tell you'. And then he left for the stage" (Ibid.).

During our interview, recounting the episode makes her emotional. "I was shaking so hard. So I was saying to myself, either I'm dead or something can happen. And I remember I went back to the big room where I could see the stage, and I saw these amazing directors and actresses – and it was so powerful. And I felt so tiny. Tiny, tiny. I was crying. I was like 'what do I hope? How do I dare'? And so I left the room because I was crying. I went back to my thirty-dollar-per-night room I was sharing with someone. And the day after, I got a phone call from one of my lawyers . . . and he said, 'We got a phone call from Universal'. He said Spielberg had called Universal saying, 'find a solution with her'. And they reopened the negotiation" (Ibid.).

Left with no other options, Löchen accepted a settlement from DTS in 1994 (Ibid.).

In 1995, at the 68th Oscar awards, Pascal Chedeville received the "Scientific and Technical Award (Technical Achievement Award) for the design of the L.C. Concept Digital Sound System for Motion Picture Exhibition." That same year, Dolby won in the same category for their SR-D system. However, whereas Dolby was awarded as a company, LC Concepts was not. Löchen was dumbfounded at the ceremony. "I was so sad. I remember I was in the room with everybody else and when the guy went to take the diploma, everybody came to my table to say 'congratulations', and I was half crying but they said, 'yeah, it's him but we know everything you did'. I said, 'Why??' Because it was a male and he was a technical guy. I mean the full crew of Dolby came on stage but they were all male" (Ibid.).

Today, Löchen resides in Paris. She says everything is fine between her and Spielberg now; he even watched her first short film, *Red Ribbon*. And if you're wondering how

Elisabeth Löchen and her grandson

everything turned out with her kids, they are fine. I even got to see her grandson during our Skype interview.

◼ Marie Louise Killick (England)

If there were one thing Marie Louise Killick couldn't stand, it was how the steel needles used on record players would degrade the shellac or vinyl underneath. It "stuck like a thorn in my flesh," she explained to the sister at Brookwood Psychiatric Hospital where she had been "admitted for observation" in 1955. Killick was calmly trying to explain who she was: an audio engineer who manufactured sound recording equipment during the war. The night before, Killick had been abducted while accompanying her eldest daughter to the hospital and bundled into the back of a car, between two men. One of the men in the front was a police officer. She was then forced to enter the ward. Eventually, her lawyer would explain that she was involved in litigation against Pye Radio, Ltd., who had infringed her patent for the truncated-tip sapphire stylus, trademarked as Sapphox in 1945. She had filed the lawsuit in 1953. Killick was eventually released from the asylum (Killick, 2018b).

Upon her release, Killick headed straight to the police station, demanding to know why she had been kidnapped and taken to a psychiatric institution. The officer told her that a newspaper was checking on the facts for a story about the case, but that when "Pye Radio denied all knowledge of you, the editor contacted us, asking to make sure your family were alright." Killick remembers her reaction to this explanation in the book *A Sound Revolution:*

> I could hardly contain my anger. "Do you mean to tell me, that on the say so of somebody from Pye, I was forcibly detained in Brookwood and my children put into care? If your officers had thought to ask me, I could have given them ample evidence of my 'action' against that firm."
>
> (Killick, 2018b)

Cynthia Killick, one of her four children, recalls their vagabond life during the lawsuit, which dragged on for five years. "We lived in a selection of houses, cottages, flats, bedsits, boarding houses, hotels, caravans, boats, houseboats and for six weeks home was a small van in a builder's yard in Oxshott, Surrey" (Killick, 2018a).

The good news is she won – the press reported claims on what she was owed, including a figure of £5 million (equivalent to $14,000,000 when reported in 1958). Even after Pye appealed the verdict, the decision was upheld. The *New York Times* on July 23, 1958, ran a story with the headline, "Woman Inventor Wins." "Mrs. Killick, an engineer, invented and patented her needle in 1945 after cutting records for the British armed forces during World War II," the article reports. "Her needle is flattened and beveled at the end so that it rides on the side of the grooves of a record and not the bottom" (Woman Inventor Wins: Briton May Get 14 Million For Her Phonograph Stylus, 1958).

Killick recalls how the idea was conceived, after the war, at night while her daughters slept:

> "The idea would not let me sleep and I got up and made notes, fearful that by the morning the idea would be lost. Over the next few weeks, I spent hours in the Patent Office researching the development of sound recording and comparing previous inventors' patents . . . my idea was this. My stylus only touched the sides and not the bottom of the groove. Mounted at its tip was a gemstone, industrial sapphire or diamond, ground to a flat, which, for technical reasons, gives better sound reproduction, while leaving the bottom and sides of the groove undamaged."
>
> (Killick, 2018b)

According to the timeline on MarieLouiseKillick.org, she filed for the patent on October 25, 1945. Pye Radio opposed the application, claiming it was not "novel or valid." Later that year, Killick also applied for a patent for a machine that would manufacture the stylus and started her own business. In 1946, Decca Ltd. offered her £75,000 for the rights to the patent, but Killick refused, partly because she wanted to grow her business. On June 18, 1948, her patent was accepted, but Pye and companies had begun manufacturing her invention. The very ugly legal battle thus begun, but Killick finally emerged victorious (MarieLouiseKillick.org, 2018).

Of her victory, Killick was reported as saying, "At last my family can have a settled home. . . . I have spent all my money in ten years of legal battles. Now I am planning a world tour to collect royalties" (Woman Inventor Wins: Briton May Get 14 Million For Her Phonograph Stylus, 1958).

Extract of Report taken from Britain's leading Gramophone Journal " The Gramophone " (August 1946 Issue), of American Test carried out on Gramophone Needles.

1. Steel needles produce the most surface noise and poorest quality ; needles show wear after only **one** playing.

2. Precious metal tipped needles produce surface noise, and show excessive record wear after 25 playings.

3. Ordinary jewel-pointed needles give best reproduction ; give surface noise, play 250 records before showing excessive wear on needle and record.

4. Diamond tipped needles show wear after 1,000 playings, but show excessive record damage after 55 playings.

5. Cactus, Fibre and Wooden needles show they are quite unsatisfactory with respect to quality and reproduction and their durability is extremely low, even incapable of playing one side before the point is seriously affected by wear.

RECENT TESTS IN THE LABORATORIES OF MESSRS. KILLICK & COMPANY CONFIRM THE ABOVE RESULTS AND CLEARLY DEMONSTRATE THE ADVANTAGES OF THE KILLICK " SAPPHOX."

KILLICK & COMPANY
SOUND SPECIALISTS,
Manufacturers of Scientific Recording Instruments, Amplifiers, Radiograms, Cutting Stylus, Discs, Playback Sapphires, etc,

117 . 118 FLEET STREET . LONDON . E.C.4

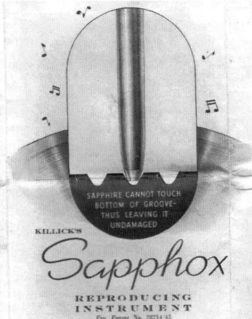

KILLICK'S

Sapphox

REPRODUCING
INSTRUMENT
Pro. Patent No. 28214/45
Supersedes Gramophone Needles

SAPPHIRE CANNOT TOUCH BOTTOM OF GROOVE— THUS LEAVING IT UNDAMAGED

KILLICK'S

Sapphox

REPRODUCING INSTRUMENT

In the world of music, technical research has, once more, perfected an outstanding achievement, and another stage in the art of Reproduction has been reached.

A NEW SCIENTIFICALLY-DESIGNED SAPPHIRE REPRODUCING INSTRUMENT called the SAPPHOX has finally overcome the numerous drawbacks inherent in the ordinary needle, which is, in fact, rapidly becoming obsolete.

Briefly, the main advantages embodied in the " SAPPHOX " patent are as follows:—

LIFE OF RECORD PRESERVED

As will be seen from the magnified illustration on the front page of this folder, THE ACTUAL SAPPHIRE, A REAL GEM HAS NO POINT. It rests lightly on the inner walls of the groove of the record in such a way that the bed of the sound track remains untouched ; consequently the wear on the record is negligible, and its life is thus considerably prolonged.

GIVES HIGH FIDELITY REPRODUCTION

The High Fidelity of the " SAPPHOX " Reproduction is maintained over a longer period than any other reproducing instrument, and the true harmonic quality is faithfully brought out. Surface noise is greatly reduced, and notes beyond the range of the ordinary needle are clearly audible. On a high quality amplifier, Straight Line reproduction from 40 c.p.s. to 14,000 c.p.s. is obtainable.

DOES AWAY WITH NEEDLE CHANGING

As a result of careful laboratory tests EVERY " SAPPHOX " REPRODUCING INSTRUMENT IS GUARANTEED TO PLAY 2,000 RECORDS BEFORE NEEDING REPLACEMENT on a Light Pick-Up. In this way the bother of having constantly to change needles is entirely eliminated.

Trailer Type for heavy-weight pick-ups.
Straight Type for medium-weight pick-ups.
E.M.I. Type for light-weight pick-ups.

Price 10/6

plus 3/6 Purchase Tax.

Pamphlet for Killick's Sapphox Reproducing Instrument

FIGURE 6.4   Diagram of the Sapphox stylus from the patent specification

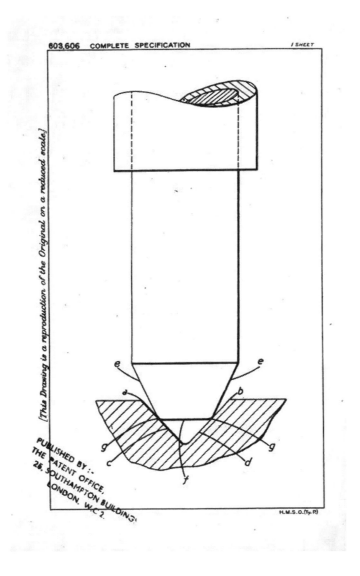

The tragedy is that she was never presented with the funds and continued to fight for the award until 1964, when she died from cancer just a few weeks shy of her 50th birthday. Her daughter, Cynthia Killick, wrote about the events in her book, *A Sound Revolution*, which includes Killick's memoirs and accounting of her invention, patent filing, and fight for justice.

Images of Marie Louise Killick, the Sapphox stylus patent specification, and the Sapphox pamphlet are provided courtesy of the family of Marie Louise Killick and the website https:// marielouisekillick.org.uk.

Marie Louise Killick, inventor of the Sapphox stylus

◼ Carol Bousquet (USA)

In 1986, during the 10th anniversary of women being allowed into the military academies, **Carol Bousquet** was assigned to serve on active duty at the US Air Force Academy. Her field was broadcasting. To celebrate the anniversary, she proposed a retrospective of women at USAFA to her chain of command. The program was approved, and Bousquet proceeded with her research (Bousquet).

She found that historically, there was tremendous pushback against women serving at all of the military academies; male prejudice against women at the academies proved to be their biggest obstacle. But once things changed, something amazing happened: "In the first few years, what they found out is that academics rose, right across the board in all four academies. They expanded their pool of intellectual greats that were otherwise closed out. So the bottom feeders are eliminated and everybody wins, which is the goal of diversity, right? Proven at the military academies, proven everywhere – and those are fair-sized institutions. Again you can see how results improve. My charge always was that we need to include the best & brightest no matter the color and gender" (Ibid.)(Bousquet).

Bousquet explains her motivation for exploring the subject of women's rights in the military. "I've always been an activist. It's in my DNA. When I see an opportunity, I try to be a feminist activist," she explains. For example, she had already been awarded five years of retroactive pay and reinstatement back into a job after winning a case she filed against a giant retailer where she once sold consumer audio equipment at age 19 (Ibid.) (Bousquet).

It's not surprising then that Bousquet would eventually chair the AES Women in Audio Committee. She had joined the AES after she began her career in the loudspeaker manufacturing industry; she even chaired the Boston AES section for several terms. She launched the AES "Women in Audio: Project 2000" in 1995 with the goal of starting a dialog on the absence of women in audio and to foster awareness of the issues (Ibid.).

But perhaps the most important event that developed her mettle started before her time in the Air Force and before her work with the AES. During the inaugural session of the "Women in Audio: Project 2000" initiative, Bousquet addressed a packed room of attendees and gave the background of how her case against the aforementioned audio retailer began in 1973.

"I began selling stereo equipment in a department store setting. I'm not a musician, however, I knew I wanted to be near the music. . . . That was what I wanted to do. I was hired in at 50 cents less an hour than the men because I was a woman. At first I said, 'that's okay. I'll make that up and I'll show them how capable I am'. And so 11 months later, they hired two men in the month of October, part time and temporary to work for Christmas. I was training them and they confided in me that they were being paid 50 cents more an hour than their top salesperson [which was Bousquet herself] that month, which was the straw that broke the camel's back. That of course prompted me to talk to management and ask what the problem was and to ask for the comparable pay. I was shot down by my manager. I explained that it was 1974 and I knew from the news there were things people could do, that there was a thing called 'women's liberation' going on and [new laws addressing discrimination] and all that. I didn't know a whole lot more than that at age 19. But, I eventually went to management and scheduled a meeting with the store manager. One Saturday morning I went to that meeting; it was the store manager and all of the departmental managers. And it was myself, a 19-year-old young woman and about 10 men. I presented my case and they laughed at me. It made me angry, and I can tell you that that anger has lasted. I mean, my career at that time was derailed. I was ultimately fired from my job for filing. I did file with the EEOC and I was fired the week after Christmas returns from my job on the floor in front of customers. I was harassed the whole time after filing, too. It was really a great social experiment and it made me very excited and of course it made it very clear that I had work to do in front of me if I was going to make it. . . . Of course I remained a fairly ardent feminist activist ever since. At any rate, I ended up 20 years later working for Ferrofluidics Corporation, working internationally with loudspeaker manufacturers, communicating to them about the technology. . . . And I was the first woman to sit on the board [of directors] of the American Loudspeaker Manufacturers Association [the industry's trade organization]. I was at an industry function with my boss (The International Autosound Championships) and we sat down at the banquet and not knowing who was sitting at the table. We learned that six of those folks were from that store. What a sweet moment on the 20th anniversary of that experience."

(Bousquet, 1995)

More details on the Women in Audio: Project 2000 initiative can be read in Chapter 1.

Dawn Birr (USA)

Dawn Birr is manager of Channel Development and Business Analysis for Sennheiser Pro Audio Solutions. Her career path was, as she puts it, a little "unorthodox." "I came in fresh out of college, with a four-year degree in psychology, and really needed a job! Sennheiser had an opening as a receptionist so I started there and worked my way through the organization for the next several years. I quickly found that I loved the company, the products and the people. A year after that I went into order entry, and a year after that I went into product management. I managed the Neumann studio brand for several years and that's where I got technical enough to be dangerous," she laughs. "I eventually moved on to live sound and I can still coordinate and talk RF [Radio Frequencies, a term associated with wireless audio systems]. I went from managing studio products to managing our premium wireless products, and for several years I managed our 3K and 5K series (Live Wireless Microphone systems). Sennheiser invested in me and trained me in RF, and I had hands-on education internally via our RF training, and I learned more about the technical side that way and also by spending a lot of time with our customers in the field" (Birr, 2019).

During that time, she got her master's in business administration from the University of New Haven. In her tenth year with Sennheiser, she was promoted to Vice President of Sales and Marketing for the U.S. Integration Systems Division. A few years later, she took on a global position as the commercial manager for audio recording before landing in her current position (Birr, 2019).

Dawn Birr, manager, Global Channel Development and Business Analysis, Sennheiser Pro Audio Solutions

She speaks excitedly about the Sennheiser product line – about their high-quality microphones, headphones, and RF systems and about their innovation in the market: there are two VR installations using Sennheiser technology, she mentions, "David Bowie is," for which Sennheiser won a 2013 "Audio Installation of the Year" award from the AV Awards (Sennheiser wins audio installation of the year award, 2013); and Pink Floyd: Their Mortal Remains at the Victoria and Albert Museum in London, in which visitors were given Sennheiser HD 2.20 headphones to wear as part of the experience. "Not only is the performance remixed in AMBEO (Sennheiser's patented 3D audio technology), there's a whopping 18 Neumann KH 420 mid-field monitor loudspeakers and seven KH 870 subwoofers in the intimate room . . . plus four video screens, resulting in an experience that comes amazingly close to recreating the feel of a live gig," according to one reviewer (Laird).

Birr has also noticed a few more women entering the industry in recent years. "I do see women coming forward more. I am encouraged by that. Even if I just use a barometer of what I've seen in the last five years at the NAMM show, or the participants in some of the panels over the last five years, I do think that there are women coming forward more and more. I think we're very lucky we have a new generation of women coming into the business and application side of audio who may have the confidence that I didn't have, and women in my era may not have had coming right out of the gate. So, I'm hopeful. I am also fortunate to work for Sennheiser – our company knows that diversity makes us stronger and I've always found a strong support system within our organization. A woman can do anything she has to do', as my grandmother would say – and as I get older, I'm finding out she was right" (Ibid.).

You can learn more about Sennheiser at sennheiser.com.

Other Women in Hardware and Software

Vanessa Mering is senior manager, Global Brand Communications, at HARMAN International. HARMAN is a division of Samsung and owns brands such as DigiTech, JBL, Lexicon, Soundcraft, Studer, and AKG. Mering is also the Northridge, California, site leader of the HARMAN Women's Network, a company-wide initiative that encourages personal and professional growth and "attracting new talent to the company" (Potter).

Ingrid Linn (Germany) is engineering project manager, Audio Hardware, at Apple. She holds a master's degree in electrical engineering/information technology from the Technical University of Munich. In an article on KVR Audio's website, she recalls her first project at Roger Linn Design. "I worked on the reverb algorithm, the compressor, as well as the tube simulator and our internal editor for the amp modeling parameters." She also helped develop the AdrenaLinn Sync plug-in, which was first written in Max/MSP, and the LinnStrument, a grid-based MIDI controller for which she helped program the pitch correction "so it always plays in tune without limiting your expressive vibratos and pitch slides" (Turl, 2014; Design, c. 2011).

Zoë James is a rising senior at the University of Rochester currently studying audio and music engineering with a concentration in electrical and computer engineering. Currently she is an intern at Telefunken Elektroakustik, and on her weekends, she is a technical specialist at Apple retail. After completing her undergraduate degree, Zoë would like to go into the world of audio electronics, one day being able to make pro audio more accessible to those who are not well off.

Zoë James

Note

1.   In 1978, musician Todd Rundgren stated, "Computers are coming on so heavy that nobody's going to bother with this intermediate technology." . . . "The economic structure will shift itself. You'll no longer go out and buy permanently recorded things, because, eventually, they do one of two things – they wear out, or you wear out" (Flanagan, 2015).

Bibliography

About Roger Linn Design. *Roger Linn Design*. c. 2011 [Online]. Available: www.rogerlinndesign.com/about.html [Accessed June 25, 2019].
Birr, Dawn. 2019. Interview with Dawn Birr. *In:* Gaston-Bird, L. (ed.).

Bosi, Marina. 1987. *Progettazione e Simulazione dell'Unita' di Controllo di un Calcolatore Superveloce per il Trattamento Numerico del Segnale Musicale*. Tesi di Laurea in Fisica, Universita' degli Studi di Firenze.

Bosi, Marina. 1988. An Interactive Real-Time System for the Control of Sound Localization. *1988 Seamus Conference.*

Bosi, Marina. 2019. Interview with Marina Bosi. *In:* Gaston-Bird, L. (ed.).

Bousquet. 1995. *Women in Audio: Project 2000*. 99th Convention of the Audio Engineering Society, October. New York: Audio Engineering Society.

Bousquet, Carol. 2019. Interview with Carol Bousquet. *In:* Gaston-Bird, L. (ed.).

Chanel, Elisabeth. 2013. *Jurassic Fight: The Steal of the Century*.

Female Pioneers in Audio Engineering [Online]. 2017. Intelligent Sound Engineering. Available: https://intelligentsoundengineering.wordpress.com/2017/08/07/female-pioneers-in-audio-engineering/ [Accessed June 6, 2019].

Flanagan, Andrew. 2015. Todd Rundgren Predicted Music Streaming 37 Years Ago. Billboard [Online]. Available: https://www.billboard.com/articles/business/6575914/todd-rundgren-predicted-music-streaming-37-years-ago.

Gaston-Bird, Leslie. 2017. 31 Women in Audio: EveAnna Manley. *31 Women in Audio* [Online]. Available: http://mixmessiahproductions.blogspot.com/2017/03/31-women-in-audio-eveanna-manley.html [Accessed March 3, 2017].

Kellner, Charlie. 1980. *The Alphasyntauri: A Keyboard Based Digital Playing and Recording System with a Microcomputer INterface*. 67th Convention of the Audio Engineering Society. New York: Audio Engineering Society.

Killick, Cynthia. 2018a. *'Life on the Run', by Cynthia Killick* [Online]. Available: https://marielouisekillick.org.uk/life-on-the-run/ [Accessed June 20, 2019].

Killick, Cynthia. 2018b. A Sound Revolution. *Lulu.com.*

Laird, James. Pink Floyd: Their Mortal Remains Exhibition at the V&A – Review. *Trusted Reviews* [Online]. Available: www.trustedreviews.com/opinion/pink-floyd-their-mortal-remains-exhibition-review-v-a-museum-2948237.

Löchen, Elisabeth. (2019.) Interview with Elisabeth Löchen. *In:* Gaston-Bird, L. (ed.).

Manley.com. "History". [Online]. Available: www.manley.com/history [Accessed November 7, 2019].

Manley Labs Wins 2016 TEC Award for Outstanding Technical Achievement. 2016. *Manley Labs* [Online]. Available: www.manley.com/news/2016/2/1/manley-labs-wins-2016-tec-award-for-outstanding-technical-achievements.

Marielouisekillick.org. 2018. *Timeline* [Online]. Available: https://marielouisekillick.org.uk/life-on-the-run/ [Accessed June 20, 2019].

Mathworks. 2019. Design an Audio Plugin. *The MathWorks, Inc* [Online]. Available: https://uk.mathworks.com/help/audio/gs/design-an-audio-plugin.html [Accessed June 26, 2019].

Mikelbank, Peter. The Dino That Roared; Small French Company Charges DTS Infringes on Its Patent. *Variety.*

Parisi, Paula French Firm Enters Digital Audio Derby. *The Hollywood Reporter.*

Pendleton, Jennifer. 1994. DTS Engaged in a Patent Fight with French Rival: Entertainment: The Dispute Centers on New Movie Theater Sound Systems Developed by the Westlake Village Digital Firm and L. C. Concept. *Los Angeles Times*, February 22.

Potter, Samantha. Vanessa Mering – Marketing Manager at HARMAN Professional [Online]. Available: https://soundgirls.org/vanessa-mering/ [Accessed September 27, 2019].

Reynolds, S. 2012. Resident Visitor: Laurie Spiegel's Machine Music.

Sato, Erisa. (2018). Interview with Erisa Sato. *In:* Gaston-Bird, L. (ed.).

Scaletti, Carla. 2017. ICAD 2017 Keynote Speaker: Carla Scaletti. 2017 ICAD Conference. The Pennsylvania State University. *YouTube.com.*

Scaletti, Carla. 2019. Interview with Carla Scaletti. *In:* Gaston-Bird, L. (ed.).

Sennheiser Wins Audio Installation of the Year Award. 2013. *Access All Areas* [Online]. Available: https://accessaa.co.uk/sennheiser-wins-audio-installation-of-the-year-award/ [Accessed September 27, 2019].

Spiegel, Laurie. 1981. Music and Media. *Kalvos & Damian!, Chronicle of the NonPop Revolution* [Online]. Available: http://kalvos.org/spieess1.html.

Spiegel, Laurie. 1986. Music Mouse – An Intelligent Instrument User Manual. 1988 ed. [Online]. Available: http://retiary.org/ls/progs/mm_manual/mouse_manual.html [Accessed September 27, 2019].

Spiegel, Laurie. 2012. *The Expanding Universe: 1970s Computer Music from Bell Labs by Laurie Spiegel* [Online]. Available: http://retiary.org/ls/expanding_universe/.

Spiegel, Laurie. 2019. Interview with Laurie Spiegel. *In:* Gaston-Bird, L. (ed.).

Stack, Robert. 1999. EveAnna Manley Acquires Manley Laboratories. *Pro Sound News Europe.*

Syntauri Corp alphaSyntauri Digital Synthsizer. c. 2017. *Encyclotronic Electronic Music Archive* [Online]. Available: https://encyclotronic.com/synthesizers/misc-instruments/syntauri-corp-alphasyntauri-digital-synthsizer-r1240/.

Syntauri Corp alphaSyntauri Digital Synthsizer. Symposium on Computers in the Arts, 1981 Philadelphia, PA, Los Angeles, CA, Piscataway, NJ, New York: Institute of Electrical and Electronics Engineers, Philadelphia Section & IEEE Computer Society.

Turl, Ben. 2014. Women in the Musical Instruments Industry: An interview with Ingrid Linn. *KVR* [Online]. Available: www.kvraudio.com/aboutkvr.php.

Woman Inventor Wins: Briton May Get 14 Million for Her Phonograph Stylus. 1958. *New York Times.*

7 Acoustics and Design

Careers in Acoustics

"Acoustics is a study, but there's a wide range of how you use that knowledge," says acoustician **Samantha Weller**. "There are acoustic scientists in the military; at Virginia Tech there were studies in acoustics for Boeing airplanes and jet propulsion; then there's underwater acoustics. You could possibly find a career anywhere you see fit if it's a passion. "Depending on where you are in the country there are certain opportunities that are available through internships. Acoustics comes into play a lot with architecture firms . . . there are a couple in Los Angeles, then there's big firms like Arup which is a huge engineering firm all across the country and world, and they have an acoustics division that does internships. So, it's really what type of acoustics you want to focus on. If it's theater acoustics then there are smaller firms that have internships, too" (Weller, 2019). Other careers in acoustics include:

▶ **Architectural acoustics:** Includes the design of concert halls, movie theaters, recording studios, even hospitals and other spaces.

▶ **Audio engineering:** Designing mobile devices or emergency alert systems.

▶ **Environmental noise:** Evaluating and minimizing noise in residential areas from roads and railways.

▶ **Bioacoustics:** Examining how man-made noise affects animals in nature.

▶ **Product design:** Making machines such as vacuum cleaners or car engines quieter.

▶ **Musical acoustics:** The study of musical instruments, or even music therapy.

▶ **Speech and hearing:** processing speech, synthetic speech, and analyzing speech for forensics purposes.

▶ **Ultrasound:** From taking images of a fetus in the womb to looking at cracks in airplanes or submarines, even destroying tumors or kidney stones.

▶ **Seismology:** the study of structures beneath the earth's surface and earthquakes.

▶ **Underwater acoustics:** The use of signals to locate fish, determine shipping routes, look at geological formations under water, or search for oil (*Where do acousticians work?*)

Tools of the Trade: Acoustics

Pen, Paper, and Tape Measure: "Distances and volume are important. Sometimes, an acoustician has been asked by a client to solve a problem, or if it's a new space, it's important to ask what the functionality of the space will be" (Weller, 2019).

Ears: Jamie Angus says, "Ears and a clicker, and an understanding of how sound works in rooms . . . going in and listening to the space, thinking about what you're trying to do, and having an understanding of what happens when sound bounces off surfaces" (Angus, 2019).

Information: "Getting information, learning it, linking it together" (Ibid.). For example, how will the space be used? What problems currently exist? Are there special conditions during which problems are revealed (time of day or traffic patterns)? Do people have differing expectations for the use of the space (lectures, rock bands, or chamber music)?

Understanding: "Understanding is a tool – it's not the same as *knowing the theory*, but *understanding* what's going on, which you might be able to do without having hard sums (equations and data). That's true for acoustics, the mix, and how you keep everything clear in the mix. How do you make sure the brass doesn't cut across the vocalist? This is about *understanding* what happens'" (Ibid.).

Software: Samantha Weller gives an overview of the tools used in her practice: "We're using CAD (Computer Aided Design), we're using Revit (building information modelling software), and for larger projects where we're designing things for theaters, we use ODEON (acoustical modeling software) and with that we use MATLAB. We also use RHINO a lot. Those are for modeling and **ray tracing** to see where sound is going to bounce off certain things. There are also mechanical systems to consider: our biggest worry. We use a project called AIM (Acoustic Information Model), which has an HVAC mechanical system calculator" (Weller, 2019).

A software package usually contains a suite of tools that include the following:

- Real-time analyzer: This is a device that can measure the relative level of signals across a spectrum of frequencies.
- Sound Pressure Level (SPL) Meter: A Sound Pressure Level meter is used to measure loudness or "volume." A commonly recommended and affordable SPL meter was made by Radio Shack, but now there are apps for mobile devices and phones that can be used, although readers are cautioned to make sure the microphone is calibrated!
- Impulse Response: An impulse response is a burst of sound, usually made by a cap gun, hand clap, or by popping a balloon. This burst of energy will contain all the frequencies of the audible spectrum and is seen as a short burst when looking at a waveform. From that impulse response, information about the decay time in the room can be calculated.
- Pink Noise: This is a signal of random noise, where each octave band of noise has equal energy. It is usually used for measuring auditoriums, lecture halls, and studios because it more closely approximates how humans hear.
- White Noise: This is a signal of random noise, where all frequencies in the audible spectrum have equal energy. It is usually used for measuring equipment to give a true frequency response that the equipment is meant to deliver. For example, an amplifier is measured using white noise, and its specifications might say, "flat frequency response from 20 Hz to 20 kHz." An engineer calibrating a studio or auditorium will use pink noise to measure acoustic sound as it is transmitted from that amplifier through loudspeakers.
- Sweep: This is a sine wave that "sweeps" through the audible spectrum.
- Phase aligned: This phrase refers to the phase of signals as they leave a loudspeaker. Because of the way low frequencies are emitted from the woofer, the time it takes to arrive

at the listener's ear may be different than the sound from the tweeter (high frequencies). This time is also affected by inductors and capacitors in the amplifier circuitry. For example, a snare drum, which contains both low and high frequencies, will sound "smeared" if the loudspeakers are not phase-aligned.

- Time aligned: Time alignment has a slightly different meaning in acoustics than it does for music. (One might "time-align" a house pair of mics with spot mics on the stage.) When an array of speakers is used in a large space, the timing of those speakers is aligned so that audience members further back from the stage do not hear an echo from the sound sources (voice or music) from the stage; the speakers should be delayed so that sound from the stage has time to travel to the people further away and more or less line up so that all of the sound arrives to the listener at the same time.

Gaffer tape: One can never have enough gaffer tape.

Fun Facts: Weighting Curves

What does it mean when someone takes an "A-Weighted" or "C-Weighted" measurement? Simply put, the curves describe a filter that is used when an acoustic sound is measured. Because of the way humans hear, especially the frequencies around speech, an A-Weighted curve filters out the frequencies (see figure below) that are not as important to intelligibility of speech or background noise. This might be important for a classroom, for example, so the lecturer can be clearly understood. C-Weighted curves are more accurate for calibrating studio monitors for music, television, and film mixing since the bass frequencies for music and effects at louder volumes are just as important as speech.

FIGURE 7.1   A-, B-, C-, and D-weighting curves

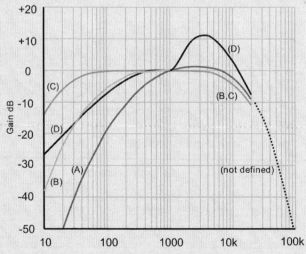

A-weighting (blue), B (yellow), C (red), and D-weighting (blk)

Source: Created by lindosland using OpenOffice Draw

▉ Jamie Angus-Whiteoak (England)

When **Jamie Angus-Whiteoak** was 11 years old, she went on a field trip to broadcast station WOR in New York with her school. "I was blown away by all the stuff. One of the engineers showed me the transmitter racks. When they interviewed me for a radio show, they said, 'So what do you want to be when you grow up, honey?' I said, 'I want to be a radio engineer!'" (Angus, 2017).

As a child, she used to love taking apart old radios and televisions and tried making her own speakers. "They didn't sound very good as you might expect, but they still sounded better than the ones on my parents' hi-fi. So I thought, 'I need to learn about this'. I read magazines and books that I could get a hold of, and I built more loudspeakers. When I was 17, I built a set of loudspeakers – that was when I'd sort of sussed the design down a bit more – and I built 4 so that I could have a quadrophonic system, which was kind of freaky."

When it was time to return to the UK and attend university, a couple of her friends asked if they could buy the speakers. Not knowing how much to ask, she asked her friends what they would pay, and the friends suggested $400 per pair. "I thought, 'Oh my God, *eight hundred dollars!*' So, I used that to buy a Nackamichi cassette recorder and things. I already liked really good quality gear" (Ibid.).

Jamie went on to study electronics and experiment with circuits and amplifiers. She studied for and got her first degree around the time the microprocessor was invented. "I decided to do a PhD thesis on designing a DSP processor. I thought that that would force me to have to learn microprocessors and DSP at the same time! I was in a Comms (Communications) department, so I also learned about spread spectrum systems and error correcting coding. It was a very stimulating environment.

"Then I worked at Standard Telecoms Labs, which no one's ever heard of, but they invented optical fibers and PCM (pulse code modulation). I ended up working in the speech coding processing group for telephone terminals. They had an anechoic room so I could test my speakers out and learn some acoustic measurements, as well as do speech coding, which links with audio coding" (Ibid.).

When Jamie had returned to the UK to go to secondary school, she was upset that she was not allowed to do music as a subject because she was doing sciences. "That angered me so much, as it did one of my colleagues (Ross Kirk) in the York University Electronics department who also had the same problem at school – he was a bass clarinetist. So, we created the UK's first Music Tech course in 1986 with Dave Malham and Richard Orton." Twenty-one years later in 2007, her son got his music tech A-level. "I was chuffed . . . I thought, 'Hey I finally got my revenge!'" (Gaston-Bird, 2017).

Some of Jamie's work focuses on the principles of diffusion and diffusors, which very simply put is the attempt to create uniformity in an acoustical space so that it sounds the same no matter where you are in the room. You may be familiar with the diffusors you see on the rear wall of a modern recording studio: a large square of light wood with many small squares, or strips, within which such squares and strips are raised at varying heights. "I realized that there was only a certain size that you could lift and put on a wall. So people put lots of them together – that wasn't a good idea," Jamie explains. "And because of the **spread spectrum** thing I knew I could maybe make it better by modulating it. So I tried it out theoretically using Excel. I remember the first time the results came out; I did a little jig down the corridor saying 'Woo-hoo!' What I thought would happen actually happened!" (Ibid.). With this revelation, she invented modulated and absorption diffuser structures. In her paper, "Using grating modulation to achieve wideband large area diffusers," she explains that narrow diffusion patterns can be improved using a mathematically derived sequence of "modulated" ("changing") structures, with the effect of achieving good diffusion using smaller diffusers to cover a wide area (Angus, 2000).

Jamie also worked with other researchers on using cylindrical diffusors in a space, and the results are published in the Acoustical Society of America with her research team as "Volumetric diffusers: Pseudorandom cylinder arrays on a periodic lattice" and "Diffusive benefits of cylinders in front of a Schroeder diffuser" (see: "Recommended Reading").

She has also done research into "Green amplification." "Maybe the switching (amplifiers) aren't as efficient as we think. And you can make a linear amp *more* efficient or *as* efficient. . . . Every home has amps and they're all consuming power, so if you can make it better you can save a little bit of energy worldwide" (Ibid.).

Jamie's long and varied career was recognized with the AES Silver Medal in 2019, making her the first woman to have earned the prestigious award. During her career,

Jamie Angus-Whiteoak

she has also invented direct processing of Super Audio CD signals, one of the first four-channel digital tape recorders, and has worked on signal processing, analogue circuits, and numerous other audio technology topics. She is an AES fellow and recipient of the IOA Peter Barnet Memorial prize for her contributions to audio, acoustics, and education.

Fun Facts: Spread Spectrum

Actress Hedy Kiesler Markey and pianist George Antheil filed patent US2292387A with the title "Secret communication system." Markey was also known by her stage name, Hedy Lamarr. In the article, "A short history of spread spectrum," the author writes that the patent was "a system for radio control of torpedoes. The idea was not new, but Lamarr's concept of frequency hopping to prevent the intended target from jamming the controller's transmissions was. Antheil chipped in with a technique for synchronizing the frequency hopping between transmitter (in a plane flying above the torpedo) and the weapon itself. It's probably no coincidence given Antheil's musical training that the system used 88 different carrier frequencies, equal to the number of keys on a piano" (Semiconductor, 2012).

The technique is known as FHSS: "Frequency Hopping Spread Spectrum" and is used by Bluetooth devices today. The idea that is when interference is detected on a channel, the device moves to another channel, or "hops frequencies" (Ibid.).

Jamie Angus-Whiteoak, who invented the modulated and binary absorption diffuser structures, summarizes how spread spectrum is used to improve the performance of diffusers:

1) "Most surfaces look like gloss paint to sound waves and so have the acoustic equivalent of 'glare' like you get from polished surfaces and mirrors.
2) Acoustic diffusers act like 'Flat' or 'Matte' paints and remove the 'Glare'.
3) For practical reasons, the size of these diffusers is limited to about 2 foot (60 cm) square, otherwise they become too heavy to lift and/or awkward to install.
4) This means that to cover a reasonable area you have to tile them and use multiple diffusers to cover a large area.
5) If they are all the same this results in the reflections of these copies being correlated and they reinforce each other, which means the wall of diffusers looks more like a multifaceted mirror, not quite what you want.
6) By modulating the diffusers with a pseudo random binary sequence you can break up the regularity and get back to the ideal 'Matte' reflecting surface.
7) Modulating in this context means choosing the diffuser that is used at a particular spot depending on whether the random sequence is a 1 or a 0. The difference may be by using a different diffuser, or by using an asymmetric one the other way up for the '1s' compared to the '0s'
8) This process is identical to what spread spectrum systems do which is to use a pseudo-random sequence to spread the signal in frequency. For diffusers the pseudo-random sequence spreads the diffusion in angle" (Angus-Whiteoak, 2019).

FIGURE 7.2 Patent US2292387A "Secret communication system"

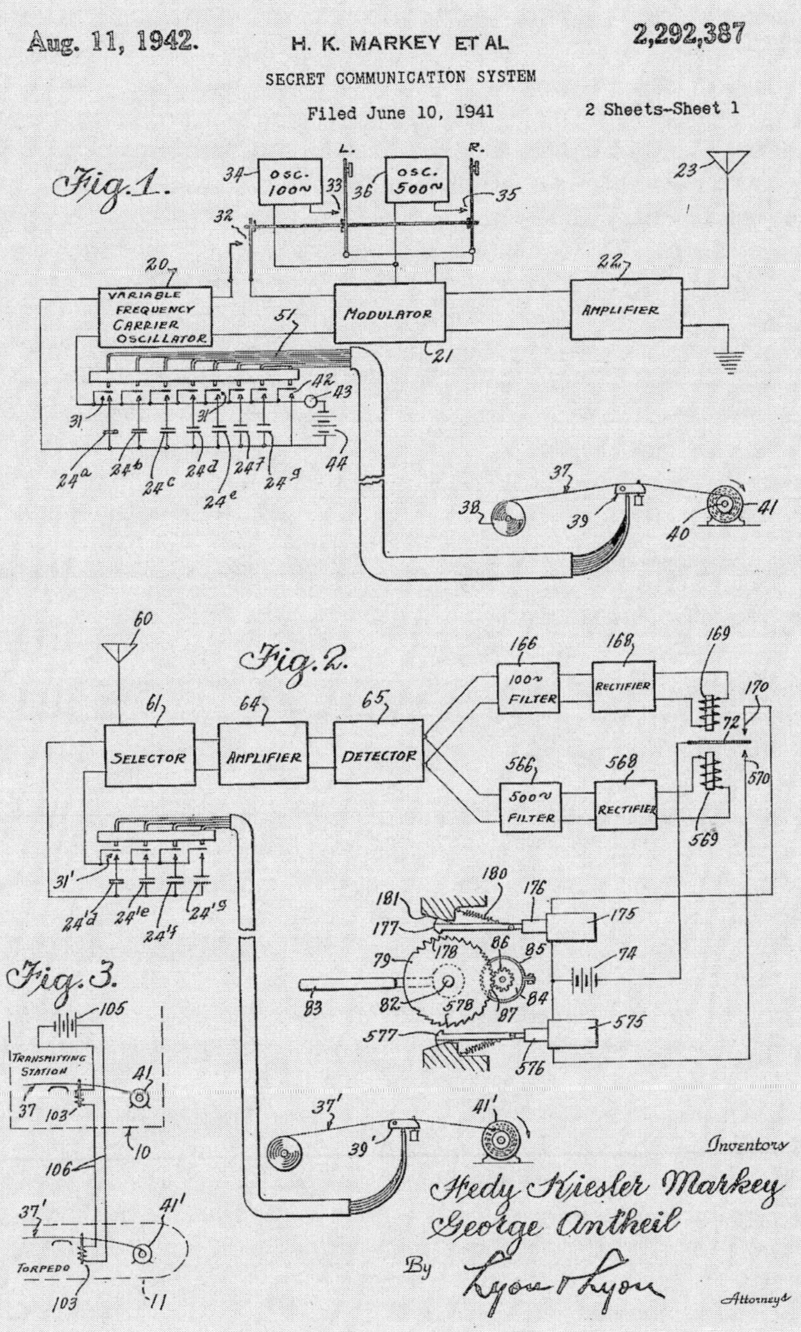

▓ Elizabeth Cohen (USA)

"If I was going to write a prescription for someone I'd say, 'live music, live music, live music . . . and *multiple genres* of live music to develop your joy, and your critical listening skills" (Cohen).

Dr. **Elizabeth Cohen's** advice carries the combined weight of her distinguished background as an audio engineer and her several leadership roles in preeminent organizations. Her prescription to be immersed in music lets us, as practitioners, be empowered by the fact that we become better engineers by partaking in the music we love, not just as passive listeners, but with an attentive ear (Ibid.).

The use of the word "immersed" is no accident, as Dr. Cohen's comment was in the context of trends in immersive audio. Fortunately, today's technology is affordable. Together with the emergence of collaborative tools, it's an exciting time for audio, she says. "I think that the delivery of tools into more and more hands at better price points . . . this whole idea of inclusion of multiple perspectives and multiple points of view is a really exciting thing. These things allow you to do things in such an easy way. I think this whole idea of collaboration – so that money goes to the hands of artists – that to me is really cool. I like the idea that artists can survive. They have multiple options for reaching people and working with each other" (Ibid.).

Her work on real-time collaboration earned her the AES Citation Award for "pioneering the technology enabling collaborative multichannel performance over the broadband internet" (Society). "The essence of music is shared communication. And that depends on instantaneous feedback," she says in an article that discusses the need for instantaneous collaboration for collaborative performance over the internet (Krieger, 2014).

Dr. Cohen holds a master of science in electrical engineering and a PhD in acoustics from Stanford, where she has been on the faculty since 1980. She was the first woman to be elected president of the Audio Engineering Society in 1995 (in its 47-year-old history at the time). She spent time at Bell Labs, starting in their internship program. "It was like Christmas every day," she recalls. "The engineers were happy to answer your questions. And as an intern you rotated through all of the labs. It paid off because gravity wave antennas at that time were just big tuning forks. And I knew how to work with tuning forks and make measurements on tuning forks . . . and I got to use 471 op amps to build cellos for Max [Matthews'] original design. They needed a set of those at Stanford. So that was my next job: building cellos" (Cohen).

She went on to other projects and later earned acclaim for her acoustical design of the Joan and Irving Harris Concert Hall in Aspen, Colorado, in 1993. She has followed the evolution of acoustics technology from huge oscillators ("that were the size of your face," she laughs) and getting measurements with a pop gun to SMAART Live (inventor Sam Berkow was one of her protégés) and the contributions of Brüel & Kjær in the area of measurements. Now, small studio owners can get self-calibrating systems, but she has a caveat: "It requires human attention to detail . . . if you don't know where to place your microphones. . . 'garbage in, garbage out'. Acoustics is a science but there's also art, and a subset of that is good taste" (Ibid.).

Her other accomplishments include, in part:

* Science and Engineering Fellow to the White House Economic Council, in charge of Arts and Information Infrastructure Initiatives

- Recipient of the Women in Music Touchstone Award
- Fellow of the Acoustical Society of America
- Fellow of the Audio Engineering Society
- Member of the Academy of Motion Picture Arts and Sciences
- Member of the Moving Image Archive Studies Program
- Member of SMPTE Ad Hoc PostProduction Committee
- Member of the Sound Preservation Board of the Library of Congress
- Vice Chair of Education for the AMPAS Science and Technology Council

(ICAD, 2017)

For aspiring, young engineers, she recommends electrical engineering, "and you really need signal processing . . . and you better be a damn good musician. You can't get away without the ear training." There are several critical listening and ear training resources available, including books, CDs, and online resources (see: Recommended Reading). She encourages students to learn about "narrative." "If you can't tell a story then you're not going to do very well. Especially in songs, you better know how to deliver that and understand the impact of phrasing, and what the artist you're working with really needs. You really need to have skills: chops in an interdisciplinary way. You need musicianship, which means you can listen critically and verbalize what you're hearing. And you need the engineering skills. And this is especially true for women: the need for competency is even greater. Things haven't changed enough . . . yet" (Krieger, 2014).

Cohen is actively involved with helping to build music and new media programs and is engaged with mentoring others. She is especially proud of her former teaching assistant: encoding pioneer, author, and AES past president Marina Bosi; and Qianbaihui Yang, an honorary AMPAS sci-tech Intern, who is now at Skywalker Sound. Inclusivity, she says, takes a systematic approach, by building up a critical mass of skilled engineers. "Being as good is not good enough. You have to be better. And so the real key is building critical mass. So you do things like make sure your grad students are placed in great jobs; or that you have women who follow you, whether it's as AES President or board members or whatever the case may be; or that you hire them for design teams. That's the only way because you don't get equity without critical mass, and one is not looking only for *equality* one is looking for *equity*, too" (Ibid.).

Samantha Weller (USA)

"You can learn a lot from a room by clapping," says **Samantha Weller**, acoustical consultant at Threshold Acoustics (Weller, 2019). In fact, many audio engineers, when walking into a new space, may clap their hands. This creates a short burst of sound followed by a decay of sound. With a keen ear, the engineer can get a sense of the reverb time in the room and whether there are anomalies like early reflections or flutter echo.

Weller focuses on theater acoustics. She began her studies at the University of Cincinnati Conservatory of Music and studied technical design and production. She also had a passion for moving loudspeakers around to get the best sound. "I was more interested in placing loudspeakers than the music coming out of them. I liked hearing how they worked in a room." Her mentor, Dr. Tim Ryan, encouraged her to consider a graduate degree in architectural acoustics. She enrolled in Virginia Tech and earned her master of architecture degree with an emphasis in architectural acoustics, a strategic decision she made based on talking to other architects. "It would be easier for me to get an architect's license," she explains. "I went a different route. It's the architectural acoustician that does AV Integration. But many of my co-workers interested in electrical engineering and mechanical engineer have degrees in acoustics" (Ibid.).

The highlighted projects in her portfolio include acoustic design for the Crested Butte Center for the Arts in Colorado, USA; Northwestern University's Kellogg School of Management; the Rubenstein Forum at the University of Chicago; and the Chamberlain Group Headquarters.

Wanting to know about how far my own high-school level knowledge of trigonometry could take me, I asked Weller about the mathematics knowledge required for the job. "Well, obviously being an architect, I love math," she confirms. "There's lots of trigonometry but the computer does handle the equations. There are some people who are good in MATLAB but that's the great thing about the company I work for now. All of us are expert in something and others are still learning. With an open office, if I have any questions there's always someone who I can ask who might have 10 more years of experience or be more knowledgeable than me . . . and vice versa."

Fun Facts: Ray Tracing

Curious to learn more about **Carol Bird** and Manfred Schroeder's contribution to the field, **Samantha Weller** takes a moment to explain for me the concept of **ray tracing**. "Ray tracing is physically looking at what sound could possibly do in a room. You would draw a point which is your source, and then from there you could draw a line from that source going out to hitting walls, ceilings and other obstacles in its way. From there, once it hits something you would take the angle it hits, mirror that angle, and that's how it will reflect off that surface. It's a rough estimate: sound is like water and it's going to go everywhere, but we can at least kind of predict how it's going to behave by what we call ray tracing and that's the path that line makes. And that's really to make sure that when we add diffusion it works to our benefit and we are not creating weird echoes and not focusing the sound where we don't want it to go" (Weller, 2019).

Orla Murphy (Ireland)

Orla Murphy isn't a "car person," but she found the task of designing audio for the acoustic environment of the automobile cabin to be exciting. "It's moving at 100 kilometers per hour down a bumpy, noisy road, but you want to make the music listening experience as good as possible, so I think it's an interesting engineering challenge. The primary purpose of a car is to get from point A to point B, but actually *music in the car* is the one enjoyment feature that makes long journeys worthwhile" (Murphy, 2019).

Her childhood experience playing viola with the National Youth Orchestra of Ireland and her love of math, science, and physics were the foundation for her university pursuits:

Orla Murphy of Jaguar Land Rover sets up a mic array in an automobile cabin

Murphy obtained a master of engineering degree (MEng) in "Electronics with Music," which is a joint degree between engineering and music at the University of Glasgow. As part of that degree she did a six-month internship with Jaguar Land Rover. "The reason why I applied was that I couldn't really pick if I wanted an electronics engineering job or music studio kind of role. I wanted a 50/50 mix, and thought if I worked for an automotive company on the sound team it would be good because I would be an engineer but have some specialism in the sound and music field" (Ibid.).

After she graduated, she returned to Jaguar Land Rover where she worked as an audio EQ (equalisation) engineer for four years. She currently works there as an engineering quality transformation manager. "I liked the idea of having a product that goes into another product . . . something you could buy, something more tangible" (Ibid.).

The entire design team has to negotiate where features go in the car – everything from speakers to mirrors, lights, storage, and dashboard features. Once that's done, Murphy's team comes in to calibrate and optimize the audio playback system. "We had specialist mic equipment and software by B&K (Brüel & Kjær) to measure the EQ within the vehicle. Our team decided the EQ for the speakers and how they combine and how the equalization changes for different road speeds; we also have inputs from mics in the car to do noise cancellation. All of that was with proprietary software. That way we could calibrate and control every speaker while we were testing. Ultimately, we are controlling the volume, phase and most importantly the EQ of each speaker and tweaking 24 to 28 speakers in the car as well as multiple mic inputs. How you combine all of those together is what calibration is all about" (Ibid.).

Critical listening is an important part of Murphy's work. Although the job entails specialist equipment and tons of engineering, at the end of the day it is your ears that tell you whether the computer's answers are correct and/or pleasing. "We did a lot of critical listening: the tools, the measurement, and the process of tweaking – all of that you can follow to a 'T,' but the car will only be 70% as good as it's going to be with that. The final bit is critical listening. The graph says one thing, but your ears say another as you sit and drive and move your head around. I was good at the first section of being methodical and determining use cases, but the second part took experience. That first stage is easy to learn. But then you're biased because you think, 'this is as good as I could get it'. Getting it to 100% means doing tests with real listeners, using fresh ears, and knowing how to tweak the sound" (Ibid.).

The audio team at Jaguar Land Rover have implemented 3D sound. Most of us are familiar with 2D sound in the car, with speakers in the doors, window frame, or dashboard. "The first implementation of 3D surround is putting speakers in the ceiling because it raises up the sound stage (for example, if you are in a concert hall, you'll notice that sound is coming from the ceiling). Delaying sound to the raised speakers also gives the illusion of space – people said the car felt bigger" (Ibid.).

Murphy has won several awards, including "Design Engineer of the Year" from the British Engineering Excellence Awards; Institute of Engineering and Technology (IET) Young Woman Engineer of the Year Award; and the Royal Academy of Engineering's "Engineers Trust Young Engineer of the Year," among others. She is also a STEM ambassador with the IET.

Fun Facts: Ear Training and Critical Listening

If you can go from saying "that sounds muddy" to summarizing "I need a cut at 200 Hz;" or if you can evolve from saying "that sounds dull" to knowing you need a slight boost between 8 to 10 kHz, then you have successfully trained your ears.

There are a few books with accompanying CDs as well as online training programs that will help you accomplish this, which you can find in the "Recommended Reading" section.

You can also find a partner and make your own game. In the "EQ game" (an activity I learned during my undergraduate course in audio technology at Indiana University), you make boosts or cuts to a band of frequencies while your partner closes their eyes. Then your partner sits at the listening position and "A-Bs" (compares) the two settings (with and without the changes) while music plays and tries to guess what changes were made. You could also play this game making changes to compression settings or delays.

You'll find that by doing this for several weeks, your way of listening to audio changes. You might even get a little faster at mixing by knowing which parameters you want to change before you reach for the controls.

◼ Linda Gedemer (USA)

In 1983, **Linda Gedemer** had just graduated high school and was offered scholarships to two colleges: Peabody/Johns Hopkins and the University of Miami. "Guess which one I chose," she says, laughing. She opted for Miami not only because of the warm weather, but because the program there had been established for some time (see Chapter 9 for a contextualization of audio programs in the early 1980s). Miami's vibrant music scene was so inspiring that when Gedemer graduated with bachelor's degrees in music engineering and electrical engineering, she wanted to travel the world and do live sound for bands. However, "when it came time for me to graduate, I called various companies around Miami and outside of Miami to try to secure an internship and was basically told, 'We don't hire women'. Flat out. Dreams dashed. But I had to do an internship, it was part of my requirement" (Gedemer, 2019).

She decided to return to her hometown of Chicago to fulfill her internship requirement but was faced with an environment of "toxic masculinity" at the recording studio at which she worked. Never intending on staying in Chicago permanently, she set her sights on Los Angeles and moved there. She found a job as a sales engineer for an amplifier company and ultimately moved on to a design/build AV contracting company that was hired to work on the emergency paging systems for the new LA metro rail stations. "That's how I went from music engineer, to audio engineer, to audio/visual control systems engineer. Along the way I had to learn about video and control systems. Then I got hired by Disney Imagineering as an audio/visual engineer and went to France to work on Disneyland Paris" (Ibid.).

In 2008, she decided to pursue a master's of science in acoustics at Rensselaer Polytechnic Institute in New York, followed by a PhD in audio and acoustical Engineering at the University of Salford in 2016. She thus became an audio/visual control systems and acoustics consultant. "I would say my main tools are AutoCAD and AFMG's EASE and EASERA software." She also uses CATT-Acoustic and Immersive Designer Pro for video projection (Ibid.).

Since 2016, she has been technical director and VR audio evangelist for Source Sound VR. That same year, she served as co-chair for the AES first Audio for Virtual and Augmented Reality (AVAR) conference in Los Angeles, along with Andreas Mayo. "I had been going to VR conferences and no one was speaking about audio, and that was the catalyst where Andreas and I had a fateful conversation: we had to do a VR/AR conference for our community." The conference was a great success: over 400 people attended the event (Ibid.).

Shortly afterwards, the AES followed up its success with another AVAR event in 2018 at the DigiPen Institute of Technology in Redmond, Washington. Gedemer again served as co-chair with Lawrence Schwedler and Edgar Choueiri (chaired by Matt Klassen). She was recognized for her work with the AES Board of Governor's Award. In addition, Gedemer holds an AES Citation Award and MPSE Golden Reel award (Ibid.).

Linda is currently a principal at the engineering consulting firm of Alfa Tech, where she heads expansion efforts in AV and acoustics.

Women in Acoustics Committee

The Women in Acoustics Committee, a subgroup of the Acoustical Society of America, has a website at womeninacoustics.org. They were established in 1995 with a mission to encourage the recruitment and retention of women in the profession and to provide networking opportunities and mentoring, among other functions. Additionally they list women who have earned medals from the ASA and who have been honored at luncheons, held twice a year.

Bibliography

AES Citation Award: Elizabeth Cohen [Online]. Available: www.aes.org/awards/?ID=1809 [Accessed July 10, 2019].

Angus, Jamie. 2000. Using Grating Modulation to Achieve Wideband Large Area Diffusers. *Applied Acoustics*, 60, 143–165.

Angus, Jamie. 2017. Interview with Jamie Angus. *In:* Gaston-Bird, L. (ed.).

Angus, Jamie. 2019. Interview with Jamie Angus. *In:* Gaston-Bird, L. (ed.).

Angus-Whiteoak, Jamie. 2019. . . . How is spread spectrum used in acoustics? *In:* Gaston-Bird, L. (ed.), *Forty Words or Less: How Is Spread Spectrum Used in Acoustics?* Facebook.com, Official Audio Engineering Society Discussion Group

Cohen, Elizabeth. 2019. Interview with Elizabeth Cohen. *In:* Gaston-Bird, L. (ed.).

Gaston-Bird, Leslie. 2017. 31 Women in Audio: Jamie Angus. *31 Women in Audio* [Online]. Available: http://mixmessiahproductions.blogspot.com/2017/03/31-days-of-women-in-audio-jamie-angus.html [Accessed March 7, 2017].

Gedemer, Linda. 2019. Interview with Linda Gedemer. *In:* Gaston-Bird, L. (ed.).

ICAD. 2017. *Keynote Speakers* [Online]. International Community for Auditory Display. Available: http://icad.org/icad2017/program-2/keynote.html [Accessed July 10, 2019].

Krieger, Lisa. 2014. *Ambitious Music Project Could Result in Super-Fast Internet Speeds.* Available: www.thestar.com.my/lifestyle/people/2014/12/10/music-at-the-speed-of-light/ [Accessed June 20, 2019].

Murphy, Orla. 2019. Interview with Orla Murphy Leslie Gaston Bird. *In:* Gaston-Bird, L. (ed.).

Semiconductor. 2012. A Short History of Spread Spectrum. *EE Times* [Online]. Available: https://www.eetimes.com/document.asp?doc_id=1279374 [Accessed June 20, 2019].

Weller, Samantha. 2019. Interview with Samantha Weller. *In:* Gaston-Bird, L. (ed.).

Where Do Acousticians Work? [Online]. Institute of Acoustics. Available: www.ioa.org.uk/careers/where-do-acousticians-work [Accessed July 10, 2019].

8

Live and Theater Sound

Careers in Live Sound

Front of House (FOH) Engineer: Responsible for mixing the sound that the audience hears at a concert, theatrical play, or musical. This can include choosing mics, dialing in equalization, adding reverb or delay (sometimes to match musical cues), as well as diagnosing problems such as feedback. Also responsible for monitoring safe listening levels (SoundGirls.org).

Monitors/Monitor Engineering: The monitor engineer mixes sound heard by the band on stage. If the drummer needs to hear the guitar in her monitor, she will let the monitor engineer know. During a show, this can be communicated with hand signals. The monitor engineer also makes sure to "ring out" the system so that microphones do not cause feedback when placed too close to a monitor. Monitors can be "wedges" on the floor or In-Ear Monitors (IEMs). Another crucial skill is a cool head – when musicians are struggling to hear each other in stressful difficult conditions where an audience is watching your every move, it will be up to the monitor engineer to keep things running smoothly (SoundGirls.org).

Sound Technician(s): This is sometimes handled by one person or more than one person on larger productions.

▶ Sound Crew Chief: Responsible for the sound crew, call times, setup and load out, logistics, managing the local crew, and other audio needs.

▶ System Engineer: Installs the sound system and setting EQ and time alignment; assists in designing the sound system for the event or tour. System engineers should have solid technical education and hands-on experience and knowledge. There is a high demand for experienced system engineers.

▶ Front of House Technician: The FOH tech is often called the "FOH babysitter," as their role is to take care of and assist the FOH engineer. Responsible for setup and maintenance of the consoles and outboard gear as well as "walk-in" music, announcements, and media feeds. FOH techs can be called on to record the performance through digital technology such as Pro Tools and should have solid experience with different consoles and outboard processing. Sometimes they mix the opening artists.

▶ Monitor Technician: Respectively, the "monitor babysitter" is responsible for taking care of the monitor engineer. The monitor tech sets up and maintains the monitor system and consoles and is often responsible for In-Ear Monitoring Systems (IEMs) and RF coordination. The monitor tech will most likely be responsible for mixing monitors for the opening artists and should have solid experience with different consoles and outboard processing, as well as different types of monitors and IEM systems.

▶ Stage Technician: Responsible for the setting up and wiring the stage for microphones, stands, and cables, as well as the patching systems. Stage techs assist other departments in the setup of the overall sound system. On festivals there will often be two or three stage techs that deal with fast turnarounds. Stage techs should be organized and provide faultless patching and changeovers.

▶ System Technician: System techs mainly are responsible for getting the sound system set up, dealing with the rigging and cabling. Like the stage tech they will assist other departments in the set-up of the overall system (SoundGirls.org).
(Reprinted with the permission of Karrie Keyes and Soundgirls.Org)

Tools of the Trade: Live Sound

In her SoundGirls article, "What's in Your Go Bag?", **Elisabeth Weidner** discusses her "Go Bag," a case full of equipment that she takes when she's called on a gig at the last minute. The case itself is sturdy, waterproof, and just the right size to be accepted as carry-on luggage. In it, she carries:

*   Gaffer tape: in varying colors and sizes, because as you know . . . *you can never have enough.*
*   Testers: Although there are a few kinds, she has a favorite. "I never leave home without my dbx CT3 Cable Tester. This is, hands down, the best cable tester I have ever had. It supports testing of DMX, Speaker Twist, XLR, DIN, RJ45, RJ11, TRS, 1/8", phono, and BNC. I still haven't told you the best part. This thing splits in half so that you can check connections that are on opposites sides of the building!"
*   Microphones: "Trust me on this," writes Weidner, "I have been the hero many times for having one of these available."
*   Cables: XLR and TRS
*   Adapters: XLR male-to-male and female-to-female; TRS (Tip-Ring-Sleeve, 1/4"), NL4 (four-pin connector), BNC, RCA barrels, and BNC-RCA. I also keep a spare cable.
*   DI Box: A "direct injection" box used for connecting unbalanced outputs from guitars and keyboards to balanced stage inputs.
*   Label Printer
*   Tools: "I pack a cordless drill/driver . . . a pack of various bits, a precision screwdriver set, an Allen set, a crescent wrench, a headlamp, and a soldering iron with all necessary accessories. Also, never leave home without your multitool" (Weidner).
(Reprinted courtesy of Karrie Keyes and Soundgirls.Org)

Jessica Paz shares with me a couple of her favorite tools. "I have Meyer Sound MAPP XT: it predicts speaker coverage. Another software is d&B audiotechnik's ArrayCalc which predicts SPL and coverage based on frequency. I use DiGiCo consoles for everything, with its 'T software'.

I can't do a theater show without it – it is the most powerful theater desk so far.

"I use Mainstage for hosting plugins: Waves, FabFilter, SoundToys, TC Electronics' VSS3 plug-ins; Waves H reverb and H Delay. I also use Logic Pro and Ableton Live for sound effects creation. I have at least 1 TB of sound effects in my library, with sounds from BBC, Sony . . . but when I'm looking for new sounds, SoundSnap.com, SoundDogs.com, and FreeSound.org are kind of like 'Get out of jail free!' "

However, she does not use gaffer tape! "As a sound designer, I don't use gaffer tape because I work in union houses where gaffer is used by FOH and deck. *They* are the masters of gaff tape" (Paz, 2019a).

Fun Facts: Safe Listening Levels

Whether you are going to a concert, working front of house, or even mixing in the studio, it's important to protect your hearing. In order to measure the levels of sound in your environment, you can use apps for iPhone or Android phones that have Sound Pressure Level [SPL] meters - but take care to calibrate the microphone!

The World Health Organization's "Make Listening Safe" campaign informs us that "85 decibels SPL (A-weighted) is considered the highest safe exposure level up to a maximum of eight hours. The permissible time for safe listening decreases as sound levels increase. For example, a sound as high as 100 dB – the level produced by a subway train – can be safely listened to for only 15 minutes each day" (Make Listening Safe).

The report also contains the worrisome statistic that music listeners set their comfortable listening levels between 75 dB and 105 dB. More concerning is the observation that nightclubs and bars can average up to 112 dB (ibid.). The Centers for Disease Control (USA) makes a similar recommendation using A-Weighting. (See Chapter 7 for an explanation on weighting curves.) (NIOSH)

Kathy Sander (USA)

"If there were women doing live sound on the national and international level, I wish I had met them," says **Kathy Sander**. In 1974, one year before **Boden Sandstrom** and **Casse Culver** founded **Woman Sound** in Washington, D.C., Sander was visiting England and invited to join Elton John's tour as a "gopher." She declined but joined the tour a year later in Texas (Pettinato). "On my first tour, 'Goodbye Yellow Brick Road,' I helped with stage, sound and lights – that is when it became absolutely clear to me that sound was the only thing for me. After a stint at Conway Recorders in LA from 1975–76, where I did everything from booking appointments to tape op to running the board for Mariachi and underground bands, it was clearer that live audio was my direction." Her favorite tours include the US Festival, Amnesty International Tours, and Live Aid. "There were no female role models when I was in live sound. Women were not valued, so my role models were often men: the ones that treated me as an equal and taught me and nurtured my skill and pushed me" (Sander, 2019).

As a child, she listened to the radio constantly – at night and while on her paper route. "My grandparents gave me their old short wave radio with a record player. I was hooked from age 13. I bought many 45s and albums and of course the *Sound of Music*. I had a friend that loved and played music so much it ignited a passionate interest that I didn't even know was possible . . . My father worked in a radio station but I only observed recording commercials" (Ibid.).

Eventually Sander left the business, became a geophysicist, and started a research firm, which she has since sold. She is now retired, but still has some advice: "Even today, luck, timing, and forward-looking technical skills can be your biggest ally" (Ibid.).

◼ Betty Cantor-Jackson (USA)

Betty Cantor-Jackson had been on a long hiatus since the death of Grateful Dead lead singer, Jerry Garcia, in 1995. It was 2011 before she would want to record live shows again. That year, upon hearing former Black Crowes lead singer Chris Robinson performing with his new band, the Chris Robinson Brotherhood, she approached them and began recording their concerts. The recordings were released as *Betty's Blends* (Melamed, 2017).

The title of the series, *Betty's Blends*, is probably meant to invoke memories of another set of recordings she made, the *Betty Boards*. These comprised over 1,000 reel-to-reel tapes she made of Grateful Dead concerts through the years. One is listed in the Library of Congress: the 1977 concert at Cornell University's Barton Hall on May 8, 1977 (2018) (Budnick, 2017). Others have been found and restored, while others were, unfortunately, lost.

Cantor-Jackson got her engineering start in 1968 in San Francisco at an "underground FM station, thinking she might want to be a DJ" (Fensterstock, 2018). The recording technician for the Grateful Dead was Bob Matthews, with whom she accepted an internship. Subsequently, from 1968–1981 she worked as their engineer and producer on albums such as *Anthem of the Sun* and the *Dead Set*, all the while recording "hundreds of reels of soundboard tape" (Ibid.).

For a while, the crew for Grateful Dead shows was all women. "(Betty) Cantor was on the board, (Candace) Brightman directed the lights, (Melissa) Cargill and (Rhoney) Gissen had . . . set up the revolutionary sound system (under Owsley's stewardship) and back at the office, several departments boasted reasonable gender parity." However, none of these women were featured in the 2017 documentary *Long Strange Trip: The Untold Story of the Grateful Dead* (Fensterstock, 2018).

◼ Brandie Lane (USA)

Brandie Lane is a sergeant first class in the US Army who leads the Audio Branch of the West Point Band. Before she joined the Army, she was head audio engineer at Sono Luminus, a classical label. She and **Leslie Ann Jones**, were the first women to win a Grammy for Best Engineered Album, Classical for *Porter, Quincy: Complete Viola Works* (2010). She also has a Downbeat award for "Best Engineering for a Live Recording" (AES). She holds a bachelor's degree from University of Miami in music engineering technology with a minor in electrical engineering, and in 2016 she earned a master's degree in management and leadership from

Sgt. Brandie Lane

Webster University. She also did the recording and mixing for the Macy's 4th of July fireworks for 2017 and 2018 (Lane).

"Both of my parents were college music professors, so a love for music was instilled in me while still in the womb," she says. "To get back at my parents for taking my drum away at age 2, I started playing percussion at age 10 and quickly developed a passion for music performance. As I grew older, I developed an interest in science and wanted to focus on a career field that allowed me to use my creative strengths in music and still involve a strong science/engineering component. The audio and recording field made the most sense" (Ibid.).

By the early 2000s, recording schools had become immensely popular, and incoming freshmen already had skills that students from decades earlier would envy. "I arrived at University of Miami simply armed with passion and a lot of determination. I struggled to catch up to a lot of my classmates who had started their own labels, set up their own dorm studios, and had been recording since high school. However, I remember my first time in the UM studio as an assistant. I brought a book to read because I figured a 5-hour session could easily become boring. However, after what seemed like 15 minutes, the band had recorded enough songs for a demo, my book stayed in my bag, and we broke down the session. I knew at that point I was in a field that would challenge me and keep my interest for a long time" (Ibid.).

Since joining the Army, she has become a section leader and is in charge of the Audio Branch of the West Point Band. "As a leader in my organization, I help facilitate and support new ideas for projects and events that educate, train, and inspire America's future leaders at the United States Military Academy" (Ibid.).

They have two live rooms/rehearsal spaces that are tied into the control room at Egner Hall. "We use the film score setup: a Decca tree with spot mics. Sometimes we add virtual instruments and MIDI and marry that with our concert band, rock band, and field music group," she explains (Ibid.). (Field music consists of instruments such as big rope drums, bugles, and fifes.)

For now, they work in stereo, but they implement surround audio for research dealing with traumatic brain injury and field recovery efforts for wounded warriors. "Some cadets at West Point are also researching battlefields and rendering 360° and virtual reality environments of battlefields, and we are putting immersive sound design and music to go with their projects" (Ibid.).

She is very interested in being asked to talk about empowering women to stay in audio. "Professionally, I look up to any female in the audio field. There are so many incredible figures including (but not limited to) Leslie Ann Jones, Ulrike Schwarz, and Agnieszka Roginska, and I'm always inspired when I get a chance to personally interact with them." Her late mother also continues to influence her. "My mother will always be my personal role model. She was lovingly referred to as a 'steel magnolia' and embodied the perfect combination of charm and a sharp tongue. No matter what, she treated everyone with respect, but did not accept anyone's excuses for not living up to their potential or not giving 100%. She passed away in 2010, but I know her legacy will live on through her thousands of students and hopefully through me" (Ibid.).

Sgt. Brandie Lane at the FOH position

Sgt. Brandie Lane in the studio

This year, her team is aligning with the Grammy Music Education Coalition to provide musical content for a platform launching in 2020. "We will be recording various arrangements from Latin artists who provide arrangements and using our concert band, rock band, and jazz band," she says (Ibid.).

You can find more information at www.westpointband.com.

Karrie Keyes (USA)

Karrie Keyes is a great listener. "Really listening to the artist to determine what they need sound wise on stage," she says, "and with SoundGirls.Org, really listening to the challenges women face in this industry" (Gaston-Bird, 2017b).

Keyes recalls that listening to the radio and music lessons were the childhood experiences that led to her love for audio. She also loved punk music and punk culture. In a bio written by SoundGirls.org cofounder Michelle Sabolchick Pettinato, Keyes recalls, "One day we went to a friend's rehearsal space, and I saw a soundboard for the first time and wondered if I could run it. Our friend, Ben, was older and in a band that got paying gigs, he laughed at me and told me girls couldn't run the soundboard. I wondered what other skills were needed to run it or if the only qualification needed was being a dude" (Pettinato).

Around 1986, Keyes was at a concert in Los Angeles by Black Flag, a punk rock band. She approached one of the roadies and expressed an interest in learning to do live sound. The roadie happened to be Dave Rat, who owned Rat Sound. That evening he invited her to hang out and learn to wrap cables, and the next day she traveled to Palo Alto, California, to do

Karrie Keyes, monitor engineer and cofounder of SoundGirls.org

another show. "By the time we returned to Los Angeles I knew what I wanted to do – sound – and my new friend Dave was going to teach me. He inspired me not to give up" (Ibid.).

Over time, Keyes began helping to build the company, and along the way he helped them achieve several milestones. "There were only six people working full time and really not being paid. Every dime went to cover operating expenses and buying more equipment. There were several milestone achievements: providing production and engineering for The Red Hot Chili Peppers, Sonic Youth, and Pearl Jam; landing the Coachella Contract; being on the first companies to sign on and purchase an L-Acoustics rig. Everyone did what needed to be done. From living in the warehouse to being without hot water for two years, to building boxes within the shop. I did lots of gigs with no sleep. I took care of the business side, HR, accounting, purchasing, and working every gig I could" (Keyes, 2019).

Eventually she became the monitor engineer for the Red Hot Chili Peppers and then for Pearl Jam. As of 2019, she has been with Pearl Jam for 27 years. Being a monitor engineer for that long, Keyes has amassed quite a bit of experience and has customized the setup to fit Pearl Jam's needs. In a radio profile of Keyes for National Public Radio, Caroline Lossneck writes, "Everything from weather and humidity to crowd size and sweat-drenched microphones can change what the musicians are getting in their ears" (Losneck, 2016).

She doesn't like being asked what it's like being a woman engineer. Instead she would prefer "to be asked the same questions men are asked. To at least be given that respect." Keyes lists some of her influences: "Angela Y. Davis and the awesome women that are leading women-focused organizations. Laura Whitmore of Women's International Music Network, Fabi Reyna from She Shreds and Mindy Abovitz of Tom Tom Magazine; women like Beyoncé, Madame Gandhi and Grimes that are taking control of their careers and art" (Gaston-Bird).

In 2017, Keyes received a She Rocks Award. The awards were founded to "designate women who display leadership and stand out within the music industry" (*The WiMN Breakfast and She Rocks Awards*, 2013). The ceremony itself used to be a breakfast conference at the

NAMM show in Anaheim, California, but in less than a decade, it ballooned to be a premier event on the scale of the *Billboard* awards. "I attended the awards for the first time in 2015," said Keyes, "and loved hearing the stories and achievements from the honorees. The sense of community that this awards show creates is very important and empowering. I am humbled to be honored. The hard work I do on a day-to-day basis for the SoundGirls community is important for women and girls. To be recognized for that work is very rewarding" (Angeline, 2013).

■ Michelle Sabolchick Pettinato (USA)

"I used to take apart my record player as a kid and try to make it run backwards so I could listen to all the 'supposed' backward messages on my rock n' roll albums," recalls SoundGirls. org cofounder **Michelle Sabolchick Pettinato**. "My dad also had a little reel-to-reel tape recorder from when he was in the military and I remember always playing with it as a child. It was broken and I took it apart to try and fix it. I was always fascinated by how stuff actually worked. I had a passion for music and science. I made plenty of mix tapes in my teens and I used to make my girlfriends come over and we'd sing along to my favorite records while I would record us with a cheesy little microphone into my tape deck" (Pettinato, 2019b).

In 1987, Pettinato attended the Recording Workshop in Chillicothe, Ohio, and then went to Full Sail in 1989. Things were tough at first; she traveled between Florida and her home state of Pennsylvania looking for work. She got her "big break" in 1992, when a friend offered her a job mixing sound for the Spin Doctors. It was a close call – the gesture came on the day she was supposed to be traveling back to Florida. "As fate would have it, Amtrak went on strike, putting my plans to travel on hold. That very afternoon my friend, John Heidenriech, called to see if I would be interested in taking over his mixing gig with a band doing a national club tour" (SoundGirls.org). At the time, the band was unheard of.

"Shortly after I began working for them, their debut album skyrocketed up the charts! That was the beginning of my professional career as a Live Sound Engineer and I never looked back" (Pettinato, 2019a).

She has been mixing FOH (front of house) on tour ever since that time, touring with Gwen Stefani, Ke$ha, and rock artists Mr. Big, Melissa Etheridge, Goo Goo Dolls, and many others (SoundGirls.org).

Most recently, she has started an online video course called "Mixing Music Live" and authored an e-book on the subject by the same name. "I am really excited about being able to help anyone who is interested in getting started in live sound by sharing my knowledge and experience. The course teaches the basic fundamental principles of live sound with a focus on what you need to know to properly operate any soundboard. I find that a lot of people just getting started or interested in live sound are overwhelmed with how much there is to learn, and they are going about it backwards by trying to learn equipment rather than the principles. If they had a solid foundation of the concepts like signal flow and gain structure, they would be able to understand what they are doing and learn new equipment faster" (Pettinato, 2019b).

"I also encourage any women working in audio and music production to become a member of SoundGirls.org. SoundGirls is a fantastic resource for women in the music business. It's not women only, is open to anyone, and there are members from all over the

Live sound engineer and SoundGirls cofounder, Michelle Sabolchick Pettinato

world. It is an incredibly supportive community that works to empower women in audio and women who might be considering a career in audio or music production" (Pettinato, 2019b). More info can be found at mixingmusiclive.com.

Characteristic of her welcoming personality, she encouraged people to greet her during her tour in 2019. "I will be mixing Elvis Costello this summer so come out and say hello!" (Pettinato, 2019b).

Fela Davis

Fela Davis is a New Jersey-based audio engineer and co-owner of both 23dB Productions and One of One Productions Studio (with Denis Orynbekov). She has worked with Me'Shell Ndegéocello, Christian McBride, Ron Carter, and Swiss Chris, among many others.

She has some advice for young women seeking careers in audio. "They can do it. Anything is possible. Denis is from Khazikstan and I'm from deep South Carolina. No one would have pegged us then as being the next audio 'anything'. Our parents weren't in it. No one expected anything from us, so anything we do is 'bonus'! You have no limit to yourself. Continue to hang and help positive people and I promise you it will return tenfold!" (Davis).

"I wanted to be a toymaker. I wanted to do something with my hands. I was always into electronics, stereos, VCRs – I was the only one who knew how to program it!" (Gaston-Bird, 2017a). Davis would read *Mix* magazine and try to learn as much as possible through high school until she went to Full Sail University in Florida. "My parents had no idea what an audio engineer was. I didn't know audio engineering was a thing. I didn't find out about it until 8th grade when I took a career placement test and it listed 'audio engineering' for my result. I thought it was something to look into" (Ibid.).

Davis experienced some isolation when starting to work in audio. "It took me a few years to feel comfortable when I first began as a stagehand. I always felt overwhelmed by how much I didn't know and wasn't sure if I could ever catch up. And I would go to concerts and hear many shows that were not impressive. I'd think to myself, 'if they can do it, then I will definitely do it!' So, I soldiered on and learned as much as I could from the engineers that sounded the best to me. That move made my confidence higher and made me realize I had a chance to really make this happen" (Ibid.).

Fela Davis and business partner, Denis Orynbekov
Source: Photo credit Eric Rhee

"Smaller sound companies needed help, so I got my foot in the door mixing for small gigs. They're not killin' it, they're not ballin', but they made a nice living. I later worked for Clair Global. Working with smaller companies gave me a chance to work one-on-one with the clients and learn how to communicate as a tech and business owner. Larger companies have account managers and they can be a little over-protective about the technicians speaking with the clients. After years of working for sound companies and venues, my dream gig showed up via LinkedIn! The 6-time Grammy award winner Christian McBride contacted me through the social media platform, and I never looked back!" (Gaston-Bird, 2017a).

"Artists don't always realize the power they have in the industry to change things," she continues. For example, front of house engineer Amanda Davis got her chance to work for Janelle Monae. "[Monae] said, 'I want a Black female engineer'. That's how many engineers of color get their starts in this industry: Quincy Jones demanded a Black engineer be in the room when he was working on 'The Wiz' back in 1978, and now Ollie Cotton is head of sound at the world famous Apollo Theater in Harlem!" (Davis).

After meeting Christian McBride, Davis wanted to get better with her studio skills. She found an ad on Craigslist that was looking for someone to help master an album for free, and that's when she met her now-business partner, Orynbekov. They have been working together for five years (as of this writing). "I was mixing at small clubs at the time, and a side hustle was to record bands live to 2-track and sell the mix for $10. I had gotten good at getting them to sound decent and started featuring the ones I liked on my YouTube channel. I started editing the music at home and adding plugins to fatten the mix for YouTube. I reached out to Denis after seeing his ad on Craigslist and liking the music he recorded" (Davis).

Fela Davis and business partner, Denis Orynbekov
Source: Photo credit Ilya Popenko

Currently, Davis and Orynbekov are working on a podcast called *The Art of Music Tech*. They also engineer podcasts for other people, for example, Joy Pratcher's *Unexpected Success* and *Forever FAB Fashion – The Art of Living Well and Beauty*, a show they produce for plastic surgeon Dr. Shirley Madhere. "Podcasting has become affordable because before, you had to pay thousands of dollars for content storage. Now with sharing software you pay ten to fifteen dollars a month and you can put up *unlimited* content" (Ibid.).

They are also endorsers of many audio products like Focusrite, Stealth Sonic IEMs, Lewitt Microphones, Ear Trumpet, Wireworld cable, Acustica Audio (". . . and more!" adds Davis.) Most recently, they have opened their own recording and production space in Fort Lee, New Jersey, called One of One Productions (with Pratcher joining them as a partner). And of course, mentoring is important to Davis. She has done mentoring workshops with SoundGirls.org, and many people reach out to her. So many, in fact, that she is using the podcast platform to reach out to them. She listens to the Napoleon Hill audiobook daily and employs the techniques she learns. "It's worth its weight in gold going to AES and NAMM [conventions] with like-minded people who want to win. That's the coal to keep the engines moving," she advises enthusiastically (Ibid.).

Carolina Anton (Mexico)

"I always want to help women because at the beginning, no one supported me." **Carolina Anton** heads up the Mexico chapter of SoundGirls.org. "When I started up SoundGirls Mexico so many women from Brazil, Argentina, and Peru were calling me from South America where it's hard to work in sound engineering and be a woman" (Anton).

Carolina Anton, live sound engineer and chapter head for SoundGirls Mexico

"I studied music; I'm a drummer, I have been playing since I was 15 years old; I played with my band a toured for maybe 3 years. Then I started studying piano and composition. Later I started studying Japanese culture and got my master's degree. I went to Japan and lived in Kyoto for two years." Anton received her master's degree from Urasenke Gakuen Professional Chado College, Midorikai (Japanese Art and Culture) in Kyoto, Japan (Ibid.).

"When I got back, I had lost touch with everyone, so I thought if I started doing sound I could meet musicians and play again." Anton says she started from the bottom in the industry. "In Mexico, there were not good sound schools so I didn't study properly. I even tried to go to England and study." Eventually she started learning on her own. "All the people who I asked said, 'No, you are a woman, you have no idea what it's like, and you're not going to be happy'" (Ibid.).

But she was determined. "I travelled to the United States and bought every book about sound I could. 'What is XLR', 'what is balanced' . . . and maybe after 5 years of trying and trying and knocking on doors someone told me to come and learn. It was an unpaid opportunity at a rental company. I started cleaning cases and doing cabling and warehouse work, and then I started learning how to use the mixers. I saw that most of these guys didn't know how to use a digital console and were afraid of digital things. I know English but they didn't. That's when I started growing" (Ibid.).

Anton started touring as system engineer for eight years and then started doing Front of House (FOH) and later did monitors for seven years. "In the middle of this situation, one very nice and good friend of mine, Fernando Guzman, went to the US and worked with SSL (Solid State Logic) doing training all over the USA. And one time he met Karrie (Keyes) when he did a training for SoundGirls. And that's when he called me and he said, 'Please come to help me because SSL wants to have a female representative' . . . because at that time I was the first woman to use the SSL Live (console) for 100 hours. So he said, 'You have to come and help me because you are important for SSL as a woman and as engineer to represent us.' So that's why he invited me, and I went to Rat Sound and met Karrie. We started talking about women and about Mexico, and she asked me if I wanted to be head up the Mexico chapter of SoundGirls and I said 'Yes, yes, yes! Let's do it!'" (Ibid.).

In 2016, Anton began heading up the Mexico chapter of SoundGirls.org, managing to bring together women within the industry for the first known time in that region.

In 2019, for the fourth consecutive year, Anton and a group of professional engineers (mostly women) had a booth at one of the largest exhibitions in LATAM: Exposoundcheck, a huge event with companies and attendees from the USA, Mexico, and Latin America. At the time of our interview, she was still in the planning stages. "It's big. More than I thought. They are going to have a conference and live sound shows and the theme is immersive sound. In Mexico there has been nothing about immersive sound for live shows . . . until now" (Ibid.).

A partial list of Anton's tour credentials includes FOH for Gloria Gaynor, Electric Forest Festival, and Every One Orchestra; she has done monitors for Kool & the Gang, Janelle Monáe, and Latin Grammy winner Natalia Lafourcade's "Hasta la Raiz" historical theaters tour; and sound system design for Zoé and Marc Anthony, among many others. She has also done DiGiCo console tech support for the Cranberries Reunion tour (2010), Slash, and Earth, Wind and Fire. She recently cofounded the 3BH Company, for whom she does integration and sound calibration for film studios, including Splendor Omnia, Pulpo Post Studio in the Dominican Republic, and others.

Carolina Anton with the SSL Live L500

"The most important thing is to be happy. No matter what. It is not a job, nothing – only to be at peace with yourself, to be happy and enjoy the path whatever the path is: sound, philosophy, teaching, whatever. That's the most important thing: to be in tune with ourselves and the people we love" (Ibid.).

Leya Soraide (Bolivia)

"One of the skills that marked my childhood would be that my whole family was dedicated to setting up sound systems for churches and small festivals, so I grew up traveling a lot, sometimes cabling and obviously helping my parents. I was curious about learning the console; I was very attracted to the number of knobs and buttons," recalls **Leya Soraide.** "My favorite role model is my mother. She is an example to follow, and she also knows audio. She taught me to solder my own cables. Another serious role model is **Karrie Keyes.** I really admire her a lot for her work with Pearl Jam and Soundgirls.org" (Gaston-Bird, 2017c).

As part of a blog series I wrote called "31 Women in Audio," I asked Soraide if she gets any questions about being a woman in audio. "I do not like to be asked if a woman can do the same job as a man, I really don't like that, and I don't like when they ask my age. I'm very aware that experience comes with age, but no one is *born* knowing how, and now I am training as a person and professional. I really hate when people ask me why women who studied sound before me did not pursue it. I really don't know." She would prefer to be asked about her interests, projects, and everything related to audio (Gaston-Bird).

Live sound engineer Leya Soraide

Leya Soraide at work

She is also cofounder of AudioCap, which organizes audio courses, and has organized courses in the city of Cochabamba-Bolivia with Cana San Martin of Argentina, Rodrigo Lopez de Mexico, and Jorge Azama of Peru. In 2016, she also ran an organization called dB Girls Bolivia to train professionals in live sound, music recording, film, and television. "In my country, there are few women who study the career of sound and even a lower percentage who pursue the career. Maybe it is because there is still machismo in our environment" (Gaston-Bird, 2017c).

Soraide recently graduated from the sound engineering program at UNITEPC in Bolivia (La Universidad Técnica Privada Cosmos). In 2019, she traveled to Mexico to take a sound reinforcement course in Mixtlan, and she now works in the city of La Paz as a monitor engineer and stage technician for Sound One Bolivia. She expresses gratitude toward those who gave her opportunities. "I am grateful to the people and companies that trusted me. I am especially grateful to my mentors: engineer Rafael Alarcon and engineer Miguel Garcia Salvatierra who both taught me the wonderfulness of audio."

Tzu-Wei Peng (Taiwan)

Tzu-Wei Peng worked as a violin teacher until she got into live sound after checking out a "behind-the-scenes" tour for the Summer Sonic Music Festival in Tokyo, Japan. "The production was fascinating," she recalls. "Knowing how many people and how much effort was involved in this festival impressed me a lot. Furthermore, I have seen a good number of women working in production over there." She started looking for entry-level work with companies in Taiwan, but most of the companies only hired men. "This experience made me realize there is no female in this field in my country. I found another woman who was interested in live production but the most she could get involved with was shadowing a small company where she knew a friend" (Peng, 2019).

The next year, she met SoundGirls.org cofounder Michelle Sabolchick Pettinato. "Her story inspired me very much, and knowing there is a community really warms my heart." She worked hard getting more experience and had an opportunity to intern with Karrie Keyes at Rat Sound Systems and meet lots of professional women working in sound. "They motivated me every day just by seeing them being a badass in the industry" (Ibid.).

However, Tzu-Weng's concerns about live sound turned from finding job opportunities to the amount of waste produced from one show. "I have always cared about environmental issues . . . seeing something that is against my values happening in my working environment almost every day made me stop and think about what I do. I love working in music and production, but I can't ignore the fact that the industry is causing so much impact on Earth." Eventually, Tzu-Weng left the sound industry and began to study material science and engineering. "I believe if we can change the material of the items we are using, especially disposable items, there are so many things that can be changed in a good way" (Ibid.).

Jessica Paz (USA)

The category for "Sound Design for a Musical" made its debut at the Tony Awards in 2008, and in 2019, Jessica Paz was the first woman to win the award for the musical *Hadestown*,

Jessica Paz, Tony Award-winning sound designer

along with her male colleague Nevin Steinberg. "It was such a surprise to me," she says in a video by Meyer Sound. "I wasn't even expecting it. I was so thrilled to just be working on the show and working with this team of people . . . the creative team, the orchestrators, everyone just brought 155% percent of themselves. And it drove you to want to do the best work possible, and that's all I was really thinking about. So when the awards started being announced I was like, 'Oh, right! This is amazing!'" (Paz, 2019b).

Paz's job begins outside of the theater. "It's a number of weeks of pre-production where I'm working on paper and in my computer to design the sound system for the space. Then there are a number of production meetings with the rest of the creative team, and then we move into the theater. I tune the sound system using software called SMAART, so that I can equalize the system in the room and do my various tests and set up of the sound system. Then it's a week or two of rehearsals followed by 3 or 4 weeks of previews in front of an audience.

"Previews last around three weeks: Our day usually begins at eight o'clock in the morning. We have rehearsal with the actors from 1 o'clock to 5 o'clock, and then we watch the show in front of an audience in the evening and I can make live changes to the sound of the show via a remote control on an iPad while I'm sitting in the audience" (Ibid.).

She keeps tabs on how the show sounds through her front of house engineer and "deck audio engineer" who maintain the show after it opens. "I'll go in once every other week and sit and watch the show and I'll give them notes about maintenance and anything that might need fixing, or give them feedback about how they're handling understudies and swing actors and what they sound like, and just make sure that the integrity of the sound of the show remains the same" (Ibid.).

The equipment list for the show *Hadestown* reveals the level of detail and accommodation for each performer, musician, and director. "We use a DiGiCo SD7, redundant Mac

Mini's running QLab for sound effects and playback; 2 Mac Mini's with [Apple] Mainstage software where we run IO (inputs and outputs) for reverbs and other processing for the vocal system and band system; QLab and Mainstage are over MADI IO into the SD7. We have two remote racks that handle input from all of the band and another for RF mics for all actors. We have each principal actor and the 'Three Fates' who are all double miked, that's about 13 actors . . . so 25 or 26 RF channels including spares and doubles. We're using Sennheiser EM3732 receivers with SK5212 transmitters. The boom mics are a combination of DPA4066's with a DPA4061 attached to it for the backup. Our 'Fates' – the ladies who sing in close harmony – use Shure IEMs (In-Ear Monitors) to hear a click track sometimes. We also have a redundant QLab system for the band to play back click tracks. Any of the songs that have automation are click tracked so automation is timed – that's so tempos don't fluctuate in the music, and they're triggered by the Musical Director with a foot pedal (he is also playing piano and conducting). The band has monitoring for each individual band member using an AVIOM system with AVIOM distribution and A360 mixers" (Ibid.).

Hadestown started as a concept album, she explains. "There's no localization: it's a stereo show. We don't add time to a person's mic input based on where they are on the stage, there are no delay zones: it's basically an album" (Ibid.).

She didn't find a so-called "traditional" route to working in theater. "I've been a bookkeeper, an assistant manager at a scuba diving store; and when I found theater it was the first time I felt like I belonged and found my calling. I'm self-taught: I've done everything from community theater to Broadway, and it feels incredible to have the Tony be an affirmation of what I've pursued and to make it that far. And to be a women in an industry and field – I know less than a dozen women doing what I do. To be the first woman is amazing, and a little sad, but hope fully there will be more" (Ibid.). Her advice to women seeking work in the field is, "Pursue what you love. Pursue what you love wholeheartedly, and don't let yourself be discouraged by the fact that the numbers aren't in our favor right now knowing that the parity isn't 50/50. The only way that that's going to change is by people pursuing what they love to do (Paz, 2019b).

Bibliography

AES Member Profile: Brandie Lane [Online]. Audio Engineering Society. Available: www.aes.org/member/profile.cfm?ID=1452368818 [Accessed July 10, 2019].

Angeline, Myki. 2013. Front And Center: Sound Engineer and Co-Founder of SoundGirls, Karrie Keyes. *Women's International Music Network* [Online]. Available: www.thewimn.com/events/the-wimn-breakfast-and-she-rocks-awards/.

Anton, Carolina. 2019. Interview with Carolina Anton. *In*: Gaston-Bird, L. (ed.).

Budnick, Dean. 2017. What's Become of the Bettys? The Fate of the Long-Lost Grateful Dead Soundboards. *Relix* [Online]. Available: relix.com/articles/detail/whats_become_of_the_bettys/ [Accessed September 27, 2019].

Davis, Fela. 2019. Interview with Fela Davis. *In*: Gaston-Bird, L. (ed.).

Fensterstock, Alison. 2018. Do You Want to Talk to the Man-In-Charge, or the Woman Who Knows What's Going On? *National Public Radio* [Online]. Available: www.npr.org/2018/03/20/593958534/do-you-want-to-talk-to-the-man-in-charge-or-the-woman-who-knows-what-s-going-on [Accessed June 19, 2019].

Gaston-Bird, Leslie. 2017a. 31 Women in Audio: Fela Davis. *31 Women in Audio* [Online]. Available: http://mixmessiahproductions.blogspot.com/2017/03/31-women-in-audio-fela-davis.html [Accessed March 1, 2017].

Gaston-Bird, Leslie. 2017b. 31 Women in Audio: Karrie Keyes. *31 Women in Audio* [Online]. Available: http://mixmessiahproductions.blogspot.com/2017/03/31-women-in-audio-karrie-keyes.html [Accessed March 12, 2017].

Gaston-Bird, Leslie. 2017c. 31 Women in Audio: Leya Soraide. *31 Women in Audio* [Online]. Available: http://mixmessiahproductions.blogspot.com/2017/03/31-women-in-audio-leya-soraide.html [Accessed March 9, 2017].

Keyes, Karrie. 2019. Interview with Karrie Keyes. *In:* Gaston-Bird, L. (ed.).

Lane, Brandie. 2019. Interview with Brandie Lane. *In:* Gaston-Bird, L. (ed.).

Losneck, Caroline. 2016. Meet the Woman Who's Been Pearl Jam's Sound Engineer for 24 Years. *Weekend Edition Sunday* [Online] [Accessed June 26, 2019].

Melamed, Dave. 2017. Exclusive: Betty Cantor-Jackson Rekindles Her Love OF Taping Through the Chris Robinson Brotherhood. *Live for Live Music* [Online]. Available: https://liveforlivemusic.com/features/betty-cantor-chris-robinson-brotherhood/.

Noise and Hearing Loss Prevention. The National Institute for Occupational Safety and Health (NIOSH) [Online]. Available: https://www.cdc.gov/niosh/topics/noise/default.html [Accessed September 27, 2019].

Paz, Jessica. 2019a. Interview with Jessica Paz. *In:* Gaston-Bird, L. (ed.).

Paz, Jessica. 2019b. Jessica Paz: Tony-Winning Sound Designer. *In:* Sound, M. (ed.).

Peng, Tzu-Wei. 2019. Interview with Tzu-Wei Peng. *In:* Gaston-Bird, L. (ed.).

Pettinato, Michelle. 2019a. *About Michelle* [Online]. Available: www.mixingmusiclive.com/pages/about-us [Accessed September 27, 2019].

Pettinato, Michelle. 2019b. Interview with Michelle Pettinato. *In:* Gaston-Bird, L. (ed.).

Pettinato, Michelle. Kathy Sander- the First of the Five Percent [Online]. *SoundGirls.org*. Available: https://soundgirls.org/kathy-sander-profile/ [Accessed July 10, 2019].

Pettinato, Michelle. Traveling the Long Road – Karrie Keyes. *SoundGirls.org* [Online]. Available: https://soundgirls.org/traveling-the-long-road-karrie-keyes/ [Accessed June 26, 2019].

Sander, Kathy. 2019. Interview with Kathy Sander. *In:* Gaston-Bird, L. (ed.).

Soundgirls.org FOH Engineer [Online] Available: https://soundgirls.org/foh-engineer/ [Accessed September 27, 2019].

Soundgirls.org. It's About the Music – Michelle Sabolchick Pettinato. *SoundGirls.org* [Online]. Available: https://soundgirls.org/its-about-the-music-michelle-sabolchick-pettinato/. www.mixingmusiclive.com/pages/about-us.

Soundgirls.org. *Monitor Engineer* [Online]. Available: https://soundgirls.org/monitor-engineer/ [Accessed June 26, 2019].

Soundgirls.org. *Sound Technician* [Online]. Available: https://soundgirls.org/sound-technician/ [Accessed July 10, 2019].

Weidner, Elisabeth. What's in Your Go Bag? *SoundGirls.org* [Online]. Available: https://soundgirls.org/whats-in-your-go-bag/ [Accessed June 26, 2019].

The WiMN Breakfast and She Rocks Awards [Online]. 2013. The Women's International Music Network. Available: www.thewimn.com/events/the-wimn-breakfast-and-she-rocks-awards/ [Accessed July 10, 2019].

World Health Organization. *Make Listening Safe.* Geneva, Switzerland: World Health Organization [Online]. Available: apps.who.int/iris/bitstream/handle/10665/177884/WHO_NMH_NVI_15.2_eng.pdf

9   Education

A Brief History of Audio Education

During the mid- to late-twentieth century, opportunities for learning the science of audio engineering (specifically music recording studios) began as on-the-job training. Young men and women who aspired to become studio engineers started out as apprentices or interns, working their way up from a "runner" to an assistant ("second engineer") and then to a lead engineer or "first engineer." In this system, because of the male-dominated nature of the industry, some women were sometimes discouraged or endured lewd advances as they sought to get their start in the business (see: sexism, Chapter 1).

Training in an academic atmosphere was hard to find for anyone wanting to work in the music industry. Students could pursue degrees in electrical engineering, telecommunications, music, mathematics, composition, or physics, but before the 1980s there were very few degree programs focused on "audio engineering," "music technology," or "recording arts" as we know them today. However, young graduates who wished to pursue careers in broadcast engineering, physics, composition, music, hardware design, acoustics, and so forth had a relatively easier time finding a career path than did those (especially women) who wished specifically to work in a music recording studio.

For the following timeline, I relied on an archived version of the Audio Engineering Society's education directory and recollections from Martha de Francisco and Wayne Jackson, my former professor at Indiana University Bloomington.

Arnold Schoenberg described the idea for training recording engineers (he called them "soundmen") in 1946, and in 1949 Erich Thienhaus, a lecturer in acoustics, founded the first known institute for the training of sound engineers at the University of Detmold in Germany. The "Erich Thienhaus Institute" (ETI) is still in operation today at the Hochschule für Musik Detmold. He also held six Tonmeister ("Sound Expert") conferences between 1949 and 1963. De Francisco offers a brief history:

> "The Tonmeister school was the very first of its kind. In 1949 – this was 4 years after World War II – German musicians were looking for a place where they could establish a music conservatory. Cities like Berlin and Hamburg were bombed out, and they found a prince in the province in Germany who offered to give them one of his palaces. And they established a music school in that palace in Detmold. These musicians from Berlin had known a person in charge of the audio recording side of the film industry in Berlin, and they thought it would be a great addition [to the program] to offer training in audio. This man was Erich Thienhaus, who decided to come provide this training. The education was designed to be half music and half technology. In this way, the Tonmeister career at the music academy was established."
>
> (De Francisco, 2019)

In the 1970s, computer and electronic music programs began to emerge. Early pioneers such as **Suzanne Ciani**, **Carla Scaletti**, and **Laurie Spiegel** pursued their studies in programs such as these, which began to branch out and become audio technology or music engineering programs. In 1971, the Recording Workshop was founded in Chillicothe, Ohio; in 1975, the Recording Institute of Detroit began, and in 1979, the Trebas Institute in Quebec, Ontario, was established.

By the 1980s, the number of recording engineering programs began to grow: South Plains College in Loveland, Texas, began its Sound Technology Program in 1980; the University of North Carolina at Asheville began a program in 1982; and the Ontario Institute of Audio Recording Technology began in 1983.

Wayne Jackson, who began the Audio Technology program at Indiana University with Dr. David Pickett, recalls that in 1982 there were only a few schools in the United States that offered degrees in music recording, including the Peabody Institute and the University of Miami. During that same time period, there were also for-profit programs such as the Recording Workshop and the San Francisco Recording Arts Institute (Jackson, 2019). In 1983 Berklee College of Music started up their Music Production and Engineering program, which **Robin Coxe-Yeldham** helped to develop (Berklee, c. 2014).

Also in 1982, a specialization in sound engineering was developed at the Technical University of Gdansk in Poland and was for many years the only sound engineering course in Poland (*About Multimedia Systems Department*, 2004–2017).

In 1986, Jamie Angus and her colleague Ross Kirk at the University of York created the UK's first Music Technology course with David Malham and Richard Horton. "We were doing DSP on Atari computers," she recalls. "CD had just come out. Back then everyone thought we were completely crazy," whereas today the initiative seems logical (Angus, 2017).

The AES' education directory reveals that by 1996, there were 21 institutions listed in the USA where students could earn a bachelor's degree in "sound recording," "music engineering," or "audio technology." That same year in Europe, the 17 programs in related fields ran the gamut, from "acoustics" to "media" or "electronic engineering" (with a focus on "signal processing"), not including the ETI. There were a handful in Australia (about five) and two in South America, both in Chile (Society, 1996).

Using that same directory, as of 2019 there are over 75 programs in the United States offering bachelor's degrees and a similar number across Europe. There are 24 programs in Latin America/South America region and 14 in the Asia-Pacific Region (Society, 2019). This is not meant to be a comprehensive list, and I did not cover the emergence of high-school and certificate programs.

However, as the number of programs increases, there is only a slight increase in the percentage of female applicants enrolling in programs. In a small number of cases, the number of women has increased, but in many other cases, it has stayed roughly the same: between 10 and 20 percent of applicants to the audio programs studied by Mathew et al. were women (Mathew et al., 2016).

In fact, in their paper, "Music Technology, Gender, and Class: Digitization, Educational and Social Change in Britain," Georgina Born and her colleagues describe a "leaky pipeline." Between the ages of five and sixteen, 40 percent of students choosing "music technology" are girls, but after age sixteen the number goes down to 25 percent, and then during A-levels, the number is 18 percent. At the point of university enrollment, the number is 10 percent. After some discussion about the learning environments and gender-based culture around technical spaces, the researchers conclude "while girls and women are no longer formally excluded from scientific and (music-) technological pursuits, they are subject to observable processes of gendered exclusion – occupationally, discursively, spatially, and practically" (Born and Devine, 2015).

Academic Standards

Educators looking for accreditation standards for audio-related courses in the USA will find recommendations in the National Association of Schools of Music handbook for 2016–2017, Appendix I.G. titled, "the NASM Standards for Studies in Recording Technology" (Music, 2017).

In the UK, the Joint Audio Media Education Support (JAMES) has emerged as an accrediting body offering a wider range of fields, including courses in music business, performance, songwriting, and composition as well as:

- audio recording and music production
- audio post-production for film and TV
- live sound
- lighting
- sound for film and TV
- film studies
- games, and
- rigging and technical theatre (J.A.M.E.S., 2018)

Careers in Higher Education

In most cases, you need a master's degree or a PhD to teach at the university level, although there are many exceptions where professional experience and/or a bachelor's degree will suffice for qualifications.

Terminal degree: A "terminal degree" is the highest degree you can get in a given field. In the US, there are such a varied number of master's degrees in music technology and recording arts that you could argue that the master's degree is a terminal degree in the field. However, there are a few that offer PhDs:

▶ NYU Steinhardt (focusing on "research methodologies in music and music education, Scientific computing, the psychology of music, digital signal theory, 3D audio or music information retrieval, foundations in mathematics and computer science, curricular development and hands-on teaching experience")

▶ Georgia Tech (focusing on "interactive music, DSP, mobile music, network music and robotic musicianship")

▶ Indiana University Purdue's School of Engineering and Technology (for "students with backgrounds in music technology, music, computer science, engineering, informatics, human-computer interaction, . . . and other relevant fields")

▶ Penn State and the University of Rochester offer a PhD in acoustics, while the CCRMA (Stanford University Center for Computer Research in Music and Acoustics) offers a PhD in computer-based music theory and acoustics. As such, when applying for a job in the United States, you *could* make the case that a master's degree is a terminal degree in the field if you are talking specifically about degrees labeled "recording arts" or "music engineering".

▶ Outside of the USA, there are a few more options. In Canada, McGill University offers a PhD in music, media, and technology. In the United Kingdom, if you want to teach at a university at the senior instructor or higher level, it's rare you would get past the application stage without a PhD. Programs that offer study at this level include the University of Surrey, University of Salford, and University of York (England); the Sonic Arts Research Centre (Belfast); Glyndwr University (Wales); and the University of Edinburgh (Scotland) (Society, 2019). In Europe and Australia, most of the PhD courses focus on acoustics, signal processing, and electrical engineering. The Iran Audio and Acoustic Institute offers a PhD in sound reinforcement in large venues (Society, 2019). *Author's caveat: I have only used one database: this is not meant to be a comprehensive list of schools and programs. You may wish to search further for opportunities in your geographical area.*

Research Assistant: Funded opportunities to study in audio-related fields exist. In the UK, applicants typically must be seeking or possess a PhD.

Adjunct Lecturer/Tutor: Adjunct lecturers (US) or tutors (UK) are invited to teach at a university on a part-time basis. Although it is a great way to get a start, it won't be enough to pay for basic living expenses, and there is controversy over whether institutions of higher education should employ adjuncts or open up more full-time positions so teachers can have a living wage. The fields of recording arts and music technology are no exception to this.

Instructor (US)/Lecturer (UK): Many times, these are full-time positions. Unlike the "tenure track" faculty, jobs may be done on a one- to three-year contract, or "at will."

Tenure Track Faculty: Assistant Professor/Associate Professor (US), Senior Lecturer (UK). In these positions, there is an expectation that the faculty member will engage in research, service, and teaching in varying proportions (for example, 20% research, 20% service, and 80% teaching). In order to advance to a tenured position, faculty must submit a dossier highlighting their work in these three areas. Tenured faculty often have a voice in curriculum development and university initiatives.

Full Professor: A full professor is also tenured (which leads some associate professors to wonder why they should go through the stress of submitting another dossier if they don't have to) (Wilson, 2012). Full professors might be invited to do more service as an assistant or associate dean and have a dossier that further establishes them as an expert in a given field.

Research and Creative Work

For faculty whose job specifies research and creative work, there are plenty of activities in which you can engage: for example, novel mic techniques, psychoacoustics, and so forth. However, when it comes to getting published, many university administrators use a different "ruler" to measure your output. Be very sure to ask for clear descriptions about the criteria required to achieve tenure at your institution.

For example, faculty in the history department may need eight peer-reviewed journal articles to be considered for tenure. What is the academic equivalent of this for music recording; is it eight studio recordings? Do these recordings need some level of peer review, such as commercial release and critical reception? Do you need to win a Grammy to prove your prowess? Without achievements such as these, how do you prove your eminence to a committee of historians, biologists, political scientists, and so on?

Apart from studio recordings, there are topics to write about and study, such as learning styles in the audio classroom and emerging technologies, and a few places to publish this research as well. Peer-reviewed examples include: *The Journal of the Audio Engineering Society*, *The Journal of the Acoustical Society of America*, *IEEE Journal*, and the *College Music Symposium*. You can also try conferences and trade publications.

Service

Service usually entails advising students, keeping office hours, advising student clubs, going to committee meetings, and hosting open house events. Women are encouraged to be thoughtful of being chosen for service because of their gender or race. Naturally, it is good to bring your experience to the discussion, but sometimes if you are the "only" one of your demographic, you may end up doing more work than your other colleagues. In the article, "Diversity Fatigue is Real," Mariam Lam writes, "Underrepresented faculty and staff members share the burden of diversity work in many visible and invisible forms: They often assume heavier workloads in teaching, advising, mentoring, and counseling, and spend more time on outreach, recruitment, training and workshops, and other service work. While their institutions benefit from collective gains in student success, those who do this work find it exhausting to do more than their fair share, indefinitely" (Lam, 2018). Not to say you should not engage in this work, just be certain that you are able to share the workload with *all* faculty who should be involved in this outreach.

Teaching

Of course, teaching is the goal! Learning how to evolve your teaching style and curriculum with the ever-changing pace of the industry has been the topic of a few conference papers.

Demonstrating proper mic placement, getting a good drum sound, how to coil mic cables, and professional studio etiquette are all part of teaching. Students want hands-on work in the studios, and depending on your program, you might be able to give individualized instruction, or students may break up into groups for their lab work.

There is sometimes a gender-based dynamic between boys and girls or men and women in the studio. If you are an educator, be sensitive to these differences and to their exceptions. *Sometimes* the male students are more eager to get in front, take charge, and show off what they know. *Sometimes* women are singled out to be the one to take care of the group's documentation. But that's not always the case. *Sometimes* it's good to put women together in groups. But not all the time! If you can, try to ask your students what dynamic they prefer. You can just as easily have women who are outspoken, assertive, and may resent being put in a gender-specific lab group.

Education-Related Volunteer and Nonprofit Work

The organizations in Chapter 2 have staff and volunteers who handle programming for young girls and women. These organizations host camps, workshops, training sessions, and mentoring opportunities, each of which offers a rewarding experience – and you can become a role model for another girl or woman finding their way into the industry.

Tools of the Trade: Education

Laptop, tablet, and mobile phone: If you are teaching audio recording, you'll need one of these to plan your classes, do demonstrations, show how various apps work, and compare audio quality. A student might want you to listen to their mix on Spotify, Soundcloud, or YouTube.

Virtual Learning Environments (VLEs): You'll probably be uploading your coursework to one of these:

- Canvas
- Blackboard
- eCollege
- open source platforms (like Moodle)

When using any of these, be sure to ask the university's online team for lots of storage! Your course will be media-heavy; you'll want to upload uncompressed music and video files, so you need tons of space.

Gaffer tape: One can never have enough gaffer tape. (Trust me – if you're teaching hands-on work, you'll need it just as much as the student crew.)

Martha de Francisco (Columbia)

"There was a woman at the Mercury label," remembers **Martha de Francisco**, perhaps referring to **Wilma Cozart Fine**. "And there was **Elaine Martone**. But in Europe, there were no women in any of the major labels: EMI, DG, Philips, Decca." That observation means that at one time, de Francisco may have been the only woman working at a classical music label in Europe as a producer and recording engineer (De Francisco, 2019).

Born in Columbia, De Francisco travelled to Germany in 1975 to pursue a diploma from the Erich Thienhaus Institute (ETI) at the Hochschule für Musik Detmold (see: "A Brief History of Audio Education" earlier in this chapter). The program consisted of music theory, score reading, ear training, mathematics, electronics, studio technology, and hands-on recording. "Tonmeisters are trained musicians," explains De Francisco. "I followed the same curriculum as a conductor might – without the conducting" (Ibid.).

De Francisco recalls that being a woman from Columbia – a developing country – made her concerned about her future in the industry. "One of the main rules in Germany was you had to leave as soon as you finished school. But I realized that after 5 years of study, I wasn't done learning everything I wanted to learn. And I was offered the job of theater sound designer at the Munich Opera. I joined for one season and a few months. Sound design for opera means amplification of orchestras from behind the stage, or voices that sound like they're coming from heaven, both of which gave me the opportunity to learn about live theater, and Munich has one of the great opera houses in the world so I could really observe important conductors, opera singers, and live theater productions, and participate in that.

"Then I was invited to join an American company, Soundstream: the first one to offer digital recording and editing. So, from 1981–1985 I was with Soundstream. This was the beginning of digital recording before the CD came out. I traveled with the recording equipment and introduced digital recording to European record labels. We recorded sessions in analog and transferred them to digital, and I took care of the digital editing at Soundstream's headquarters in Germany. There was a time when digital recording needed to be developed. So I was one of the pioneers of developing digital editing" (Ibid.).

After the launch of the compact disc, Soundstream did not go on. De Francisco was offered a job at Philips Classics in Holland as a producer, engineer, and editor based on a referral from her professor at Detmold. "This was one time where being a woman came into play. I was booked to record a famous piano trio for a session in Switzerland. I accompanied my former teacher – now colleague – Volker Straus on the sessions, but the musicians didn't know that their producer was leaving after the first day and leaving them with a young woman. During the first 30 minutes of recording, one of the musicians was not at ease with the notion of having a woman producer . . . until he and the others noticed I knew what I was doing and could do it well, and then they were happy.

"Much later, [Straus] told a beautiful story about me when I was a student. He learned that a younger student – me – was trying out their own recordings and had come up with her own mic setup. He tells that he was shocked when he heard my recording sounded better than his, and amazed that a young person could do something that sounded better in the same conditions. He said, 'I have to watch out for this young woman', and he was pleased about that" (Ibid.).

Among her recording artists are Claudio Arrau, the Beaux Arts Trio, Alfred Brendel, Christoph Eschenbach, Nelson Freire, John Eliot Gardiner, Matthias Goerne, Håkan Hardenberger, Heinz Holliger, Lang Lang, Gustav Leonhardt, Jan Lisiecki, Neville Marriner, Truls Mørk, Anne Sophie Mutter, Kent Nagano, Jessye Norman, Simon Rattle, Sviatoslav Richter, Lara St John, Daniel Taylor, the Montreal Symphony Orchestra, the Philadelphia Orchestra, and the Wiener Philharmoniker (AES, 2015).

De Francisco was later offered a position as associate professor at the Schulich School of Music of McGill University in 2003, where she has been ever since. She is a member of the Centre for Interdisciplinary Research in Music Media and Technology (CIRMMT). Her research includes surround-sound techniques, music recording with virtual acoustics, studies on piano brightness, and the aesthetics of recorded music. She was the producer/Tonmeister of the acclaimed research and production project "The Virtual Haydn," an extensive practical study on acoustics and interpretation (Ibid.).

"I started new as an educator and researcher, I didn't have a PhD, but I had the experience of recording 400 CDs for the global market as well as international and European awards. As such I was able to enhance the classical recording side of McGill's university program.

"A year after I started teaching there, the recording students started winning the AES Student Competitions. It was an explosion of prizes. Upon my arrival at the university I was unable to find a student who could help me as an assistant engineer, but now there are ten people who help, and some of our graduates have gone on to occupy important positions in classical sound recording; others I know are teaching at universities around the world. And when I meet *their* students, I recognize my style of recording in their work. And *that* is a wonderful feeling!" (Ibid.).

Robin Coxe-Yeldham (USA)

Robin Coxe-Yeldham earned the title "First Lady in Audio Education in America" as she accepted an **AES "Granny"** award in 1995. Prior to that, she earned the "Dean of Faculty" award from the Berklee College of Music in 1991. She helped to start the AES' "Women in Audio: Project 2000" (Mix-Staff, 1999). **Cosette Collier**, in an interview, supposed that Coxe-Yeldham might be responsible for the statistic that "five percent of all working professionals in audio are women," which was frequently cited before the USC Annenberg Study was released (Collier).

However, Coxe-Yeldham would not live to see the fruits of her efforts: she died of cancer in 1999. The professional audio world mourned her loss, but her legacy continues as a special award. Each year, the Berklee College of Music gives the Robin Coxe-Yeldham Award to an outstanding female student in their Music Production and Engineering Department.

In fact, **Jett Galindo** was a recipient of Berklee's Robin-Coxe Yeldham award, and **Lisa Nigris** cites Robin as an influence, saying, "She was an amazingly talented engineer and educator. She seemed to be able to balance work, home life, everything, with ease. I was privileged to have taken some courses with Robin while at Berklee" (Gaston-Bird, 2017). Grammy award-winning producer Tony Maserati says he still uses techniques he learned from her (Smith, 2016).

Coxe-Yeldham was born in Indianapolis in 1951. She was first drawn to graphic design and designed album covers in 1981. She began teaching at the Berklee College of Music in 1982 and helped establish the Berklee Department of Music Production and Engineering in 1983 (Mix-Staff, 1999). Wanting to know more, I contacted Don Puluse who was her Department Chair at Berklee. When he was hired at Berklee to serve as the first chair of the newly created Music Production & Engineering (MP&E) Department, he "inherited" the faculty, "and Robin was one of them," he recalled. "Robin taught all of the hands-on courses. She was thorough. All of the (equipment) manuals came from Japan and she would rewrite them." These manuals were in English, not Japanese, but the translations were often poor, so Robin would go through and correct them (Puluse).

"She was close to students; not just a teaching relationship, a long-term relationship. She started the Berklee HUB program," he says, which is a networking tool for Berklee MP&E alumni. "They had almost 50 students internationally. In music production there were more than a handful of women students, and they were generally our top students and influenced by Robin." Using the network, recent graduates could find other Berklee alumni and connect with them. "Someone moving to a city for the first time would have an introduction to someone," Puluse explains (Ibid.).

Puluse recalls that a producer had come to campus for a National Academy of Recording Arts and Sciences (NARAS) day at Berklee and convinced Rush Titleman to do producing for a group of musicians in front of a live audience. Coxe-Yeldham was the engineer for that. "It's really significant," Don explains, "most people don't allow observers in a studio session" (Ibid.).

She was an advocate for women in audio and held a panel at Berklee. "She said, 'I'm going to call it 'successful alumni'. It was a panel of four women who got out in the field and were working. And the audience was packed. The words 'woman' or 'girl' were never mentioned – and it was great," he says. Coxe-Yeldham also worked to introduce young girls to aspects of the audio and music business professions (Ibid.).

Berklee's current MP&E Chair, Rob Jaczko, was a student of Coxe-Yeldham. "I took her synchronization class. She had complete and total command of the tools. She was a great educator: balanced, articulate in her presentation. She had the demeanor to push you harder. And when it came down to learning the DTX synthesizer she was an absolute wizard. Education is performance art coupled with technology. She practiced what she preached, and she was a great teacher. And when you work with people like that it makes an impression (Jaczko, 2019).

"Robin kept the name Coxe-Yeldham when she remarried, but was officially Coxe-Skolfield," Puluse says. Simaen Skolfield was the name of her second husband. "They left Boston and moved to Florida where Robin was diagnosed a short while later" (Puluse).

Robin at Studio B, owned by Wayne Wadhams
Source: Courtesy of Carl Beatty and Dakota Yeldham

■ Theresa Leonard (Canada)

"I've spent the past 30 years working with some incredible artists in the recording studio as well as in the audio program at the Banff Centre in Canada. I've worked with many producers and been a mentor to countless engineers and musicians, but being an educator has given me the most satisfaction. As a woman, educator, producer, and facilitator, I've helped people achieve their goals and attain career heights. I've helped them connect the dots to get to where they are" (Leonard).

For over 20 years, **Theresa Leonard** was director of audio at Banff Centre in Canada. She is currently the director of the Edgar Stanton Audio Recording Center for the Aspen Music Festival.

Leonard was the third woman to have served as president of the Audio Engineering Society in 2004–2005 and has worked on Juno Award-winning and other award-nominated albums.

Leonard grew up in Canada. After losing her father at a young age, for much of her life her family was herself, three brothers, and her mother, a single parent from whom Leonard says she inherited a strong work ethic. She had an interest in sound even as a child and studied piano and enjoyed music. She recalls, "I always had an ear for it – music, recordings, and sound. It was always important to me that things sounded well." Leonard earned bachelor's degrees in music and education and was unaware of audio as a career until a colleague mentioned the program at McGill University. She ended up enrolling there as a graduate student: "I had to do a lot of prerequisite course work because I hadn't taken all these courses, but then I got into the program. At the time they only took four people . . . so once again in my life it was me and three guys." Wieslaw Woszczyk was a mentor for Leonard at McGill, who she says was a great example for her. "I saw how he stood up for audio, which was a well-respected area at McGill. I always took that with me" (Ibid.).

After earning her master's degree in sound recording at McGill, Leonard was recruited for a work-study position at The Banff Centre and was also hired to work in post-production for a television show. During that time, she heard about an opening for a music recording engineer at the University of Iowa School of Music, where she worked for four years until 1995. She was contacted for a position at Banff that year and decided to return (Ibid.).

While at Banff, Leonard developed a number of initiatives. With her background in music, audio, and education, she helped to expand the audio program so it would welcome more international applicants. "When I went back to Banff in 1995, they decided they were going to call [the program] 'Music and Sound.' Someone had met Steven Temmer who had written the Tonmeister Technology book, and they liked the idea but weren't sure where to go with it. That became my role," she says (Ibid.).

The Banff Centre had an established history of delivering both classical and jazz music programs. Leonard says the audio program was built around those areas because there were so many concerts to record and a lot of musicians were seeking demo recordings. "I also instituted 'The Recording Residency,' with pricing so that we wouldn't be competing with local, commercial projects in Banff," she says. "We were able to bring in projects that made sense for the Banff Centre and that helped to fund the studio by bringing in professional work" (Ibid.).

Leonard saw an opportunity to recruit not only international students but also faculty: "The level of musicianship at Banff – being an international arts school – is very, very high, and we had to be extremely professional." In the new audio program, she started to bring in faculty to suit different programs: classical producers, jazz producers, and even an independent ("indie") band residency. Students applied from Tonmeister programs in Germany as well as from France, South Korea, Poland, Canada, and America, among others. Leonard was mindful to recruit locally as well; one of her most well-known students is Grammy Award winner Shawn Everett (Alabama Shakes). Notable women like **Piper Payne** also attended Banff and worked with Leonard. "Piper was one of the senior people to help train and organize some of the workflow for younger artists coming through," she says. Leonard was influential for **Poppy Crum**, whom she mentored while working as a recording engineer at the University of Iowa. Crum was a work-study in Banff audio, went on to study at McGill and further, and is now the chief scientist at Dolby Laboratories. "We had a really rich faculty at Banff, including professional seminars and conferences, so that the students learned from many others. It's different from a university where students learn from the same faculty. With the work-study program, the students learned more from each other, and sometimes even more than they did with the faculty I brought in," she says.

Leonard says one reason for this is the students she chose for the program came in with such varied backgrounds: "They didn't all come from universities; some came from the 'real world' and wanted to home in on certain skills. We'd also get older people wanting to 'brush up.'" Many of these students from the Banff Centre audio program have advanced toward astonishing success stories in their careers (Ibid.).

Leonard says although the audio program was based in music and sound, students did sound support for artists, including a sound and vision residency and sound for film or video. "We did audio post production. Live sound (reinforcement) happened mostly in the theater program, and we had a great rapport with them." It was important to Leonard that there were opportunities for students to work in video, so the sound department grew to be called "Film and Media." Leonard says, "Our students did a lot of service, but in service there is great art" (Ibid.).

There was also a "Women in the Director's Chair" series at Banff every year. Leonard had women come in as audio faculty to support this: "I always had an equal number of women as faculty if I could. I didn't have to try, really, I just did it. One thing I will say about Banff Centre was that there were always pretty equal numbers of women and men in the audio program." Leonard says that since many of the students that applied to Banff came from artistic backgrounds, like music schools, and in Europe it seemed to be more common for those programs to have women (Ibid.).

Leonard also started the AES section in Alberta, Canada. "Through my work with the AES in Banff, I got heavily involved in running a few events at AES internationally. That's how I got involved as a Governor at AES, then Education chair, then President." During her time as AES president, Leonard did some international outreach. "I had a lot of students from Latin America, South Korea, Japan, China, and Mexico, in addition to Europe and North America. I was not only helping grow the initiative of Latin America within the AES, but I also met amazing potential students who came to Banff." Leonard says she's a big communicator, and with an educational background, this helped her to facilitate and strive to understand more fully the structure of AES. "We had a good team of people making sure things got done. I met a lot of people, and it was nice to be President in that mostly male dominated society" (Ibid.).

Being in a leadership position and working on inclusivity issues was enlightening for Leonard, but it wasn't easy. She drew upon her own experiences to help guide her. "Having grown up in a family with brothers and having worked with a lot of men in audio, I know issues with people can happen. It could easily be either women or men who aren't supportive. I do think women are underrepresented and we have to see ourselves reflected in these higher positions and keep

Theresa Leonard, director of the Edgar Stanton Audio Recording for the Aspen Music Festival

taking the high ground and supporting each other. At the time, in AES, I saw there were many who didn't see that as a priority. But it did take men and women to make that happen. I learned that it is important to talk about it, because sometimes people don't see it" (Ibid.)

During her tenure, Leonard sought to change the dynamic between students and professionals at AES events. "You don't have to make something students will understand: students are bright, brilliant people. I tried to bridge the gap between students and professionals – I wanted to help change that" (Ibid.).

Leonard has recently taken on the role as director of the Edgar Stanton Audio Recording Center for the Aspen Music Festival and School in Aspen, Colorado. She is thrilled to keep growing in the audio area both professionally and educationally and will continue her work as a music producer and audio program consultant.

Lisa Nigris (USA)

There are many opportunities for audio engineers to work in university programs focused on music performance. These schools need live recording and sound reinforcement for their concert series and recitals. Music conservatories host dozens of events each year: solo performances, ensembles, symphony orchestras, and operas. At the New England Conservatory (NEC), it is the role of the "Recording and Performance Technology Services" department (although there are a few different names, such as "Audio/Visual Services" or "Recording

Services," for example) to take care of the mic setup, recording, and delivery (via streaming or recorded media) of these shows.

Lisa Nigris is the director of Recording and Performance Technology Services at NEC. "But there's no audio engineering program," Nigris explains. Her student workers are performance or composition majors, "so they have to learn a new field by working in my department" (Nigris).

Nigris remembers her own musical childhood. "I started taking piano lessons around the age of 7 and voice lessons at 12. There was a bit of a revelation when my folks gave me a (Yamaha) DX7 and I discovered that manipulating sounds interested me more than performing songs. My earliest experience with recording was holding a mic in front of my TV speaker trying to capture Kiss performing on Don Kirshner's Rock Concert. I learned about the need for isolation in recording that day . . . among other things" (Ibid.).

"In high school, my voice teacher mentioned that Berklee would be a good place to go to explore the various fields within the music industry. That was good advice. . . . While at Berklee, we had to record a sound alike project. I chose the Beatles' tune, '8 Days a Week'. It was a humbling experience. Sounded awful, but I learned a lot about listening, and capturing the desired performance from an artist." She recalls she was one of three other women in her class. "A little later in my college career, I was interning at Blue Jay and realized that the repetition of studio work was not for me. I thrived on the pressure of live sound and live concert recording. This was a major revelation that guided my career" (Ibid.).

Nigris worked at the Aspen Music Festival as director of Audio Production. "Nine years ago, I got a call from my boss saying 'Aspen is going to call you. They can *borrow* you but they can't *have* you.' The festival needed help getting their recordings accepted by National Public Radio. It was only supposed to be a one- to two-year gig," but she ended up working there for eight years (Gaston-Bird).

Nigris has since been working at the New England Conservatory, which has built new concert halls, bringing the total to eight. "We were one of the first colleges to have a Dante system in place to solve a problem in Jordan Hall: the lines themselves were corroding. . . . We're using Dante everywhere and this past September (2018) we rolled out a new service in seven of the eight halls for recital videotaping and streaming. In addition to a new dorm and library space, [one] building includes a black box theater, large recording studio, orchestra rehearsal space, and a small stage area in the dining commons" (Gaston-Bird).

Nigris and her team decided to implement Steinberg's Nuendo as their DAW of choice. The decision was based on the ability to process files in less-than-real time, which was important for a busy concert season. "most [Pro Tools users] are doing short pop tunes . . . but when you're doing a *concert* that's an enormous chunk of time" (Nigris, 2019), (Fortner, 2019).

Valeria Palomino (Mexico)

Valeria Palomino is a recording engineer and educator in Mexico. She shared with me her observations about teaching, new music, music technology, and gender. "This educational atmosphere regarding genders is for both men *and* women. It's equal in that sense. We have misunderstood many things for many years – women and men – about our healthy relationships in the professional arena. It's not something *you teach them* or *they give us*. I think we all have been making lots of mistakes about it. And we need this conversation and content of your book [Women in Audio]. It's really going to make a difference. That's important. If we don't connect really well between everyone, we are not going to advance" (Palomino, 2019).

In the classroom, she defines and assigns tasks. "I notice that it's important in the studio to have very clear positions defined for the process of the recording. These two to three guys are responsible for miking techniques; the other 3 guys set up the console, and the other 3 guys are on the software" (Ibid.).

She talks about "guys," but there are girls there, too. "It's always around 10% girls. Every 20 people, I have 2 girls more or less." When asked about how gender comes into play in the classroom, she replies, "The difference is they are always on your eye because they are so few. If you have just one or two students in your group who are women, you notice. The same thing happens at the AES conferences: you notice the women, because there are only two or three. And that's terrible but it's impossible to avoid."

Her female students have asked her to speak about the issues. "The best definition of misogyny I have is this attitude of, 'I forgive you for being a woman, but only if you are perfect'. It's very sad but it is true in many contexts. So, some women believe they cannot fail in this industry because it is male-dominated. I travel to different schools and give lectures and a group of girls asked me to speak about gender, so I've been reading about it. And they agreed with me when I talked about perfectionism. For example, [when girls] start to do dangerous things like carrying too much heavy equipment. If you are small person, you cannot compete with that – and you shouldn't be sorry about it. The first person that has to accept that is yourself. You always have something else to give. Don't be so anxious to demonstrate you can do anything. I try to calm down women *and* men: they can fail, and they can be different, and that's nice because differences are good for a project" (Ibid.).

Although she loves teaching, for the moment, Palomino is focused on making albums and recordings, especially new classical music. "I love the recording and divulgation of new music; even though it's exciting to record the great, old, traditional composers. I love when we create the custom and habit of recording live music – modern music – and try to be in the middle of that connection between audience and new music, which is not easy. This is not a natural process in this century. And the musicians and the projects are open to exploring different mic techniques and quality of the sound.

"You need to record with a high class of microphone. That doesn't ever change. I prefer to use a Decca tree with 3 omnis; [placed] high in a hall with good acoustics, especially the ones built for chamber music and small orchestras" (Ibid.).

Among her credits, Palomino has recorded the music for *Frida* (directed by Julie Taymor, 2002 Oscar winner for Best Music), *Zurdo* (directed by Carlos Salces, 2003, Ariel winner for Best Music), *The Crime of Father Amaro* (directed by Carlos Carrera, 2002, nominated for an Oscar for Best Foreign Film), and *Walkout* (directed by Edward James Olmos, 2006), among others.

She studied piano and arts at the Universidad Nacional Autónoma de Música (UNAM). "In Latin America 26 years ago I started to do that. There were not universities in which you could study this formally. The audio I had to learn on the street. There was no internet! Now you can search everything, and you have it in a minute. Back then, you had to find a book, or a friend would give you an article. This was in 1994. And you had to believe what your mentors told you.

"I have always said 'yes' to new professional experiences. Sometimes students go to university for something specific. Students want to focus on this or that. But have your mind open to new opportunities. Audio is changing every 8–10 years and the opportunity to work in anything presents itself. For example, many of them started wanting to record great bands in a great studio, but that kind of way of doing things is dying. I make productions, live sound, recording, I do many things because you have time to learn a lot of things. And our careers are long. You have to be open. You might not get a job in that specific area but at least you explored it. This is advice for life" (Ibid.).

Liz Dobson

Dr. Liz Dobson used to love playing with cassettes. When she was around eight years old, she would "bounce down" performances (because she couldn't play every instrument at once, of course), and she thought it made a cool sound. "Tape for me was like playing with Lego. I would record my family, I would even play tape backwards." Looking back, she puts that in context of how a child's environment becomes a kind of privilege. "If we didn't have that," she explains, "I wouldn't have had those avenues of interest." She could even try coding because she also got to play with an early computer, the ZX Spectrum Computer. During our chat, she quickly performs a search for one so I can see it. "Ah! There it is, the one with the squidgy keyboard," she announces (Dobson).

"You would have dinner while [your program] was loading. You could type in code and make little melodies. In the 80s when MIDI came out, I convinced my family to get me a MIDI keyboard" (Ibid.).

Eventually Dobson would major in classical music composition and performance and later earn a master's degree in composition for film from the University of Wales, but she was always drawn to music technology. "In 1994–1998, there were some MIDI computers and

Liz Dobson, principal enterprise fellow, lecturer in music technology, and National Teaching Fellow at The University of Huddersfield

S1000 Akai samplers in our studio, and there were modules in music technology. Music tech in uni was rare. I took a required module which was pretty basic, like connecting a MIDI keyboard to a computer which these days you would be expected to know before you started a degree program. . . . A lot of what I do is self-taught, just because the education system in this country wasn't set up to teach that" (Ibid.).

She initiated and later cofounded the **Yorkshire Sound Women Network**, wrote the Audio Equity Pledge (#audioequitypledge), and coordinated the industry collaboration around its development. The pledge is a list of recommendations for "supporting women in sound" and a "resource for people working in education, industry, audio production and other related fields." You can read more about it by following the link on her blog at drlizdobson.com/blog.

Currently she is developing a multimodal methodology for researching music technology engagement in extracurricular settings and managing a series of workshops called "Go Compose" with Sound and Music (a national charity for new music). She is also a Community Interest Company (CIC) director for the Yorkshire Sound Women Network and is doing a residency at Q-02 (a "subsidized workspace for experimental contemporary music and sound art").

Cosette Collier (USA)

Cosette Collier is a professor of audio production at Middle Tennessee State University (MTSU). She graduated from the University of Memphis with a BFA in commercial music/recording engineering and an MA in communication/film and video production. She began

Cosette Collier, professor of audio production at Middle Tennessee State University

teaching at MTSU in 1993. She is also mentioned extensively in Chapter 1 for her role in the AES Women in Audio: Project 2000 initiative.

Collier has specialized in mastering, audio restoration, and audio forensics. More recently, she has begun working on a history of the demise of Nashville's historic Music Row as well as a textbook on basic audio and is researching virtual and augmented reality and Dolby Atmos.

Bibliography

About Multimedia Systems Department [Online]. 2004–2017. Gdańsk, Poland: Gdansk University of Technology. Available: https://sound.eti.pg.gda.pl/profile_en.html [Accessed July 10, 2019].

AES Education Directory [Online]. New York. Available: www.aes.org/education/directory/ [Accessed June 20, 2019].

Angus, Jamie. 2017. Interview with Jamie Angus. *In:* Gaston-Bird, L. (ed.).

Berklee. c. 2014. *Celebrating Berklee's Faculty Pioneers* [Online]. Boston, MA: Berklee College of Music. Available: www.berklee.edu/berklee-today/spring-2014/celebrating-berklees-faculty-pioneers [Accessed June 20, 2019].

Born & Devine. 2015. Music Technology, Gender, and Class: Digitization, Educational and Social Change in Britain. *Twentieth-Century Music*, 12, 135–172.

Collier, Cosette. 2019. Interview with Cossette Collier. *In:* Gaston-Bird, L. (ed.).

De Francisco, Martha. 2019. Interview with Martha De Francisco. *In:* Gaston-Bird, L. (ed.).

Directory of Educational Programs [Online]. New York: AES. Available: https://web.archive.org/web/19961023233750/www.aes.org/education/europe4.html [Accessed June 19, 2019].

Dobson, Liz. 2019 Interview with Liz Dobson. *In:* Gaston-Bird, L. (ed.).

Fortner, Stephen. 2019. *New England Conservatory Chooses Steinberg Nuendo* [Online]. Available: https://www.prosoundweb.com/new-england-conservatory-chooses-steinberg-nuendo/ [Accessed September 27, 2019].

Gaston-Bird, Leslie. 2017. 31 Women in Audio: Lisa Nigris. *31 Women in Audio* [Online]. Available: http://mixmessiahproductions.blogspot.com/2017/03/31-women-in-audio-lisa-nigris.html [Accessed Friday, March 17, 2017].

Jackson, Wayne. 2019. Interview with Wayne Jackson. *In:* Gaston-Bird, L. (ed.).

Jaczko, Rob. 2019. Interview with Rob Jaczko. *In:* Gaston-Bird, L. (ed.).

J.A.M.E.S. 2018. *James HE & FE Course Accreditation Overview* [Online]. Available: www.jamesonline.org.uk/accreditation/accred_overview/ [Accessed July 10, 2019].

Lam, Mariam. 2018. Diversity Fatigue Is Real. *The Chronicle of Higher Education*, 64.

Leonard, Theresa. 2019. Interview with Theresa Leonard. *In:* Gaston-Bird, L. (ed.).

Martha de Francisco [Online]. 2015. New York: Audio Engineering Society. Available: [www.aes.org/events/139/presenters/?ID=3741] [Accessed July 10, 2019].

Mathew, et al. 2016. *Women in Audio: Contributions and Challenges in Music Technology and Production.* Audio Engineering Society Convention 141.

Mix-STAFF. 1999. *Current* [Online]. Available: www.mixonline.com/recording/current-373920 [Accessed July 10, 2019].

National Association of Schools of Music Handbook 2016–17. National Assocation of Schools of Music.

Nigris, Lisa. 2019. Interview with Lisa Nigris. *In:* Gaston-Bird, L. (ed.).

Palomino, Valeria. 2019. Interview with Valeria Palomino. *In:* Gaston-Bird, L. (ed.).

Puluse, Don. Interview with Don Puluse. *In:* Gaston-Bird, L. (ed.).

Schoenberg, et al. 1987. *Arnold Schoenberg Letters*, University of California Press.

Smith, Ebonie. 2019. Masters Behind the Mix: A Behind-the-Scenes Glimpse into Music Making With Engineer/Producer Tony Maserati. *Atlantic Records/All Access/From The Studio* [Online], June 8. Available: www.atlanticrecords.com/blog/22641 [Accessed December 12, 2016].

Wilson, Robin. 2012. Why Are Associate Professors so Unhappy? *Chronicle of Higher Education*.

10 Games, Immersive Sound, and Audio for Virtual, Augmented, and Mixed Reality

Tools of the Trade: Game Sound

Game Engines: To build an entire video game from the ground up, you need a game engine. Unreal Engine (www.unrealengine.com) and Unity (unity.com) are popular ones: both are free to download, and both have tutorials for beginners. Both Unreal and Unity allow users to do sound design. If you want to release a game commercially on these platforms, you will have to pay for a license. **Winifred Phillips** also says, "there are proprietary tools used by individual publishers and developers who make their own game engines. My thing is to give the teams the tools they need – whatever the engine" (Gaston).

Middleware: If you don't want to build a game engine and you want specialized tools for audio, you may want to try Wwise (www.audiokinetic.com/products/wwise/) or FMOD (www.FMOD. com/). Both are free to download, and both come with tutorials. An article by Anne-Sophie Mongeau compares FMOD, Wwise, Fabric, and Unity. She finds that Wwise is less expensive, but FMOD is "designed like a DAW (Digital Audio Workstation)," with several important caveats: "Games are not a linear medium," she cautions, "and sound integration is not sound designing or editing" (Mongeau, 2016).

Programming Languages: Knowing a computer programming language is definitely useful. Unity is based on C# ("C sharp"), and there is an open-source framework called "MonoGame" (www. monogame.net), which also uses C#. C++ is also used (Unreal uses C++), but at least one educator encourages new programmers to use a different language to begin learning how to code because of its steep learning curve. Java is popular for Android apps and a bit easier to learn. Javascript and Python are also used. Each of these languages has many online tutorials for self-paced learning.

Tools of the Trade: Immersive Audio

In immersive audio, loudspeakers are placed all around the listener. Various arrangements include:

- Quadrophonic (two speakers in front and two in back),
- 5.1 (three speakers in front, two in the rear, and one subwoofer),
- 7.1 (either three front, two side, and two rear; five front and two rear (cinema); or three front, two "height" channels, two rear channels, and a subwoofer).

There are also 10 channel setups (10.1 with one subwoofer or 10.2 with two subwoofers), 22 channel setups (or 22.1, 22.2, etc.), and just about everything in between.

In order to assign sound to all of these channels, a **Digital Audio Workstation** needs to be flexible with routing audio. Here are some options:

- Reaper: Many engineers working with immersive sound use Reaper because the user can assign any number of loudspeakers they like and choose how to pan sounds from one speaker to the next.
- Nuendo (Steinberg): This is a popular choice that allows many popular loudspeaker configurations from mono to 10.2 for multichannel audio at a lower price point than Pro Tools.
- Pro Tools (Avid): The latest version of Pro Tools Ultimate allows for surround and Dolby Atmos mixing; however, Pro Tools First and Pro Tools do not offer this feature.
- Logic Pro X (Apple): This offers multichannel configurations up to 7.1.
- Max/MSP (Cycling '74): Because of the versatility of Max/MSP, a user can set up behaviors to be assigned to various loudspeakers. This can be used in an interactive scenario, for example.

There are also multichannel versions of **plug-ins** that you can buy, some more affordable than others. A partial list includes:

- NuGen Audio: The Upmix, Downmix, SEQ-S, and ISL Limiter plug-ins from NuGen Audio work in surround formats up to 7.1. Upmix takes a stereo signal and can expand it to 5.1 or even 7.1.4 (.4 refers to four height channels) for **Dolby Atmos** or **Auro 3D.**
- iZotope: The Insight metering plug-in and RX7 Advanced plug-ins (noise reduction, de-click, etc.) feature multichannel support up to 7.1 channels.
- Waves 360°: This plug-in bundle includes the C360 Surround Compressor, Durrough Surround meters, the R360 surround reverb, the S360 Surround Panner, and M360 Surround Manager and Mixdown, which performs **bass management**.
- Audio Ease: Their Altiverb and Indoor plug-ins are capable of multichannel sound, with an impressive array of impulse responses. With Indoor, you can move a virtual microphone inside of a space to create a sense of distance from near to far or even around a virtual wall.

Microphones for immersive audio can be set up as room mics, or a single ambisonic mic can be used to capture sound in 360 degrees. Such microphones include:

- Sennheiser AMBEO
- Soundfield
- Zylia
- Røde NT-SF1

Agnieszka Roginska is editor of *Immersive Sound: The Art and Science of Binaural and Multichannel Sound,* a textbook which goes into further detail about many of these concepts and which is listed in the section, "Recommended Reading".

Tools of the Trade: Binaural Audio

Microphones: Binaural audio is meant to create a better approximation of what the "ear hears" by placing microphone capsules inside each ear. One example of this is the Neumann KU100 microphone, nicknamed "Fritz" (see photo). Roland makes the CS-10EM, a pair of binaural microphones that double as in-ear monitors.

Plug-ins can then be used to process binaural audio with an averaged **HRTF** (see: Fun Facts, "Head Related Transfer Functions") so listeners can have a common experience. Two examples include:

- Facebook 360 Spatial Workstation (formerly Two Big Ears): A suite of plug-ins for processing and publishing 360-degree audio.
- AMBEO Orbit: A panner that can be used with the Sennheiser AMBEO mic or other binaural signals to pan sounds in a 360-degree space.

Zoë James holding the Neuman KU 100 microphone

▮ Other Tools

Headsets: A number of headsets can be used, which is important when mixing audio matched to picture within a VR/AR/MR space. They include the Oculus Rift (VR), HTC Vive (VR), Valve Index, and Magic Leap One.

Gaffer tape: One can never have enough gaffer tape.

Women in Game Sound: Industry Statistics

Statistics from the "GameSoundCon Game Audio Industry Survey" conducted in 2016 and 2017 reveal that the percentage of women in the composing and sound design industry has risen from 3.5 percent in 2014 to 12.7 percent in 2017 (Schmidt). However, there is a gender-based pay gap, and women are earning about 83 percent of what their male counterparts earn.

"High-level Summary of Results (2016):

- A greater percentage of women reported income from game audio (10.4%) than prior years (7% in 2015, 3.5% in 2014).
- Even after accounting for the lower average number of years of experience in the game audio industry, women make less than men.
- The difference in total income is equivalent to approximately 2.1 years of experience; That is, the 'cost' of being female in game audio is approximately the equivalent of having 2.1 fewer years of experience in the game audio industry" (Schmidt, 2017).

"Among the findings (2017):

- Women game composers and sound designers are up to 12.7% of the industry
- Up from 10.4% in 2016 and 7% in 2015
- Average Salary (employee): $69,848
- Women, on average earn 83% of what men earn" (Schmidt, 2017).

Winifred Phillips (USA)

"There was a whole arcade of retro games in the Thomas Jefferson building at the Library of Congress. . . . It was a lot of fun, in those marvelously fancy government rooms lined in mahogany paneling, to see people playing Pong and Pitfall" (Phillips).

In 2019, **Winifred Phillips** was invited to give a lecture entitled "The Interface Between Music Composition and Video Game Design." The lecture took place during an event hosted at the Library of Congress called Augmented Realities: A Video Game Music Mini-Fest. Part of the mission of the Library of Congress' music division is to "spotlight different kinds of music, in the effort to protect and support copyright and intellectual property," Phillips explains (Ibid.).

Phillips was invited because she composed music for titles in the *LittleBigPlanet*, *God of War*, and *Assassin's Creed* franchises, among many others. Hers would be the "very first video game music composition lecture ever given at the Library of Congress" (Phillips). She is also the author of *A Composer's Guide to Game Music* (The MIT Press), which earned her a Global Music Award for an exceptional book in the field of music (Society, 2015). In addition to her Library of Congress lecture, Phillips has also given presentations for such prestigious conferences and organizations as the Society of Composers and Lyricists, the Game Developers Conference, the North American Conference on Video Game Music, the Ludomusicology Conference, the Montreal International Game Summit, and the Audio Engineering Society convention.

Winifred Phillips at the Library of Congress

At the start of her professional career, Phillips wasn't pursuing a career in video games. At first, she worked on a series for National Public Radio called *Radio Tales*, which featured adaptations of famous literary works such as *War of the Worlds* and the short stories of Edgar Allan Poe. She was hired to join the *Radio Tales* series as its music composer and sound designer. "But I was always playing video games," she recalls. Her decision to make a career change occurred during some time off, while in the midst of playing the original *Tomb Raider* video game. "When you are in the ballroom in the mansion during the tutorial," Phillips recalls, "you can turn on the stereo and listen to the game's music. The music is isolated from game play, and nothing else places any demands on your attention. And that is when it first occurred to me that I could be writing music for games.

"I started doing research. I wanted to see how I could reach out to developers and game studios, to let them know who I am and what I've done. Around that time, I made a connection with the folks at Sony Interactive Entertainment America. They were putting together a composer team for *God of War*," Phillips explains. "And so *God of War* became my first project as a game composer . . . along with the *Charlie and the Chocolate Factory* game for High Voltage Software, which is another game project I landed at the exact same time." Phillips adds, "The game was a tie-in with the Johnny Depp film, and the director Tim Burton himself approved the musical score I composed for the game. So, it was exciting for me to know that he'd listened to my work and that he'd thought it was right for the world he was creating for *Charlie*" (Ibid.).

While giving an overview of the art and technology of creating music for video games, Phillips pauses to emphasize the difference between linear and dynamic music. "Linear music is traditional music composed for film, television, the symphonic concert hall," Phillips explains. "It is music that is designed to proceed with a definitive, unalterable structure over the course of time. That's why we call it linear: because it follows a line." Phillips then adds, "Dynamic music is structured so it can alter itself, depending upon what's going on in the game. Dynamic music is flexible. It's structured in parts – in components – in bits and pieces. These bits and pieces can then be manipulated by the game engine, allowing the music to change in reaction to the shifting circumstances of the game. The goal is to make the music sound as though it were written in a linear fashion . . . so that the listener can't really tell that it's been structured dynamically," she explains. "Instead, the music just seems to be reacting to what the gamer is doing in a very cinematic and organic way. That's the ultimate goal of dynamic music composition" (Ibid.).

While Phillips' description of dynamic music calls attention to the unique demands of video game composition, there are some interesting similarities between composing for games and composing for film and television. As an example, a composer might choose to create an orchestral composition for strings, brass, and percussion or perhaps a rock composition for drums, guitar, and keyboards. For the sake of convenience, these instruments might be grouped together in "busses" or "stems," to provide the audio engineer with greater efficiency during the mixing process. Then, these separate instrumental recordings would invariably undergo a stereo-mixing process, which would allow the separately recorded instruments to play together in a single stereo audio master. This isn't necessarily the case when preparing dynamic music for video games (Ibid.).

Phillips takes a moment to differentiate the terminology of audio engineering from that of dynamic game music construction. In film and television, the music editor might use the "busses" to create "stems" to group the instruments together for mixing purposes. But in video game music creation, it's necessary to have "layers" instead. The meanings of the two terms are similar, but not interchangeable. As Phillips puts it, "there's a little bit of a difference in terms of the philosophy behind it. When you are stemming, you might stem individual instruments," (for example, a strings stem, a brass stem, or a percussion stem) "because you want fine control. You're not thinking about how they sound on their own because they're not going to play on their own – you're just getting them separated so that you have fine control over each one," Phillips says. "With vertical layering, it's possible that these things could play by themselves, or they could play in subsets depending upon what's going on" (Ibid.).

To better describe the concept of vertical layering, Phillips discusses an example from her own repertoire. "I composed music for quite a few *LittleBigPlanet* games for Sony Europe

Winifred Phillips

with Media Molecule, Sumo Digital, Tarsier Studios, Double 11, Supermassive Games, and other developers. In most of the *LittleBigPlanet* games, there would be 6 simultaneous layers of music individually recorded for each dynamic music composition. The six layers would play simultaneously in the game – in the same way that stems play together in a DAW (Digital Audio Workstation) – but the game engine could alter the dynamics of each layer. The game engine could turn layers on and off. Essentially, the game engine could act like a virtual mixing engineer. This would allow the music to be interactive, according to that vertical layering system" (Ibid.).

Looking back on her career path, Phillips reflects for a moment on her childhood as an avid gamer. "For me it was all about video games as a passion," she recalls. "I was always a geeky girl gamer! It was a wonderful way to escape reality. I used to love to pretend that the video games I played were actually a 'heads up display' – that I was controlling something that was actually happening in another world." Phillips laughs. "It's wonderful to live in the times that we do now . . . to see virtual reality come into existence. I think of the girl that I was, and how I imagined all these wonderful, futuristic things that are now actually happening in my lifetime. We're stepping into virtual worlds and interacting with them in ways I never could have dreamed."

Eiko Ishiwata (Canada)

Eiko Ishiwata is the music director for Snowcastle Games, who recently released the title *Earthlock: Festival of Magic*. She is also the main composer at Flying Carpet games, whose titles include *The Girl and The Robot* and the upcoming release *Hiboka* (scheduled for 2020) (Ishiwata, 2019).

She recalls that at age 7, she was given a soundtrack by her mother. That night, "instead of practicing, she composed her first piece." When she was 11, she had saved up her allowance to buy a four-track Tascam Portastudio tape machine. "I taught myself how to use the Portastudio and tried to write music similar to [rock band] Faith No More by playing all the parts, besides drums which I programmed in my Yamaha W7 Synthesizer" (Ishiwata, 2019).

Her mother came to St. John's Newfoundland, Canada (where Eiko was born) from Japan. She "was always a rebel against the social gender constrictions of Japan and taught me to be very independent and individualistic," Ishiwata recalls (Ibid.).

Not only is Ishiwata a freelance composer, she is also an indie game developer. She is excited about the development of her game, *Valiant Mirror*. "I realized I want my own creative vision to create my own worlds: sound and every other aspect. I'm in early development but plan to lead a team once the company has been established." She is also planning to have a symphony orchestra perform music for the game, perhaps leading to a tour. In fact, her song "Akari" from the game *The Girl and the Robot* was performed in 2017 during the Montreal Video Game Symphony by Dina Gilbert and the Orchestra Métropolitain (vimeo.com/248154049). "Akari" was also performed with the Orchestre Symphonique de Québec in Quebec City (Ibid.).

She is also in talks with other orchestras for performances for the music from *Valiant Mirror*. "We are hoping to create a lot of hype through PR for the music and our goal is to one day have a touring Orchestra similar to *Distant Worlds*. The plan for the Studio Ishiwata is

Eiko Ishiwata
Source: Photo credit Karolina Turek

to open up a location in LA once the first game is released and possibly one in Japan" (Ibid.). You can learn more about Ishiwata at eikoishiwata.com.

Ayako Yamauchi (Japan)

Ayako Yamauchi is a game sound designer in Los Angeles. She is most concerned about equitable working conditions, lamenting the fact that in Los Angeles, companies hire very young people and ask them to work day and night for minimum wage. She is trying to change the culture of companies asking people to work for free (Yamauchi).

She notes that at the American Federation of Musicians conference, more people from Europe raised their voices about why payments for streaming services weren't negotiated while in Japan, she says, musicians' unions negotiated with Google, YouTube, and Spotify. "It's not perfect, but at least they're making an effort," she says. "In the US, people are giving up. Video games work is non-union" (Ibid.).

She currently works for the Formosa Group editing and mastering dialog and has contributed to such titles as *God of War*, which won a BAFTA award in Best Sound, and *Uncharted: Lost Legacy*, which won Game Audio Network Guild (G.A.N.G.) awards for Best Dialog, Best Audio Mix, and Best Cinematic/Cutscene Audio. "America is super dominated by 'star culture' . . . people are afraid to be unique," she says. But, she says, "we all love music and sound. I am a music lover and artist." She is especially proud of the music video "Eternally," which she composed, produced, and engineered and which garnered several festival awards (Ibid.).

When trying to break into the game industry, some people told her that her lack of credits on game titles was holding her back. However, Dave Collins, chief engineer of A&M Records, listened to her work and hired her as a second engineer. "He didn't judge me even though I didn't have many credits," she says, and he was encouraging. Gina Zdanowickz, who teaches Game Audio for Berklee College of Music online, is also someone she credits for mentoring her (Ibid.).

Fun Facts: Ambisonics

Ambisonics is a technology used to create 360 degrees of sound around the listener using a single microphone with multiple capsules. Michael Gerzon and Peter Craven first developed the microphone in the 1970s, although the concepts date back to the 1930s (Malham).

There are different "orders" of Ambisonics: A "zero order" Ambisonic system is basically an omnidirectional pattern. A "first order" Ambisonic system is comprised of three components: 1) a figure-of eight polar pattern oriented to the left, 2) one to the right, and 3) one up and down. A signal processor can add and subtract levels from each lobe in order to derive other polar patterns. As the orders increase, the patterns become more directional, and some flatten.

By using higher order ambisonics, it is theoretically possible to record a sound from anywhere within a space and upon playback make it appear as though the sound is originating from that same point in space.

FIGURE 10.1 Visual representation of the Ambisonic B-format components up to third order (starting with zero).

Dark portions represent regions where the polarity is inverted.

Note how the first two rows correspond to omnidirectional and figure-of-eight microphone polar patterns.

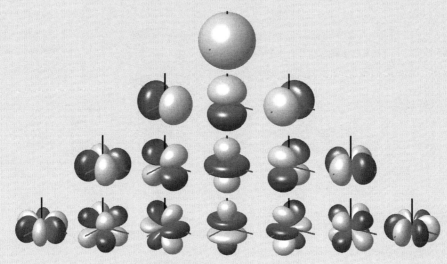

Source: Photo credit Dr. Franz Zotter, zotter@iem.at

Emily Ridgeway (Australia)

"Sound is fluid in a way. The way it gets into your ears. One of the main goals of VR audio is a sense of natural immersion, and that doesn't necessarily always mean spatial accuracy," says **Emily Ridgway.** As the award-winning audio director for *BioShock*, Ridgway's opinion carries a good deal of weight. She now works at Valve Corporation, which recently released the Valve Index, a complete VR kit that comes with a headset, controllers, and room-tracking hardware. "The audio component is fairly different from regular headphones," she explains. "We designed the audio to output via two 'ear speakers' that hover off the ears by about one centimeter each side. This lets sound flow around the user's own ears and head as it would in real life. The result is that virtual audio feels more natural than traditional isolating headphones" (Ridgeway, 2019).

Originally from Australia, Ridgway has been working at Valve since 2013 where over the years she has helped contribute to the Index VR, the HTC Vive, The Lab (Valve's first VR experience), and *Counter Strike: Global Offensive* as a writer and sound designer (Ibid.). It was the overlapping importance of sound directionality within e-sport giant CS:GO and VR that required Ridgway to become an expert in matters related to 3D sound, acoustic simulations, and Head Related Transfer Functions (HRTFs).

The study of HRTFs is almost a field of its own. The principle is that the shape of the ear and ear canal gives a unique, measurable frequency response, where certain frequencies are blocked, attenuated, or even boosted depending on the direction of the sound. This "transfer

Valve Index

function" is as individual as each person's ear shape. I couldn't help but ask Ridgeway about the effects of someone with a large amount of hair (such as myself) adding to the HRTF frequency response curve and how that might affect the VR experience.

She explains that "sounds are absorbed and reflected off our head and ears just like any other geometry." For example, if you are wearing a heavy winter coat with a hood, your sense of where sounds are coming from may change temporarily. If you then take the coat off, you have "changed your geometry" again; so as Ridgeway explains, "it makes sense that our brains could adjust to different HRTFs over time." The number of reflections in a room space can also make it difficult to localize sounds, too. She continues: "like calling your cell phone when it's lost. I can rarely pinpoint it immediately just using the sound. We have to hunt around. So we're not great at localizing sound in real life at the end of the day" (Ibid.).

Ridgway's ability to effectively convey a story using audio tools is proven, in part, by her work on *Bioshock*, for which she won the Game Developers Choice Award in the categories of Best Audio and Writing. "If you are creating an entertainment experience then it makes sense to have emotional goals. In games we have all these simulations and dynamic interactions, but it's what the player *feels* that becomes the primary goal for me," she reminds us (Ibid.).

In her early school career, she didn't care much for the music software being used. "I would write on manuscript paper and have my friends convert it into Cubase for me." By the time she started her career, however, she had to learn to get comfortable in a technical domain quickly. "I started in game audio in 2004, and there were no nice audio interfaces like Wwise and FMOD. Studios typically wrote their own engines." One audio engine was RAD Game Tools. "The sound designer would use text files to assign volume, pitch, looping – and other basic parameters to thousands of sounds in the game. Working like that was something I had to get comfortable with very early on" (Ibid.). In 2007, as dedicated game audio tools rose into prominence, Ridgway used FMOD Designer to craft the music and sound effects at Tim Shafer's studio, Doublefine, where she worked for four years.

Since joining Valve, Ridgway has extended herself even further, learning how to program in C# and C++. "The Lab VR was all created in Unity, and that enabled me to experience the joy of directly prototyping sound behaviour and interactions without having to bother other people on the team. That level of control and autonomy was addicting, and I think the end result was better for it" (Ibid.)

"My advice: Learn to enjoy learning," she says. "It's been so important to my success to get outside my comfort zone and learn new methods and areas of study that I was almost afraid to delve into, like programming and hardware. Every time I learned something new and acquired a new skill, I became a more unique professional. I don't think about trying new things as succeeding or failing but rather slowly improving and growing over time" (Ibid.).

Fun Facts: Head Related Transfer Functions

Not everyone hears things the same way, and that's because our ears are shaped differently. For audio engineers creating content for headphones and earbuds who want to create the illusion of sound coming from behind you or above you, it's important to pay attention to Head Related Transfer Functions, or HRTFs.

To build a database of HRTFs for many different shapes of ears and heads, engineers ask many different people to come to a laboratory where they place binaural microphones in the person's ears and then play sounds from loudspeakers placed in locations around the person at ear height, then above the person, and even below the person. They play pink noise and measure the frequency response at each angle for each person.

As you can imagine, the more people participate, the larger the database, and the more engineers can create an average profile. Then when you play back audio in your game or VR experience, you can either choose a single user's HRTF profile from the database, or use an average of everyone's profile. You can even create your own HRTFs.

Khris Brown (USA)

As a child, **Khris Brown** was a big fan of *Starlog*, a science-fiction magazine that began publishing in 1976. She also subscribed to the Star Wars fan club newsletter, *Bantha Tracks*. "There was a cellophane record as an insert and it was about the sound effects for *Star Wars*. I grew up in Northern California and that's where the effects were created" (Brown). When she was in the fourth grade (around nine to ten years old), she programmed her first video game in BASIC on a PET computer (Wawro). When she was 12 years old, she got to meet Ben Burtt, sound designer for the *Star Wars* films. "He was recording arrows," she recalls, in a barn before Skywalker's tech building was built. Her father was an engineer involved in construction at Skywalker Ranch. "I freaked out when I found out Ben Burtt was there and I pushed my dad to apply for the job," she remembers. Her mother volunteered in literacy teaching and environmental science (Brown).

She probably wouldn't have dreamed that years later she would be the "guardian" of Ben Burtt's private sound effects library. "I had all the raw, original recordings of light sabers and the servos from Cadillacs that were C-3P0. Each sound saber had to be unique to whatever color it was and Ben put me in charge of it. Penguins became Geonosians and cars became Anakin's pod" (Ibid.).

Before earning the dream gig managing Burtt's library, she had studied environmental science, but then dropped out to "temp" at Lucasfilm Games. "There were two sound designers. I stayed late and arrived early to prove I was doing good work." In those early days, audio crews were cobbling things together. "I was using Sound Designer in the beginning. Then there was Region Munger in the 90s. We had Waves plug-ins before their 'plug-in suites'. We were just 'duct taping' things together," she says of the various software her team had to use (Ibid.).

"This year [2019] it will have been 29 years since I started. I've done voice, audio directing, and have worked for Electronic Arts, Ubisoft, and Oculus." She runs her own company, Vox Point LLC, which does everything from budgeting the voice production process to story consultation, casting, and recording. She is also the voice director at Double Fine Productions (Ibid.).

From time to time, she meets someone who isn't familiar with her or her work. " 'Do you play games?' That's the first question I get. No, I just gave my life to this industry . . ." She doesn't finish, interrupting herself with an incredulous laugh (Ibid.).

"In games, what I feel is particularly engaging is the way in which the player moves through the world, and the additional opportunity for the player to meet strangers and create teams and to work through – similar to the rebirth of *D & D* [*Dungeons and Dragons*] – work with building guilds with strangers, work with finding out: what are these peoples' different skills and abilities? What are their different talents? So we are all always focusing on creating a collaborative future together which I think is essential for the world" (Fine, 2018).

Khris Brown recording at Activision

Khris Brown with Tim Schafer and Wil Wheaton

Brown has some perspective on being a woman and starting out in game audio. "Often I think we are invisible. There are stories of people sleeping under desks and having no money. It's not okay that it is that way. But those things DO exist. . . . Learn about innovations and inventions; having vision and seeing new ways of doing things is a part of the career opportunity that's available" (Brown).

Fun Facts: TASCAM Portastudio

The **TASCAM Portastudio** was a line of products offering four-track recording on a cassette deck. All units featured a built-in mixer; some offered EQ, and the top-of-the line models even had balanced inputs. In the early 1990s, it was an affordable way for musicians to create demo tapes or assemble song ideas. Although there were some eight-track reel-to-reel machines available and TASCAM's ADAT format was adopted in smaller studios, the user-friendly device along with inexpensive cassettes (a bit more expensive if you wanted higher grade "metal" cassette tape) made the device a forerunner to the "bedroom studio."

Pam Aronoff (USA)

There weren't a lot of women in the synth program at the Berklee College of Music (Boston, Massachusetts) in 1985–1986 when **Pam Aronoff** was beginning her studies. "There was Nancy, who I still know, and one other girl. Then I went to the electroacoustic program at Cal Arts (Santa Clarita, California) and studied with Mort Subotnick, and there was just Becky and I," out of about 14 students total (Aronoff, 2019).

About 20 years later, Aronoff was invited to teach at Loyola Marymount University in Los Angeles. Out of 24 students, only 3 were women – not a big change for such a long

Pam Aronoff and classmate Alessa Sindoni at Berklee

stretch of time. But in 2019, she is finally starting to see improvement in the gender balance of the classes she teaches. "It's almost half," she estimates. "Or maybe four women to six men. That's really reassuring to me" (Ibid.).

Her career path began during the early days of digital recording. "We had Sound Designer, we didn't have Pro Tools yet. You would hook up a 4-track cassette tape to record acoustic material that would also be laid to the Sound Designer stuff," she recalls. "By the time I got to Cal Arts (California Institute of the Arts), we had a DAT player" (Ibid.).

Gary Manfredi was a friend of hers at Cal Arts who was working at Image Smith in San Francisco, a company working on a new *Peanuts* title and needed someone to help do sound. "They didn't know anybody at all doing sound for any of this . . . So Gary tapped me and I said, '*No*, I can't do your *audio* for your *game*,'" she had said rather disparagingly. Now she laughs at the irony of her dismissal. "I said, 'I'm doing my final thesis', but he begged me. The truth is I wish I could find that early sound design. I was at a very experimental school, and my concepts were very out there. I think that first game must sound really wacky" (Ibid.).

After she graduated with a MFA in electro-acoustic music composition from Cal Arts, Scott Gershwin (who also attended Berklee) invited Aronoff to design at Soundelux. "He asked me for a fireball and a bungee jump. And my idea for a fireball – because it was so abstract – was [to use] my friend Becky singing opera. And he said, 'very interesting . . .' basically, 'you're interesting but you're not ready for prime time,'" Pam laughs (Ibid.).

Nonetheless, she found a job on her first day in Los Angeles. "Day one. I came into town, I said I was a sound designer. I had one name from a guy at Philips, and next door to them two guys had started a company called Viridis. And they were doing the 'edutainment' stuff: *Eco East Africa* was the first game I worked on. I walked in the office, and they asked, 'what do you do'? And I said, 'I'm a sound designer'. And they said, 'You're hired'. And day one I was thrown into the fire, just figuring it out. That was the CD-ROM era" (Ibid.).

Later on in her career, she was invited to work at Disney Online. "I gave my friend Gary this little cassette tape of weird stuff – not something that you would apply for a gig with," recalls Aronoff. "And he handed that to the head of the online department there, and the next thing I know I'm getting called by Disney and I actually was like, 'Nah. No. I'm too edgy for you! Have you listened? It's, like, *so dark!*' And they said, 'We're trying to give Disney a new spin.'" She ended up working at Disney Online after they agreed to her terms: a 30-hour workweek (Ibid.).

She went on to become the 27th person hired at Stan Lee Media as Queen of the Audio Department. ("They indulged my silliness," Pam says. "Technically nowadays it would be considered Audio Director.") "At that point Stan Lee had his own company . . . that was an amazing experience. And everyone wanted to be a superhero. So you had Michael Jackson and Mary J Blige – *everybody*, in on a daily basis. And they had all the faith in me." Michael Jackson wanted to be Spiderman (Vanguardia, 2019), and Mary J Blige had her own superhero designed by Stan Lee (StanZine, 2000).

"I've been all over the place," says Pam. "Viridis, Disney, Soundelux, Stan Lee, back to Soundelux, Flagship, Back to Soundelux, PopCap . . . everywhere."

In 2004 Aronoff, along with her sound team, won the Motion Picture Sound Editors (MPSE) Golden Reel Award for Best Sound Editing on *T3: Rise of the Machines*. Her other Golden Reel nominations include Best Sound Editing for *Crysis*, *Area 51*, *Lineage II: The Chaotic Chronicle*, *The Chronicles of Riddick anime*, *Lost Planet*, and *Getting Up*. Her other awards include two Game Audio Network Guild (G.A.N.G.) nominations for *Van Helsing*, *Lost Planet*, and *Plants vs Zombies*. Her film credits include *Below*, *Chronicles of Riddick*, *South Park*, *Pure Danger*, and several more (moxymusic, 2019).

Since 2004 she has been owner of moxymusic, specializing in music composing and post-production for video games, film, and radio. You can find out more about her work at moxymusic.net.

Fun Facts: Wwise

Wwise (pronounced "wise") by Audiokinetic is a software tool used to create audio for games. It has a more intuitive interface than those found within a game engine; rather it uses a plug-in to achieve integration with the game engine (Unity, Unreal Engine). Wwise also features internal plug-in tools such as spatialization, convolution reverb, granular synthesis, and more. If you want to learn the software, they offer several tutorials and even an online certification process. More information can be found at www.audiokinetic.com/learn/certifications/.

Anastasia Devana (Russia)

What if there were no digital audio workstations, no Pro Tools, no Logic . . . nothing? How would you work with an audio signal? In a way, that's the question developers at Magic Leap are answering as they enter into the new world of spatial computing, which blends virtual content into the real world. "Because audio works in a very different way in this medium, we don't have those tools yet," says **Anastasia Devana**, audio director at Magic Leap. "We actually do everything from content creation to implementation. We have people on our team who are sound designers, who are composers, who are implementers who can work with Wwise and FMOD and other middleware, and people who can work in Unity and programmers who can write even lower level code so we can have interactive audio content in Lumin OS. We like to control the audio process beginning to end. And the big reason for that is – for this particular medium – it's so new. It's a brand new way of doing audio" (Devana).

As the audio director at Magic Leap, Devana is involved with a wide variety of projects. "At Magic Leap, we make both hardware and software. . . . We make our own operating system, we make core apps; but we also make creative content, so we have Magic Leap Studios that is part of Magic Leap." Even with all of these areas of operation, the goal is simple: "We work really closely with our internal software team and our algorithms team and a hardware team to make sure that our platform is the best platform for audio content creators" (Ibid.).

In 2019, Devana was the keynote speaker at the Audio Engineering Society's conference on Immersive and Interactive Audio in York, England. She took some time to define for me some of the basic terminology needed just to hang on to conversations about "game engines" and "APIs" and developer tools. "In Unity, for example, you just open an engine and play sound using Unity. But it connects to the underlying audio API. You can think about it like an onion, because there are layers," she explains. "So, at the very core of an engine is the code that actually plays audio. That's what we call 'low level': super-obtuse machine language no regular human can understand. And then there's a layer on top of that, the 'C' API (Application Program Interface) for example. It's a set of commands that the engine understands, and it's also a little bit more understandable for a normal human being. That's the common interface: 'this is how you play something'. But then on top of that, then it can be exposed in a language that other engines would understand, like Unity; and a separate one for Unreal; and a separate one for something else" (Ibid.).

So where does Wwise and FMOD fit into all of this? "They're the tomatoes. You can chop it up and sauté it with the onion," she jokes. Although game engines like Unity and Unreal have audio features, they're not very extensive. "In game or interactive audio, sometimes you want to do more complex things, or you want the sound designer to have flexibility without needing to learn a programming language. What something like Wwise does is allow a sound designer to use a more approachable interface that can design all this complex behavior without needing to code. And then Wwise will export the sound bank, and then you can plug it into Unity and then the [Unity] programmer will hook up a few calls [to the sound bank]; but yeah, this is like the side dish" (Ibid.).

And then there's Lumin OS, Magic Leap's operating system, which features Lumin Runtime. "Right now, Lumin allows you to run multiple apps; that's what we call a 'landscape application' where you can have multiple independent applications running at the same time side by side." Magic Leap has several core apps that come with the platform, including a gallery, Avatar Chat, and browser. "Basically, you can have a browser over here, and you have your gallery over there and you can be chatting with a friend over there and you can see all of that in your space at the same time" (Ibid.).

"If you are looking at some pictures and talking on the Avatar Chat you can share your gallery app with a friend, and you can both look at it at the same time – both in their homes.

"So even the technical pipeline is not the main challenge. The real question is how do we design sound that is not only spatial and feels accurate to the user's space, but is also contextual and appropriate to the user and their surroundings, and to their shared experiences" (Ibid.).

Magic Leap press photo showing the gameplay environment
Source: Courtesy of Magic Leap

Devana got her start in music at an early age and sang in choirs for a long time. "I've also played in metal bands – totally compatible with choirs," she laughs. "But I never thought that music would be my career. So, I studied business administration for a couple of years." After moving to Los Angeles, she still felt unfulfilled, so she decided to pursue graphic design. Then she discovered Adobe Dreamweaver, which is a software program used for web design. The interface shows the user what the web page will look like, but there's another view where you can see the underlying code. "I was totally fascinated by it, like, 'Holy smoke! You can write some code, and the visuals show up, and that's amazing!' So right then and there I decided that that's whated I want to do: web design. And this was really funny because I've never done it." However, she was able to find an entry-level job and over time made a successful career (Ibid.).

But after some time, she became restless again. She decided to enroll in the UCLA extension program to earn a certificate in film scoring . . . and then it was back to the bottom of the ladder. "It was pretty challenging, because I went from senior developer, making really good money, to making no money and fetching coffee for Hollywood producers.

Anastasia Devana, audio director of Magic Leap

"And then I went to my first GDC (Game Developers Conference). And I thought, 'Oh! Wait a minute: *games*. It's technical and I can use my programming skills and my composition skills, and I can do something really cool with music for games and interactive systems.' It was around this time in 2015, when VR started to take off (again)" (Ibid.).

For one of her projects, Devana was hired to create and implement 3D sound. At that time there were several plug-ins on the market that performed 3D panning, but there wasn't any information about how they compared with each other. "So, I chose a plug-in for that project. . . . And then I decided to write an article and do an in-depth analysis of the plug-ins that were out there, so I compared the features and the performance and then made a little Android app and I put it on the App Store. Then suddenly I became the 'spatial audio person' because nobody has done anything like that. Companies started reaching out to me to review their plug-ins" (Ibid.).

As she began networking and establishing contacts, someone recommended a new start-up company, "Magic Leap," to her (Ibid.).

"It's been really awesome. Very, very challenging, but also really interesting. You'll learn something new every day, and do something that nobody's done before . . . or try to, you know. It's pretty cool" (Ibid.).

Fun Facts: Virtual, Augmented, and Mixed Reality

What's the difference between virtual reality, augmented reality, and mixed reality?

Virtual Reality: A user wears a headset with an eyepiece, "goggles," or perhaps a specially-mounted mobile phone attached. The user experiences a "virtual world," which replaces the "real world" around them. The user can look around and sometimes navigate to travel through this virtual space.

The author wearing a VR headset consisting of a mobile phone

Augmented Reality: Games like *Pokémon Go* are augmented reality experiences. A digital image is overlaid on top of the real world. In some museum installations, users can aim their phones at special images, which are replaced with an animation when carefully placed. There are toys such as "Parker: Your Augmented Reality Bear." This is a teddy bear with a special symbol on its belly. A child can aim a tablet or phone at the bear and see an animation of different "ailments" affecting the bear.

Mixed Reality: In this environment, the graphics are integrated into the room. Content could be occluded by, say, a table, so it appears as if this object is behind a table or even on top of it. Magic Leap makes content for mixed reality experiences.

Victoria Dorn (USA)

Victoria Dorn is an audio-focused software engineer at Sony Interactive Entertainment (SIE). She is a graduate of Berklee College of Music where she studied sound design and composition for video games. She began her journey at SIE as an audio engineering intern in the Research and Development department in 2013. After a few years of working with PlayStation, she was inspired by the developers in R&D to jump into the world of software engineering and earned a computer science degree from Oregon State University (Dorn, 2019).

Dorn is currently part of the R&D team at SIE working on spatial audio for the PS4 platform, where she strives to help game audio teams create the best sounding immersive audio experiences. Since working at SIE, Dorn has contributed to many projects, including the PSVR and the Platinum Wireless headset (Ibid.).

"My day-to-day work is usually composed of various tasks revolving around 3D audio, be it 3D audio developer outreach, 3D audio testing, or working on pieces of tech that are actually living within the PlayStation's 3D audio library," Dorn explains. "One of the tasks I am most passionate about is the task of developer outreach. I work on a team that is developing functionality that is being used by a wide array of audio developers from all different sizes of studios making all different types of games. We want to make sure that what we are creating works well with as many teams' audio workflows as possible. Some days, I might be chatting with different audio teams who are creating PlayStation games, learning

Victoria Dorn, Software Engineer at Sony Interactive Entertainment

of functionalities that they might want added into our library, and some days I might be taking those ideas and creating functional prototypes so that my team can understand if these features would actually give the desired results in our environment.

"I also get to have some fun coming up with creative solutions as to how developers can most effectively use our library within their game since we provide mix and match options for developers to use. For example, one studio might use all object-based audio and one might use objects encoded with ambisonics audio, and yet another studio might use a bit of both. I find it quite interesting trying to figure out the best and most effective way for a studio to use 3D audio.

"In addition to communication with developers through our regular PlayStation channels, I try to provide developers with up to date information on our library through various presentations at internal and external conferences including: GDC (Game Developers Conference), GameSoundCon, and CEDEC (Computer Entertainment Developers Conference)" (Ibid.).

Dorn is also involved with other exciting projects. "Ambisonics experimentation and recording, HRTF shenanigans, user testing, and 3D audio library side test design and creation. Overall, I find it quite awesome that I have a job that lets me bring together both my love for game audio and my passion for creating powerful and useful tools to help games tell the best aural stories possible" (Ibid.).

▨ Sally Kellaway (Australia)

"Sound is more than the things you hear: it's also the way you experience something and the way you are made to feel," **Sally Kellaway** reminds us (Kellaway).

Specializing in audio design for immersive and interactive media, Kellaway got her bachelor's degree in Brisbane, Australia, from the Queens Anne Conservatorium Griffith University, studying music technology with a focus on audio engineering and sound design. "I'm a very particular person," she explains. "I know what I like and don't like; what I'm interested in and not interested in." As such, she preferred to work alone or with her small cohort in her degree program. There were opportunities to collaborate with film and animation students, but "I don't think they had enough respect for my work when they changed things at the last minute." She recognized that students in other programs had different expectations, but they just didn't mesh with hers. She was also choosy about how she spent her time. "There is a music industry in Australia, but the scale isn't there, and I didn't want to work late nights doing live sound" (Ibid.).

"I had an Xbox and a friend with a PlayStation, so I had exposure to pretty cool games that were coming out at that time like *Red Dead Redemption* and *Bioshock*. And the games had heavy theming, and the audio directors that worked on those games did incredible work to make those games holistic packages of audio that could take you somewhere . . . and not just to have you sit there and think 'wow, that was an experience', but to be part of that world" (Ibid.).

Especially with *Bioshock*, with sound design by Australian audio director **Emily Ridgeway**, "there are subtle things that I really latched onto like the construction of the ambience and environment and how you could just move through it and it just felt like you were there, it sounded like you were there. It was incredible. So, I wanted to learn how to do that" (Ibid.).

Her university was starting a game design course, so she walked up the road to the fine arts campus and worked on her first game. It wasn't just sound design, she explains, "but sound design in a place you can interact with." Kellaway went on to learn about interactive music systems, learning from Mick Gordon (composer on *Doom 3*). "He made a concept about a game in the Australian outback. . . . I wasn't allowed to use traditional instruments and it all had to be done from sound design elements. It was a horror game, and the question was, 'How do you make someone feel creepy without a zombie in your face?'. It had a lot of impact on me; it was a real lesson going from super short, fun, student projects to something deeper and finding out how to use those systems, and how to make people *feel* using those systems." Her path led her to discover binaural audio, and she experimented with the binaural panner in Logic (Ibid.).

Capitalizing on this newfound interest, she decided to pursue a master of design science from the University of Sydney. From there she was involved in VR games and a 3D tech company. Kellaway now specializes in audio design for immersive and interactive media. She currently works as the senior audio designer at Microsoft, doing audio consultation and editing for machine learning, audio design, and location recording, among other duties. She also works with mixed reality and artificial intelligence. "Technically, I could work on games if I wanted, but I am working in enterprise space now. . . . I see this as a huge opportunity to learn mixed reality and machine learning, and [discovering] what does audio mean to that? Finding that, and helping build that, is my picture of what my career is right now" (Ibid.).

Making the transition from games to other fields is something that she says could be quite natural for game audio designers because of the growing and expansive technology they are exposed to. "They could really help create new technologies for audio . . . by doing audio in interesting ways" (Ibid.).

She also worked for FMOD for two and a half years and has worked with Wwise in addition to developing with game engines like Unity and Unreal. "It's a project-by-project basis on what the project needs," she explains (Ibid.).

Her plan for the next three years as she continues at Microsoft is to continue helping establish their vision. "I'm really excited to engage because it's audio and mixed reality in a different way: inputting signals to make weird sounds for example . . . the world and mixed reality is going to be very different so it's hard to predict what is going to happen" (Ibid.).

Fei Yu (China)

Fei Yu is a Chinese award-winning music producer, music supervisor, recording engineer, and music editor. She used to work for China Film Group as recording engineer for three years and earned a master's degree at McGill University. She is the first Asian to win the AES Student Recording Competition Gold Award twice. After graduating from McGill, she started her own company named "Dream Studios" and expanded her work to different fields, including producing music for game and different movie projects. "The game world is like living in heaven," she says. "Some triple A games have big budgets; they let you record a full orchestra just for a few minutes of music."

Yu was music supervisor for *Revelation (Original Soundtrack)*, which became one of the first game scores to be released by the film music label Varèse Sarabande and won the BSOSpirit Jerry Goldsmith Award as well as the Tracksounds Genius Choice Vote for Best

Fei Yu, recording engineer

Score: Video Game. In addition, it was nominated for an International Film Music Critics Association Award, a Tracksounds Cue Award, a Reel Music Award, and a Game Audio Network Guild Award and was named one of the top ten scores of 2015 (game and film) by Cinematic Sound Radio (Acree, 2016).

Like **Winifred Phillips**, she notes the difference between scoring for film and scoring for games. "One really important thing which is different from game music, is the film score needs to work with dialogue really well. With film scoring there is also the need to deal with picture changes. . . . But I like working in both industries; game music gives me more freedom. Film scoring makes me know how to collaborate with others."

Bibliography

Acree, Neal. 2016. *Revelation Online Wins BSOSpirit Jerry Goldsmith Award* [Online]. Available: www. nealacree.com/news/2016/1/19/revelation-online-nominated-for-reel-music-award [Accessed September 27, 2019]

Aronoff, Pam. 2019. Interview with Pam Aronoff. *In:* Gaston-Bird, L. (ed.).

AES New York 2015 Presenter or Author: Winifred Philips [Online]. New York. Available: www.aes.org/ events/139/presenters/?ID=3995 [Accessed June 14, 2019].

Brown, Khris. 2019. Interview with Khris Brown. *In:* Gaston-Bird, L. (ed.).

Devana, Anastasia. 2019. Interview with Anastasia Devana. *In:* Gaston-Bird, L. (ed.).

Dorn, Victoria. 2019. *RE: Victoria Dorn – Software Engineer, Sony Interactive Entertainment*, June 18. Type to Gaston-Bird, L.

Ishiwata, Eiko. 2019. Interview with Eiko Ishiwata. *In:* Gaston-Bird, L. (ed.).

Kellaway, Sally. 2019. Interview with Sally Anne Kellaway. *In:* Gaston-Bird, L. (ed.).

Mongeau, Anne-Sophie. 2016. Audio Middleware Comparison – Wwise/FMOD/Fabric/Unity 5. *Audio // Blog: Game Audio and Digital Art Blog* [Online]. Available: https://annesoaudio.com/2016/06/12/ audio-middleware-comparion-wwisefmodfabric/ [Accessed September 27, 2019].

Moxymusic. 2019. *Moxymusic* [Online]. Available: http://moxymusic.net [Accessed June 14, 2019].

Phillips, Winifred. 2019. Interview with Winifred Philips. *In:* Gaston-Bird, L. (ed.).

Phillips, Winifred. Video Game Music Composer Gives Lecture at the Library of Congress. *Gamasutra* [Online]. Available: gamasutra.com/blogs/WinifredPhillips/20190417/340788/ [Accessed June 14, 2019].

Ridgeway, Emily. 2019. Interview with Emily Ridgeway. *In:* Gaston-Bird, L. (ed.).

Schmidt, Brian. 2017. *GameSoundCon Game Audio Industry Survey 2017* [Online]. www.gamesoundcon. com. Available: www.gamesoundcon.com/single-post/2017/10/02/GameSoundCon-Game-Audio-Industry-Survey-2017 [Accessed June 14, 2019].

Stanzine. 2000. *Check Out Mary J. in Action* [Online]. Stan Lee Media. Available: https://web.archive.org/web/20000815082909/www.stanlee.net/maryjblige/video.html [Accessed June 14, 2019].

Vanguardia. 2019. Stan Lee reveló que Michael Jackson quería ser un superheroe de Marvel. *Vanguardia* [Online]. Available: https://vanguardia.com.mx/articulo/stan-lee-revelo-que-michael-jackson-queria-ser-un-superheroe-de-marvel [Accessed June 14, 2019].

Wawro, Alex. 2014. Q&A: Veteran Director Khris Brown on the Secrets of Great Voice Acting. *Gamasutra* [Online]. Available: www.gamasutra.com/view/news/221127/QA_Veteran_director_Khris_Brown_on_the_secrets_of_great_voice_acting.php [Accessed June 14, 2019].

Women Behind the Games at Double Fine: Khris Brown. 2018. Directed by Fine, D.

Yamauchi, Ayako. 2019. Interview with Ayako Yamauchi. *In:* Gaston-Bird, L. (ed.).

About the Author

Leslie Gaston-Bird was born in Dayton, Ohio, in 1969 to Frances and Berdell Gaston. Her father had an Ampex reel-to-reel tape recorder, and she remembers singing into it as a small child and eventually learning how to thread tape around its path with her tiny fingers. She played piano throughout her childhood, studying first with "Mrs. Wadsworth" at age six and then at age eight with Phyllis Katz who taught her classical piano. At age 11 she was the solo accompanist for a seventh-grade musical, *The Game*. At Chaminade Julienne High School, she played in Jazz Lab Band. Steve Weingart, himself a prodigal jazz pianist and high school senior at the time, taught her basic jazz voicings during her freshman year. She also accompanied the school chorus and played percussion in the marching band (xylophone and snare) and double bass in the string ensemble. She earned high scores in several regional competitions on piano.

In 1987 Gaston-Bird enrolled in the Audio Technology program at Indiana University Bloomington. Her professors were Wayne Jackson and David A. Pickett. At the time, the school had no digital editing tools; everything was done to analog tape, and she learned how to do razor-blade editing. She recorded operas, such as *The Tales of Hoffman*, and did live sound for the IU vocal ensemble The Singing Hoosiers during a summer tour. She recorded a number of bands including the Joyce Brothers, Planet Ranch, the I.U. African-American Choral Ensemble, and Mojo Hand. She also worked as a radio board operator for WFIU-FM, an NPR affiliate station, where she saw weekly job announcements for "broadcast/recording technicians" at National Public Radio. Inspired by these opportunities, she decided to take a chance and apply.

Determined to succeed, after graduating in August 1991, Gaston-Bird moved to Washington, D.C., and by December she was hired at NPR as a broadcast/recording technician. As one of 40 technicians, she rotated shifts as the "drive engineer" for *All Things Considered*, *Morning Edition*, *Talk of the Nation*, and *Performance Today* and also worked in "record central" taking in phone and satellite feeds from reporters around the globe. When the digital audio workstation Sonic Solutions debuted, she was one of five engineers introduced to the technology and worked on a piece for *National Geographic Radio Expeditions*. She also played briefly with an all-female reggae/metal band, Medusa Complex.

Having to work weekends and overnight shifts was exhausting. Gaston-Bird sought to work consistent hours and accepted a job at Colorado Public Radio in Denver, Colorado, in 1995. As the only recording engineer on staff she did location recording, mixing, and editing. Digital Audio Tape (DAT) recorders were the norm at the time, although some reporters still used Marantz cassette recorders. She started with an obscure digital audio workstation called Arrakis but made the case to switch to Pro Tools in 1998.

In 1998 she met the all-female rock band Ezmeralda and joined them as their bass player. She engineered two albums for them, *Half Gramme Holiday* (2000) and *Immortal*, the latter also available as a surround-sound audio-only Blu-Ray disc (2014). (It remains perhaps the first and only, all-female performed and engineered 5.1 Blu-Ray audio disc.)

Also in 1998 she accepted an invitation to help the Colorado Symphony Orchestra under the direction of Marin Alsop. The CSO sought to get their performances broadcast on *Performance Today*, and Gaston-Bird's expertise from working at NPR came into play. She worked with NPR to recommend and implement better microphone techniques and a recording setup, and for the next few years, she would record the CSO for broadcasts on CPR and NPR. Gaston-Bird's work as a recording engineer was featured on the CD release, *The Colorado Symphony on Colorado Public Radio*. In 2000, along with Kelley Griffin and Dan Drayer, Gaston-Bird won a Radio and Television News Directors Association Edward R. Murrow Award (Large Market Documentary) for *A Columbine Diary*.

As the technology continued the transition to digital, radio automation systems were being implemented. Gaston-Bird completed training for the NexGen Prophet system as a master user. During this transition, reporters began to record and edit their own pieces. She trained the journalists on Cool Edit Pro (which became Adobe Audition). The move also enabled the entire library of classical music to be digitized, and conversations about data compression and normalization led her to realize there was more to learn about this new digital audio world.

She left CPR in 2002 to pursue a master's degree in recording arts at the University of Colorado Denver. Her mentors there were Roy Pritts and Richard Sanders. Both of them encouraged Leslie to become active in the Audio Engineering Society.

That same year, she started working at Post Modern Company in Denver doing sound restoration for old film soundtracks in the Sony/Columbia Pictures archives. She did thousands of hours of editing and noise restoration using Pro Tools and CEDAR software.

Meanwhile, for her thesis, she assembled a crew with director donnie l. betts and Future Jazz Project to create a style of production called "Music Video Vérité" in 2003. Having been a fan of music videos since the 1980s, her goal was to combine the immediacy of live performance by recording the audio alongside the video – even as locations changed within the video. She successfully defended her thesis and entered the video into an AES student competition in 2004, winning in the category "sound for picture."

Her academic successes led her to apply for an opening for an assistant professor job at CU Denver, and she joined the faculty in 2005. Her research interests focused on multichannel sound. In 2006 she was the principal investigator for the article, "Comparison of 3 Multichannel Codecs," which she presented at the Verband Deutscher Tonmeister conference in Leipzig, Germany, and published in the AES *Journal*.

In 2011 she received a Fulbright Award to undertake research and teach at the University of York with her sponsor, David Malham. Her goal was to study ambisonic decoding for Blu-ray Discs using the BD-J (Javascript) format. She had to shorten the length of the project in anticipation of the birth of her second child in 2012 but successfully authored a 5.1 performance decoded from her ambisonic recording of Norwegian musician Christian Wallmrød's ensemble. She was tenured in 2012 and promoted to associate professor.

In 2017 she moved with her family to Brighton, England, and resigned from her tenured position at CU Denver. That year, she also became a Motion Picture Sound Editors (MPSE) member as well as a member of the Association of Motion Picture Sound (AMPS). In 2019 she was welcomed to the Recording Academy. She now freelances and has works as sound supervisor for documentaries and feature films, including Doc of the Dead, Insomnio,

Enough White Teacups (directed by Michelle Caprenter, which was nominated for a Heartland Emmy Award; *Three Worlds, One Stage* directed by Jessica McGaugh and Roma Sur); *Leap of Faith* (a documentary of *The Exorcist* directed by Alexandre O. Philippe) and *A Feral World* (directed by David Liban). Her company is Mix Messiah Productions and can be found at MixMessiahProductions.com.

AC-3:   Acoustic Coding 3, a data-reduction algorithm based on the principles of psycho-acoustic perception and used by Dolby Digital.

ADR: Additional Dialog Replacement or Automatic Dialog Replacement.

AES: Audio Engineering Society.

Automation: Can refer to "automated programming" in radio, where a computer fires off each event (announcers, music, or commercials) or to "volume automation" (or "plug-in automation," etc.), used in digital audio workstations.

AoIP: Audio over IP, a protocol that specifies how audio is transmitted using wireless computer networks. Signals are generally sent over CAT-5 ethernet cables. Examples are Dante, Cobranet, and AES-67, all of which support high sampling rates and unlimited channel capacity.

API: Application Program Interface, usually in the context of game programming.

AR: Augmented Reality.

Atmos: *See "Dolby Atmos"*

Auro 3D: A tool for authoring and playing back multichannel audio for cinema.

Bass Management: Used in surround sound mixing, the practice of employing a subwoofer to play back sound from the "Low Frequency Effects/Extension" channel and to extend the range of the main surround channels.

BD-J: Blu-ray Disc Java.

Beam-Forming: A signal processing technique used to make acoustic signals directional, sometimes with a very narrow radiation pattern.

Beat Cypher: A live performance of improvised rap music.

Binaural: A method of recording and reproducing sounds by using the physical properties of the ear to shape and contour the sound, hopefully achieving a more realistic sensation.

CD: Compact Disc.

Convolution: A signal processing technique used to capture an impulse response in a room, which can then be "deconvolved" by, say, a reverb plug-in to emulate the sound of the room.

Dante: A protocol for Audio over IP.

DAB: Digital Audio Broadcasting.

DAT: Digital Audio Tape.

DAW: Digital Audio Workstation.

DDP: Disc Description Protocol, used in creating compact discs.

De-Esser: A process that reduces sibilance, or the harsh "s" sound in speech.

Deliverables: The final, master files that are used in distribution.

Dictaphone: A tape recorder, usually associated with transcribing speech.

Dolby Atmos: A format for authoring and playing back multichannel audio for cinema and home.

DJ or DeeJay: Disc Jockey. Historically, a radio announcer would play disc-shaped records. Today, of course, this can refer to radio announcers or performers who may or may not play CDs or records.

DRM: Digital Rights Management, a way of watermarking a file so its copyright information is protected.

DP: Director of Photography: supervises the camera crew and may also run the camera.

DSD: Direct Stream Digital, a high-fidelity format using a data rate of 2.8 MHz with one bit.

DVD-A: Digital Versatile Disc, Audio: format for high-resolution stereo and surround sound.

DVD-V: Digital Versatile Disc, Video: format for video.

Flangeing or Flanging: The practice of applying pressure to a reel of tape as it travels past the heads. By affecting the speed, the sound appears to "wobble."

Foley: Adding sound effects to a film using props.

Granular Synthesis: A way of "chopping" a sound file into very small segments, which can be altered and played back in creative ways.

Haptic: Sensory feedback from a device, such as a cell phone's "vibrate" function.

HRTF: Head-Related Transfer Function.

IFB: "Interruptible Fold Back".

IPS: Inches per second.

Max/MSP: A software program used to create and manipulate sound.

MADI: Multichannel Audio Digital Interface, a protocol used to carry several channels of audio over a single cable.

Microphone: A microphone captures sound and converts it into electrical energy. The changes in voltage are analogous to the changing pressure of the sound wave as it arrives at the microphone.

MIDI: Musical Instrument Digital Interface.

MOOC: Massively Open Online Course.

MPEG: Motion Pictures Expert Group, an organization that oversees standards for audio and video compression and transmission.

MR: Mixed Reality.

MUSICAM: Masking Pattern Universal Sub-band Integrated Coding and Multiplexing, a form of data compression.

NAMM: National Association of Music Merchants, hosts of a large annual convention in Anaheim, California.

Op-amp: Operational Amplifier, usually a small chip or integrated circuit that is used to provide amplification in a device.

Oscillator: A device that generates an audio signal at a single frequency.

Oscilloscope: A device that measures and displays audio signals or DC voltage.

PA system: Public Address system, usually any setup where there are loudspeakers, from a small nightclub to a stadium concert, for playback on a film set, or for lecture halls and classrooms.

PCB: Printed Circuit Board.

PCM: Pulse Code Modulation.

QLab: Software used to play back sound effects, usually in theater.

Quadrophonic: Also known as "Quad sound," uses four channels to be played on as many loudspeakers: left, right, left rear, and right rear.

Red Book: One of the formats used to play back compact discs. A red book CD plays back audio at 44.1 kHz with 16-bits of resolution. (Other formats include orange book, which is a recordable CD or CD-R).

RF: Radio Frequency(ies).

RTA: Real-Time Analyzer.

SACD: Super Audio CD, an audio format for high-resolution stereo and surround sound.

Sample Rate: In digital audio, once a sound is represented as changing levels of voltage (using a microphone, for example), those voltage levels can be measured several times a second, and a digital number is assigned to each voltage measurement. The number of times per second the measurement is taken is referred to as the sample rate. One example of a sample rate is "44,100 times per second." This is written as 44,1000 Hertz (Hz), or 44.1 kHz (kilohertz). Other examples are 48 kHz, 96 kHz, or 192 kHz.

SEAMUS: The Society for Electric-Acoustic Music in the United States.

SPL Meter: A device that measures sound pressure level.

SPL: Sound Pressure Level, the energy created by sound waves.

Test Pressing: This is usually created in the first stage of making a vinyl record. Before the manufacturer creates thousands or millions of records to distribute, a test pressing is made so the creative team can "sign off" on the recording.

Video Game or Videogame: Both spellings are acceptable, the latter being used in the "early days." Video games can be played on a computer, a game console (Playstation, Xbox), and on mobile devices.

VR: Virtual Reality.

Zine: Abbreviated version of the word "magazine," often published by one author for a niche audience as opposed to a mass-distributed publication with staff of several journalists.

Recommended Reading

Links to these works and more can be found at the companion website for this book, WomenInAudioBook.com

Recommended Reading: Books

Brown, L.M. & Gilligan, C. 1992. *Meeting at the Crossroads: Women's Psychology and Girls' Development*. Cambridge, MA: Harvard University Press.

de Bruin, L.R., Burnard, P. & Davis, S. (eds.). 2018. *Creativities in Arts Education, Research and Practice*. Leiden, The Netherlands: Brill | Sense.

Corey, Jason. 2016. *Audio Production and Critical Listening: Technical Ear Training*. Burlington, MA: Taylor & Francis.

Densmore, F. 1970. *The American Indians and Their Music*. New York: Johnson Reprint Corp.

Hill, N., 1883–1970 1944. *Think and Grow Rich: Teaching, for the First Time, the Famous Andrew Carnegie Formula for Money-Making*. Meriden, CT: The Ralston Society.

Hinkle-Turner, E. 2006. *Women Composers and Music Technology in the United States: Crossing the Line*. Brookfield, VT: Ashgate Publishing, Ltd.

Hughes, R.J., Angus, J.A., Cox, T.J., Umnova, O., Gehring, G.A., Pogson, M. & Whittaker, D.M. 2010. Volumetric Diffusers: Pseudorandom Cylinder Arrays on a Periodic Lattice. Melville, NY: *The Journal of the Acoustical Society of America*, 128(5), 2847–2856.

Laird, P.W. 2006. *Pull: Networking and Success Since Benjamin Franklin*. Cambridge, MA: Harvard University Press.

Massy, S. & Johnson, C. 2016. *Recording Unhinged: Creative and Unconventional Music Recording Techniques*. Milwaukee, WI: Hal Leonard.

Moulton, D. 1995. *Dave Moulton's Golden Ears: The Revolutionary CD-Based Audio Eartraining Course for Musicians, Engineers and Producers*. Sherman Oaks, CA: KIQ Productions.

Phillips, W. 2014. *A Composer's Guide to Game Music*. Cambridge, MA: MIT Press.

Pogson, M.A., Whittaker, D.M., Gehring, G.A., Cox, T.J., Hughes, R.J. & Angus, J.A.S. 2010. Diffusive Benefits of Cylinders in Front of a Schroeder Diffuser. Melville, NY: *The Journal of the Acoustical Society of America*, 128(3), 1149–1154.

Prior, et al. Women in the US mUSic indUStry. Berklee Institute for Creative Entrepreneurship Available: www.berklee.edu/sites/default/files/Women%20in%20the%20U.S.%20Music%20Industry%20Report.pdf [Accessed September 27, 2019].

Rodgers, T. 2010. *Pink Noises*. Vancouver: Duke University Press.

Roginska, A. & Geluso, P. eds. 2017. *Immersive Sound: The Art and Science of Binaural and Multi-Channel Audio*. New York, NY: Taylor & Francis.

Williams, M. 2004. *Microphone Arrays for Stereo and Multichannel Sound Recordings: Designing Microphone Arrays for Stereo; The Variable M*. Segrate MI, Italy: Editrice Il Rostro.

Links

#

Back issues of *Billboard Magazine*. www.americanradiohistory.com/Billboard-Magazine.htm.

Back issues of *Starlog Magazine* (one of Khris Brown's favorites). https://archive.org/details/starlogmagazine.

Ear Training with Sound Gym (offers a free introduction, then paid membership). www.soundgym.co/

Galindo, J. 2019. Mastering for Vinyl: Tips for Digital Mastering Engineers. *Mastering* [Online]. Available: www.izotope.com/en/blog/mastering/mastering-for-vinyl-tips-for-digital-mastering-engineers.html.

Podcasts from Sound Women. http://audioboom.com/channel/soundwomen.

The Recording Academy's Producers & Engineers Wing. 2004. Recommendations For Surround Sound Production. www.grammy.com/sites/com/files/surroundrecommendations.pdf.

She Shreds Magazine, for Female Guitarists and Bass Players. https://sheshredsmag.com.

Sound is Fun: http://soundisfun.com

Smith, S. et al. 2018. Inclusion in the Recording Studio? Gender and Race/Ethnicity of Artists, Songwriters & Producers across 600 Popular Songs from 2012–2017. Annenberg Inclusion Initiative. *USC Annenberg*. http://assets.uscannenberg.org/docs/inclusion-in-the-recording-studio.pdf.

The Story of Tom Tom Magazine, a Site for Female Drummers. https://tomtommag.com/2019/05/the-history-of-tom-tom-magazine/.

Women in Rock Project: Dedicated to Telling the Story of Women Who Played Rock and Roll in the 1950s and 1960s. www.womeninrockproject.org.

Filmography

Haniya Aslam. *Main Irada*. www.youtube.com/watch?v=wbXUFGWVZxs&feature=youtu.be.

June Millington. *Play like a Girl*. www.youtube.com/watch?v=gevsWISf04g&feature=youtu.be.

Making Waves, The Art of Cinematic Sound. www.makingwavesmovie.com

Linda Briceño. *Solitude*. www.youtube.com/watch?v=30wECQ9cqg8.

1821: Sophie Germain publishes *Recherches sur la théorie des surfaces élastique*.

1843: Ada Lovelace's additions are featured in "Sketch of the Analytical Engine, invented by Charles Babbage Esq., By LF Menabrea of Turin officer of the military engineers, with Notes from the Translator." These notes would earn her the title of "first computer programmer."

1891: Margaret Watts-Hughes publishes in *The Century Magazine* in 1891.

1907: Frances Densmore records music of Native America tribes from 1907–1930.

1928–1929: Blanche Sewell, Viola Lawrence, and Barbara McLean are editors for Hollywood's first "sound films."

1928: Aletha Dickerson produces "Selling' that Stuff" for the Hokum Boys on Paramount Records.

1929: Barbara McLean performs an early dialog edit.

1934–1937: "Synchrophone" record label has a female "recording expert," Ursula Greville.

1935–1936: Helen Oakley Dance goes to Detroit and records in Chicago.

1939: Mary Weller publishes "Vibrations of free square plates: part I. Normal vibrating modes," in the Proceedings of the Physical Society.

1940s: Ethel Gabriel promoted to record producer at RCA in New York. She is the first known female producer at a major record label.

1945: Marie Louise Killick patents the Sapphox phonograph stylus.

1947: Evelyn Blanchard (married name Evelyn Palladino) engineers "Smoke! Smoke! Smoke! (That Cigarette)" by Tex Williams at Radio Recorders in Hollywood. She may have been engineering prior to this.

1947: Mary Shipman Howard starts releasing her own commercial recordings under the MHR label (classical music).

1948: Bebe and Louis Barron start their recording studio and exploration of electronic music.

1950: Lillian McMurry produces her first recording session with St. Andres Gospelaires for Trumpet Records (Jackson, Mississippi). That same year, she creates Trumpet Records to release blues records.

1950: Memphis Recording Service (owned by Sam Phillips) opens in Memphis. Marion Keisker was the studio assistant, manager, and "Jane of All Trades."

1951: Wilma Cozart Fine is hired by Mercury Records to run the label's small classical music department.

1953: Elvis records his first demo, "My Happiness," at Memphis Recording Service. Marion Keisker says she was the engineer.

1953: Lillian McMurry engineers her first session at DRC studio in Jackson, Mississippi, with Trumpet artist Jerry McCain.

1953: Dorle Soria launches Angel Records with her husband, Dario.

1953: Vivian Carter Bracken founds Vee-Jay records with her husband, James Bracken.

1953: Dorle Soria coproduces 500 albums with her husband (Dario Soria) for Angel Records, a subsidiary of EMI.

1954: Ruth White starts recording music. Her first recording project was filmmaker Paul Burnfords' short film, *Rhythms of the Freight Yard*, which she also scored. In 1955, she started doing recordings for the Los Angeles City school district, which needed folk music and other recordings.

1955: Ethel Gabriel produces Perez Prado's *Cherry Pink and Apple Blossom White*. It becomes a number-one hit and is credited with starting the mambo craze in the USA.

1955: Ruth White started the production company Rhythms Productions to produce records and educational materials for the Los Angeles city school district. She did many albums of folk music as well as albums for the physical education department.

1956: Cordell Jackson starts Moon Records to release her own music and self-produces *Beboppers Christmas*.

1956: Wilma Cozart Fine named vice president of the Mercury label.

1958: Johnnie Matthews starts the Northern Recording Company in Detroit. She was the second known African American woman to own and operate a record label. (Vivian Carter Bracken and her husband, Jimmy, were the first known.) She was one of the artists on the label.

1958: Florence Greenberg, a 45-year-old housewife in New Jersey, starts Tiara Records. She signs her first artist, the Shirelles, after an audition in her living room.

1958: Daphne Oram joins the newly created BBC Radiophonic Workshop, known for their experimentation with music and sounds for radio (and later for television). Maddalena Fagandini joined a year later.

1959: Florence Greenberg creates Scepter Records after selling Tiara Records (with the Shirelle's contract) to Decca Records. Scepter was a successful independent label through the 1960s.

1959: Ethel Gabriel launches Living Strings for RCA and becomes the first woman to receive a RIAA Gold Record as record producer (1959) for Henry Mancini's *Peter Gun Soundtrack*.

1960s: Ethel Gabriel creates the *Living* easy-listening albums that were sold by the millions: *The Living Voices, The Living Guitars, The Living Brass, The Living Marimbas*, and *The Living Organ*. The Living Strings won a Grammy in 1968, and *Living Voices: Wish Me A Rainbow* won a Grammy for Best Performance by a Chorus in 1967.

1973: Olivia Records is formed.

1973: *Virgo Rising,* a record featuring all woman performers and crew, is released by Thunderbird Records (Reno, NV).

1975: Leslie Ann Jones is hired as the first female recording engineer at ABC Studios in Los Angeles.

1977: Leslie Ann Jones becomes the first female engineer at the Automatt Studios in San Francisco.

1980: Pamela Peterson presents "History of Women in Audio" at AES and AES has first "Women in Audio" panel.

1980: Lenise Bent becomes first woman to win a platinum album for *Autoamerican*.

1980: Suzanne Ciani is the first known woman hired to score a major Hollywood feature, *The Incredible Shrinking Woman* with Lily Tomlin.

1980: Suzanne Ciani designs sounds for Xenon Pinball and is credited as the first human female voice in a game.

1981: Marianna Sankiewicz-Budzyńska was elected vice rector for education at the Gdańsk University of Technology.

1982: Sylvia Robinson is the producer for the Sugarhill Gang's "Rapper's Delight."

1983: Joanna Nickrenz is the first woman to win the Grammy for Classical Producer of the Year. The same year she is nominated for Best Classical Orchestral Recording for *Del Tredici: In Memory of a Summer Day*, which she shared with Marc Aubort.

1984: Kay Rose is the first woman to win an Oscar for Sound Editing for *The River*.

1986: Laurie Spiegel releases Music Mouse software.

1986: June Millington establishes the Institute for Musical Arts, a women's retreat, studio, school, and, eventually, summer camp, which she founded with her partner, activist Ann F. Hackler.

1986: Cecelia Hall receives an Oscar nomination for sound editing on *Top Gun*.

1989: Janet Jackson is the first woman nominated for a Grammy for Producer of the Year, Non-Classical with Jimmy Jam and Terry Lewis for *Rhythm Nation 1814*. That same year she won the Grammy for "Best Music Video" and was also nominated for Best R&B Vocal Performance, Female, and Best Instrumental Arrangement Accompanying Vocals.

1990: Cecelia Hall is the second woman to win an Oscar for Best Sound Editing along with George Watters II for *The Hunt for Red October*.

1991: Mariah Carey is the second woman nominated for a Grammy for Producer of the Year, Non-Classical for *Emotions* with Walter Afanasieff and Best Pop Vocal Performance, Female.

1991: Gloria Borders is the third woman to win an Oscar for Sound Editing on *Terminator 2: Judgement Day* with Gary Rydstrom

1993: Judith Sherman is second woman to win a Grammy for Classical Producer of the Year (and the first women to win on her own). Altogether, Sherman has 5 wins in this category.

1994: Gloria Borders earns an Oscar nomination in the category "Best Sound Editing" with Randy Thom for *Forrest Gump*.

1995: Anna Behlmer is the first woman to be nominated for an Oscar in the Sound Mixing category (*Braveheart*), which is shared with Andy Nelson, Scott Milan, and Brian Simmons.

1995: Elizabeth Cohen becomes first woman to serve as president for the Audio Engineering Society.

1996: Anna Behlmer shares an Oscar nomination in the Best Sound Mixing Category for *Evita* with Andy Nelson and Ken Weston.

1997: Anna Behlmer shares an Oscar nomination for Best Sound Mixing for her work on *L.A. Confidential* with Andy Nelson and Kirk Francis.

1997: Marina Bosi becomes AES president elect from 1997–1998 and president from 1998–1999.

1998: Trina Shoemaker was the first woman to win a Grammy for Best Engineered Album for Sheryl Crow's *The Globe Sessions*, sharing the award with Tchad Blake and Andy Wallace.

1998: Sheryl Crow is the fourth woman to receive a Grammy nomination for Producer of the Year, Non Classical for *The Globe Sessions*.

1998: Pud Cusack is nominated for an Oscar in the Best Sound Mixing Category for *The Mask of Zorro*, which she shares with Kevin O'Connell and Greg P. Russell.

1997: Paula Cole becomes the third woman (and first woman on her own) to be nominated for a Grammy for Producer of the Year, Non Classical.

1999: Leslie Ann Jones becomes the first female chair of The Recording Academy's Board of Trustees.

1998: Anna Behlmer is nominated for an Oscar Sound Mixing *Thin Red Line*, along with Andy Nelson and Paul Brincat.

2001: Anna Behlmer is nominated for an Oscar Sound Mixing for *Moulin Rouge!* along with Andy Nelson, Roger Savage, and Guntis Sics.

2003: Women's Audio Mission is founded in San Francisco by Terri Winston.

2003: Anna Behlmer is nominated for an Oscar Sound Mixing for *Last Samurai* and *Seabiscuit*, along with Andy Nelson and Jeff Wexler.

2003: Lauren Christy (who with Graham Edwards and Scott Spock are known as The Matrix) is the fifth woman nominated for a Grammy Producer of the Year, Non-Classical.

2003: Theresa Leonard elected as president-elect (2003–2004) and president (2004–2005) of AES.

2005: Anna Behlmer is nominated for an Oscar Sound Mixing for *War of the Worlds*, along with Andy Nelson and Ron Judkins.

2006: Anna Behlmer is nominated for an Oscar Sound Mixing for *Blood Diamond*, along with Andy Nelson and Ivan Sharrock.

2007: Darcy Proper is the first woman to win a Grammy for Best Surround Sound Album for her work as mastering engineer on Donald Fagen's *Morph The Cat*, sharing the award with Elliot Scheiner and Fagen.

2007: Karen Baker Landers is the fourth woman to win an Oscar for Best Sound Editing for *The Bourne Ultimatum*, which she shares with Christopher Boyes.

2007: Ebonie Smith launches Gender Amplified.

2009: Imogen Heap is the second woman to win in the category Best Engineered Album, Non-Classical for her album, *Ellipse* (and the first woman to win on her own).

2009: Gwendolyn Yates Whittle is nominated for an Oscar in the Best Sound Editing category for *Avatar* (shared with Christopher Boys)

2009: Anna Behlmer is nominated for an Oscar in the Best Sound Mixing category for *Star Trek*, along with Andy Nelson and Peter J. Devlin.

2010: Gwendolyn Yates Whittle is nominated for an Oscar in the Best Sound Editing category for *Tron: Legacy* (shared with Addison Teague).

2010: Lora Hirschbirg is the first woman to win an Oscar for Best Sound Mixing (*Inception*), which she shared with Gary Rizzo and Ed Novick. As of 2019, she is the only woman to have won in this category.

2010: Leslie Ann Jones and Brandie Lane are the first women to win a Grammy for Best Engineered Album, Classical for their engineering work on *Quincy Porter: Complete Viola Works* by Eliesha Nelson and John McLaughlin Williams, sharing the award with Kory Kruckenberg and David Sabee.

2010: Society of Women in TeCHnology (SWiTCH) holds its inaugural event.

2011: Deb Adair is nominated for an Oscar for Best Sound Mixing for *Moneyball*, Sound Mixing with Ron Bochar, Dave Giammarco, and Ed Novick.

2012: Karen Baker Landers wins an Oscar for Best Sound Editing for *Skyfall*, which she shared with Per Halberg.

2012: Bridget Shield becomes first woman president of the Institute of Acoustics.

2013: SoundGirls.org is founded by Karrie Keyes.

2013: Beats by Girlz was founded by Erin Barra.

2013: Anna Behlmer and Lora Hirschberg mix *World War Z*, marking the first known time two women have mixed a major Hollywood motion picture.

2013: Virginia Read is the first woman to win an Australian Recording Industry Association (ARIA).

2014: Becky Sullivan is nominated for an Oscar for Best Sound Editing for *Unbroken* (shared with Andrew Decristofaro).

2014: The Royal Academy of Engineering (est. 1977) elects Professor Dame Anne Dowling as president of the Royal Academy of Engineering.

2015: Linda Briceño becomes first woman to win a Latin Grammy for Best Producer, Non-Classical.

2015: Yorkshire Sound Women's Network founded.

2016: Renée Tondelli receives an Oscar nomination for *Deepwater Horizon*, which she shares with Wylie Statemen.

2016: Ebonie Smith receives a Grammy certificate for her work as assistant engineer on *Hamilton, the Original Broadway Cast Album*.

2016: Ai-Ling Lee and Margaret Latou are the first all-woman team to be nominated for best sound editing at the Oscars for *La La Land*.

2016: Ai-Ling Lee is the first Asian woman to be nominated for a sound editing Oscar.

2016: Ai-Ling Lee becomes the first woman to do both sound mixing and editing *and* to be nominated for an Oscar both Sound Editing and Sound Mixing for the same film, La La Land, with Andy Nelson and Steve Morrow.

2017: Nadja Wallaszkovits elected as president-elect (2017–2018) and president (2018) of AES.

2017: Mary H. Ellis is nominated for an Oscar in the category Best Sound Mix for *Baby Driver*, with Julian Slater and Tim Cavagin.

2017: Suzanne Ciani is the first woman to be awarded the Moog Innovation Award.

2017: Abhita Austin establishes the Creator's Suite.

2018: Ai-Ling Lee and Mildred Latrou Morgan are nominated for an Oscar for Best Sound Editing for *First Man*. Ai-Ling Lee and Mary Ellis are nominated for Best Sound Mixing for the same film (with Jon Taylor and Frank A. Montaño).

2018: Nina Hartstone is the fifth woman to win an Oscar for Best Sound Editing for *Bohemian Rhapsody* with John Warhurst.

2018: Linda Perry becomes sixth woman to be nominated for a Grammy for Producer of the Year, Non-Classical. As of 2019, no woman has ever won in this category.

2019: USC Annenberg Study released.

2019: Agnieszka Roginska elected as president-elect (2019) and president (2020) of AES.

*Special thanks to **April Tucker** who contributed to many of these entries. Oscars and Grammy information verified at awardsdatabase.oscars.org and Grammy.com.*

Appendix 2
Women's Audio Organizations

This list of organizations for women and girls in music technology was compiled by Liz Dobson and used with her permission. The original page can be found at drlizdobson. com/2018/02/18/feministsoundcollectives, where you can click on each of the groups to navigate to their page. The list will continue to evolve as more groups are formed and found. Groups wishing to add themselves can get in touch with Dobson on the site.

TABLE A2.1 List of organizations for girls and women in audio. (An asterisk [*] denotes groups whose entries on the website are pending as of this writing.)

Name	Location	Description
Analogue Ladies	Online	This group is for ladies of all kinds who LOVE analog synthesizers!
Association of Canadian Women Composers	Canada	The Association of Canadian Women Composers (ACWC/ AFCC) is the only professional association of women and women-identified composers and musicians in Canada. It actively supports music written by Canadian women. The ACWC promotes its members on this website, and publishes a bi-annual Journal highlighting activities and articles of interest. *Included because this association clearly includes sonic arts and audio-based practices.*
Audible Women	Australia	Audible Women is an online directory for women who make some kind of art that can be listened to. It is open to women who make sound, sound art, noise and music (acoustic or electronic) with a bit of an experimental and exploratory bent – interpret that as you will.
Audio Girl Africa*	Nigeria	Audio Girl Africa www.audiogirlafrica.com
Babely Shades	Ottawa/ MTL/ Toronto	We are a collective of people of colour from the Ottawa arts and music scene who seek to bring visibility and awareness to local artists and issues against marginalized groups. Our end goal is to dismantle the white supremacy in our communities in order to allow more folks of colour to participate in local arts.
BAE Collective*	US NY based	The collective BAE is a celebration of conscious female creativity and collaborative feminine energy.
Beats By Girlz	North America	is a non-traditional, creative and educational music technology curriculum, collective, and community template designed to empower females to engage with music technology. *Chapters providing education, supported by online resources*

(Continued)

TABLE A2.1 (Continued)

Name	Location	Description
The Creator's Suite*	US NY based	A hub where all artists are seen for their creativity and feel comfortable to be their authentic selves. This community is The Creator's Suite! A place for us to share resources, develop our craft, and collectively amplify our voices
Discwoman	US NY based	Founded by Frankie Decaiza Hutchinson, Emma Burgess-Olson and Christine McCharen-Tran, Discwoman is a New York-based platform, collective, and booking agency – that showcases and represents talent in electronic music. Started as a two-day festival in September 2014 at Bossa Nova Civic Club Discwoman has since produced and curated events in 15+ cities – working with over 250 DJs and producers to-date.
Electronic Girls	Italy – Venice	Label, collective and projects
Electronic Ladies	Germany	What is Electronic-Ladiez.Net?
		Electronic-Ladiez.Net is a small local network of women who make electronic music for those who like electronic music.
		Our focus is on composition, production and performance and less on "classical" DJing.'
Fair Play Network	France	FAIR_PLAY is a network designed to promote the visibility to the practices of female identifying and trans artists in the fields of sound art, experimental, electroacoustic and alternative music and related arts and techniques.
Female Frequency	North America	Female Frequency is a community dedicated to empowering female, transgender & non-binary artists through the creation of music that is entirely female generated.
female:pressure	International	female pressure is an international network of female, transgender and non-binary artists in the fields of electronic music and digital arts founded by Electric Indigo: from musicians, composers and DJs to visual artists, cultural workers and researchers.
Gender Amplified	US NY based	Gender Amplified is a nonprofit organization that aims to celebrate Women in music production, raise their visibility and develop a pipeline for girls and young women to get involved behind the scenes as music producers. The movement also connects passion for music with technical skills that can be used in a wide range of scientific and arts based fields, areas in which Women are traditionally underrepresented.
Girls Make Beats*	US Miami, Los Angeles	Girls Make Beats (GMB) Incorporated is a 501(c)(3) nonprofit organization that empowers girls by expanding the female presence of DJs, Music Producers, and Audio Engineers.

Name	Location	Description
Girls Pressing Buttons	Online	Girls Pressing Buttons is a global initiative to engage girls and women of all ages in creating music. Through community, sharing of experience and information and promoting artists and similar causes, we hope to bring visibility to the movement and those involved through sharing information, artists, methods, and hosting workshops. Contact us if you'd like to be a contributor.
Girls Rock Camp Alliance	International	The Girls Rock Camp Alliance is an international membership network of youth-centered arts and social justice organizations.
Girls Rock London	UK London	part of a movement of rock camps that take place all over the world, united by a desire to achieve gender equality in the music industry and to ensure that all girls and women get the chance to make music. The aims of the project are to empower girls and women – regardless of previous musical experience – to write and perform music, and to build self-confidence.
Go Girls Music	US Houston – Texas	GoGirlsMusic.com is the oldest and largest online community of indie women musicians. Our mission is to promote, support and empower indie women in music. *Now closed but check out the message.*
Her Noise	UK	Her Noise Archive is a resource of collected materials investigating music and sound histories in relation to gender bringing together a wide network of women artists who use sound as a medium. *Includes events and guest curators*
Her Sound Her Story	Australia	Various significant outputs including Her Sound Her Story Film The initial concept for Her Sound, Her Story came from observing the constant conversation about gender inequality and imbalance within the music industry and the arts in general. Believing in the power of art and it's capacity bring people together as well the importance of storytelling, the project was born.
The Institute for Musical Arts	US Massachusetts	The institute's nonprofit mission is to support women and girls in music and music-related businesses. Rooted in the legacy of progressive equal rights movements, IMA's development is guided by the visions, needs and concerns of women from a diversity of backgrounds and has grown from the need to nourish ourselves and each other.
The International Alliance for Women in Music	International	. . . an international membership organization of women and men dedicated to fostering and encouraging the activities of women in music, particularly in the areas of musical activity such as composing, performing, and research in which gender discrimination is an historic and ongoing concern.
In The Loop	Canada	A free workshop series introducing women to the basics of electronic music creation, recording and history.

(*Continued*)

TABLE A2.1 (Continued)

Name	Location	Description
Konstmusiksystrar [Contemporary Sisters]	Sweden	Art music sisters are a network of composers and artistic artists that define themselves as women or transgender and are at the beginning of their creation. On our website we have an ever increasing list of our members. Art music sisters act as a safe environment where our members can meet without a doubt, share experiences, create music and change the current structures within the art scene.
Ladyfest	International	Ladyfest is a community-based, not-for-profit global music and arts festival for feminist and women artists. Individual Ladyfests differ, but usually feature a combination of bands, musical groups, performance artists, authors, spoken word and visual artists, films, lectures, art exhibitions and workshops; it is organized by volunteers.
Ladyz in Noyz	International	Series celebrating women experimental/fringe/noise/sound artists/musicians. increasing gender diversity & visibility w/in music.
		- international inclusive collective network that furthers those with similar missions and individuals alike.'
LARM	Sweden	LARM culminated in a manifestation of women's activities in the art of sound with the LARM Nordic Sound Art Festival at the Culture House in Stockholm March 30 – April 22, 2007. LARM seems to grow and maintain the archive, and by contacting the LARM group by mail, interested artists may submit their presentation.
Listening to Ladies	International	Listening to Ladies is a podcast which first aired on September 26th, 2016. The episodes feature excerpts from interviews with composers who are women, interwoven with samples of their work. Interviewees include established, emerging and under-recognized composers from the USA, Canada, Argentina, Israel, Iran, Scotland, England, and Australia.
Malta Sound Women Network	Malta	[We] aim to bring like-minded women together; to share knowledge and skills in music and sound technology, sonic arts, production, audio-electronics, and anything to do with using kit to create sound!
Mint	Berlin	Mint is a platform for women – DJs and producers – working in electronic music, founded by Zoe Rasch and Ena Lind in 2013.' 'Mint's goal was to promote visibility amongst female DJs and acts. The platform has been providing female artists with opportunities to showcase their talent, develop their skills and expand their network. *Mint is now defunct since 2017.*
Normal Not Novelty	UK London	Networking space for producers, writers, engineers, DJs meeting at #NormalNotNovelty at Red Bull Studios London

Name	Location	Description
New York Women Composers, Inc.	US New York	The mission of the New York Women Composers, Inc. is to create performing, recording, networking, and mentoring opportunities for its members, and to work for the betterment of all women concert-music composers. We believe that continually focusing attention on music by women composers will hasten its full inclusion in the concert repertoire.
Omnii	UK London	London collective for female and non-binary producers/engineers/sound manipulators
Opensignal	US	Opensignal is a collective of artists based in providence RI with a concern for gender and race in electronic music/art. stay tuned for more recordings, conversations, and performances!
The Other Woman	UK London	A music network dedicated to showcasing the best in new, upcoming and current female talent. It began as a radio show on London's finest arts community radio station Resonance FM in 2006, as a reaction to working at the BBC and being disillusioned with radio playlists, that featured the same kind of female artist. Hearing this homogeneous sort of female vocal got really boring, and so the search began for edgy, alternative music by kick ass women.
Polyphones	Paris France	Polyphones is a space of visibility created for women in the field of sonic and musical experimentation. This is a network born out of the reality of the Parisian scene with national and international offshoots. . .
Producer Girls	UK London	. . . founded by London-based producer and DJ, E.M.M.A together with producer pals P Jam, Dexplicit, Ikonika and Nightwave to encourage more young women to take up electronic music production. The workshops give girls who have an interest in learning music production tips, expert advice and free software to help them get started.
Sonora Red de Ingenieras de Sonido de Colombia	Colombia	Support the professional development of sound engineers, through the realization of cultural projects that include workshops, conferences, tutorials and events; oriented to generate possibilities of study and employment that increase the representation of women in the sound engineering industry.
Rise	UK	Rise is a not-for-profit, advocate group for gender diversity within the broadcast manufacturing and services sector.
Rock n' Roll Camp for Girls Los Angeles Part of the Girls Rock Camp Alliance	US LA	We strive to nurture self-esteem and self-expression in girls, in a world that doesn't always give girls permission, space, or tools to do so. Our programs are designed to promote collaboration, build confidence, and celebrate unique and diverse voices.
Seattle Sound Girls	US Seattle	Seattle Sound Girls is a non-profit organization dedicated to helping women and girls to develop confidence and a strong foundation of technical skills needed to excel in the fields of live music production and audio engineering.

(Continued)

TABLE A2.1 (Continued)

Name	Location	Description
The Seraphine Collective	US Detroit	Seraphine Collective is an inclusive, supportive and active community of feminists designed to foster creative expression and camaraderie among marginalized musicians and artists in Detroit.
She Rock She Rock	US Minnesota	She Rock She Rock is a Minnesota nonprofit dedicated to empowering girls, women, trans and gender nonconforming folks through the art of music.
Shejay	International	The world's first and largest global network and agency for Female DJs
She Said So	International	shesaid.so is a curated network of women with active roles in the music industry. Our vision is to create an environment that supports collaboration, creativity and positive values.
Sisu	UK London	We are an artistic community providing a platform to primarily showcase female creativity. We also work with male artists who support female empowerment, to provide a gender-neutral creative commonality.
The Society for Women in Technology (New York University)	US New York	The Society of Women In TeCHnology (SWITCH) is a graduate student run club for women technology students at NYU and is open to all graduate and undergraduates.
SONA	UK Sheffield	Born out of regular Sheffield meet-ups of the Yorkshire Sound Women Network, SONA is a Sheffield-based group which fosters skills, creates space, generates networks, and forges collaboration to support women in learning and experimenting with sound and music practices.
Sound Girls	US based international organisation, with chapters worldwide.	SoundGirls.Org's vision is to inspire and empower young women and girls to enter the world of professional audio and music production while expanding opportunities for girls and women in these fields, and to share resources and knowledge through cooperation, collaboration, and diversity.
Sound Women	UK	When we started in 2011, our mission was to build the confidence, networking and leadership skills of women in radio and audio. *Sound women is now closed however they provide further links via their website.*
Studio XX	Canada	Studio XX is a bilingual feminist artist-run centre that supports technological experimentation, creation and critical reflection in media arts. XX assists in the independent production and diffusion of art created by artists who identify as women, queer, trans, and gender fluid in the field of contemporary technological practices.

Name	Location	Description
Syrphe	Asia, Africa and Latin America	. . . a platform focused on music and events in the field of electronic music, noise, avant-garde, contemporary classical, electro-acoustic, industrial, experimental, sound art (. . .) specially from Africa and Asia but not exclusively. Syrphe tries to establish new connections and exchanges between musicians, promoters, galleries, venues, magazines, radio stations from all over the world and tends to spread above all awareness about Asian and African composers.
Titwrench Collective	Denver	We believe in promoting feminist and queer perspectives, creating access for all ages of people to experience the power of experimental art and music in a communal setting and to amplifying underrepresented voices. We strive for accessibility and inclusivity in all of our events. We believe in the power of art to effect social change, foster community and transform lives for the better.
WOMB 2 Women of Music Business	UK London	A new inclusive female-promoted event in London Monthly at Number 90, Hackney Wick.
Women in Music	International	Women in Music is a non-profit organization with a mission to advance the awareness, equality, diversity, heritage, opportunities, and cultural aspects of women in the musical arts through education, support, empowerment, and recognition. Our seminars, panels, showcases, achievement awards, and youth initiatives celebrate the female contribution to the music world, and strengthen community ties.
Women in Music Canada	Canada	Established in 2012, Women in Music Canada (WIM-C) is a registered non-profit organization comprised of influential members of the Canadian music industry. The organization is dedicated to fostering gender equality in the music industry through the support and advancement of women.
Women in Music Tech	US Georgia Tech in Atlanta	Women in Music Tech (WMT) is an organization which aims at (1) attracting more women into the Music Technology Program at Georgia Tech Center for Music Technology (GTCMT) to get closer to parity, and (2) providing both women and men with a safe and respectful environment at GTCMT.
Women in Sound Women on Sound	UK and International	A network of networks linking individuals, groups and organisations to each other, promoting knowledge exchange that shapes our current understanding of sound and technology.
Women on Air	US Georgia – Atlanta	Women ONAIR was created as a networking platform for all women working in television & radio broadcasting.
Women Produce Music	International	An artist & producer led non-profit org & network promoting & supporting the activities of music-makers, producers & engineers through a series of initiatives.

(Continued)

TABLE A2.1   (Continued)

Name	Location	Description
Women's Audio Mission	US San Francisco	The only professional recording studio in the world built and run by women – to attract over 1,500 underserved women and girls every year to STEM and creative technology studies that inspire them to amplify their voices and become the innovators of tomorrow. WAM's award-winning curriculum weaves art and music with science, technology and computer programming and works to close the critical gender gap in creative technology careers.
Women in Acoustics*	US	Women in Acoustics Committee was created in 1995 to address the need to foster a supportive atmosphere within the Society and within the scientific community at large, ultimately encouraging women to pursue rewarding and satisfying careers in acoustics.
Women in Vinyl*	International	Empowering women working in the music industry to create, preserve and improve the art of music on vinyl.
Yorkshire Sound Women Network	Yorkshire and International	Yorkshire based network set up to bring like-minded women together; to share knowledge and skills in music and sound technology, sonic arts, production, audio electronics and, well, basically anything to do with using kit to make sound!

Index